Beginning
AutoCAD 2005

Other titles from Bob McFarlane

Beginning AutoCAD 2002　　ISBN 0 7506 5610 7

Beginning AutoCAD 2004　　ISBN 0 7506 6451 7

Modelling with AutoCAD 2004　ISBN 0 7506 6433 9

Beginning AutoCAD 2005

Bob McFarlane
MSc, BSc, ARCST
CEng, FIED, RCADDes
MIMechE, MBCS, FRSA

authorised publisher

Taylor & Francis Group

LONDON AND NEW YORK

First published 2005 by Newnes

2 Park Square, Milton Park, Abingdon, Oxfordshire OX14 4RN
52 Vanderbilt Avenue, New York, NY 10017

Routledge is an imprint of the Taylor & Francis Group, an informa business

First issued in paperback 2019

Copyright © 2005, R. McFarlane

The right of R. McFarlane to be identified as the author of this work has been asserted in accordance with the Copyright, Designs and Patents Act 1988

All rights reserved. No part of this book may be reprinted or reproduced or utilised in any form or by any electronic, mechanical, or other means, now known or hereafter invented, including photocopying and recording, or in any information storage or retrieval system, without permission in writing from the publishers.

Notice:
Product or corporate names may be trademarks or registered trademarks, and are used only for identification and explanation without intent to infringe.

British Library Cataloguing in Publication Data
A catalogue record for this book is available from the British Library

Library of Congress Cataloguing in Publication Data
A catalogue record for this book is available from the Library of Congress

Typeset by Integra Software Services Pvt. Ltd, Pondicherry, India

ISBN 13: 978-1-138-42922-2 (hbk)
ISBN 13: 978-0-750-66720-3 (pbk)

Contents

	Preface	vii
	Acknowledgements	xi
Chapter 1	Using the book	1
Chapter 2	The AutoCAD 2005 graphics screen	2
Chapter 3	Drawing, erasing and the selection set	14
Chapter 4	The 2D drawing aids	23
Chapter 5	Saving and opening drawings	29
Chapter 6	Standard sheet 1	36
Chapter 7	Line creation and co-ordinate entry	38
Chapter 8	Circle creation	46
Chapter 9	Object snap	50
Chapter 10	Arc, donut and ellipse creation	60
Chapter 11	Layers and standard sheet 2	65
Chapter 12	User exercise 1	83
Chapter 13	Fillet and chamfer	85
Chapter 14	The offset, extend, trim and change commands	90
Chapter 15	User exercise 2	99
Chapter 16	Text	101
Chapter 17	Dimensioning	108
Chapter 18	Dimension styles 1	114
Chapter 19	Modifying objects	126
Chapter 20	Grips	137
Chapter 21	Drawing assistance	145
Chapter 22	Viewing a drawing	151

Chapter 23	Hatching	157
Chapter 24	Point, polygon and solid	174
Chapter 25	Polylines	179
Chapter 26	Modifying polylines	187
Chapter 27	Divide, measure and break	193
Chapter 28	Lengthen, align and stretch	196
Chapter 29	Obtaining information from a drawing	202
Chapter 30	Text fonts and styles	208
Chapter 31	Multiline Text, Text Tables and Fields	218
Chapter 32	The array command	233
Chapter 33	Changing properties	241
Chapter 34	User exercise 3	250
Chapter 35	Dimension styles 2	252
Chapter 36	Drawing with various units and paper sizes	264
Chapter 37	Multilines, complex lines and groups	270
Chapter 38	Blocks	281
Chapter 39	Wblocks	294
Chapter 40	Attributes	301
Chapter 41	External references	309
Chapter 42	Pictorial drawings	314
Chapter 43	Model space and paper space	322
Chapter 44	Templates	333
Chapter 45	CAD standards	343
Chapter 46	The AutoCAD Design Center	347
Chapter 47	Toolbars and tool palettes	359
Chapter 48	Sheet sets	364
Chapter 49	'Electronic' AutoCAD	369
	Activities	377
	Index	405

Preface

AutoCAD 2005 is the latest release of (probably) the best-selling and most widely used PC-based CAD software package. The package is very similar to previous releases but incorporates several new and innovative features. These new features, combined with the traditional AutoCAD interface, will increase the users' draughting skills and improve productivity. Some of these new features include:

Drafting tools

1. Intuitive table creation
2. Fields that can be updated
3. Display of overlapping objects
4. Mark-ups for design review
5. Backgrounds for multiline text and dimensions
6. New notation symbols
7. Hatch object trim
8. Hatch tolerance for areas with gaps
9. Both the DDEDIT and ATTEDIT commands can be used to edit attributes
10. Reverse arcs and calligraphy style for revision clouds
11. Backgrounds for 3D scenes
12. Control of display as you adjust clipping planes
13. Vertical text.

Drawing management

1. Tree view organisation of sheets
2. Quick sheet creation
3. Sheet view management
4. Linked labels and callouts
5. Automatic updates when organisation or content changes
6. Plot stamp
7. Sheet list tables
8. Sheet sets archives.

Drawing output

1. Named sheet selections
2. Electronic transmittals.

Plot and publish tools

1. Background plotting
2. Simplified Plot dialogue box
3. Page setup enhancements
4. Enhanced DWF format
5. Enhanced publishing.

Productivity tools

1. Layer management
2. Maximised viewports button
3. Scaled text in OLE objects
4. Midpoint between two points
5. Object snap off for hatches
6. Object zoom
7. Relative path for image files
8. Reference type and uniform scale settings for external references
9. Time-tracking Express Tool
10. Applications on the computer's taskbar.

Sheet Set Manager

Supports the way the user manages projects and serves as a single organisational interface to the design data that the user must assemble for project teams and clients. Views from various drawings can be grouped as sheets in a sheet set, allowing the user the ability to process and package them as a unit.

Tool palette enhancements

1. Tools by example.
2. Command tools
   a) Frequently used commands can be set up as tools and organised on tool palettes
   b) They can then be customised by setting properties such as layer and linetype.
3. Tool palette organisation.

Many of these new features will be discussed in this book.

Note the following:

1. This book is intended for:
   a) new users to AutoCAD who have access to AutoCAD 2005
   b) experienced AutoCAD users wanting to upgrade their skills from previous releases
   c) readers who are studying for a formal CAD qualification at City and Guilds, BTEC or SQA level
   d) training centres offering CAD topics
   e) undergraduates and post-graduate students at higher institutions who require AutoCAD draughting skills
   f) industrial CAD users who require both a text book and a reference source.
2. The objective of the book is to introduce the reader to the essential basic 2D draughting skills required by every AutoCAD user, whether at the introductory, intermediate

or advanced level. Once these basic skills have been 'mastered', the user can progress to the more 'demanding' topics such as 3D modelling, customisation and AutoLISP programming.

3 As with all my AutoCAD books, the reader will learn by completing worked examples, and further draughting experience will be obtained by completing the additional activities which complement many of the chapters. All drawing material has been completed using Release 2005 and all work has been checked to ensure there are no errors.

4 Your comments and suggestions for work to be included in any future publications would be greatly appreciated.

Bob McFarlane

an advanced level. Once these basic skills have been mastered, the user can progress to the more demanding topics such as 3D modeling, customisation and AutoLISP programming.

5. As with all my AutoCAD books, the reader will learn by completing worked examples and further draughting experience will be obtained by completing the additional activities which compliment part of the chapters. All drawings that the user has completed using release 2000 and all work has been checked to ensure there are no errors.

6. Your comments and suggestions for work to be included in any future publications would be greatly appreciated.

Acknowledgements

It would not have been possible for me to complete the various exercises and activities in this book without the inspiration from all other AutoCAD authors. It is very difficult to conceive new ideas with CAD and I am very grateful for the ideas from these other authors. A special mention must be given to Dennis Maguire and his book 'Engineering Drawing from First Principles using AutoCAD' published by Arnold.

Autodesk® and AutoCAD® are registered trademarks of Autodesk, Inc., in the USA and/or other countries. All other brand names, product names, or trademarks belong to their respective holders.

Windows® is a registered trademark of the Microsoft Corporation.

Chapter 1

Using the book

The aim of the book is to assist the reader to use AutoCAD 2005 with a series of interactive exercises. These exercises will be backed up with activities, thus allowing the reader to 'practice the new skills' being demonstrated. While no previous CAD knowledge is required, it would be useful if the reader knew how to use:
a) the mouse to select items from the screen
b) Windows concepts, e.g. maximise/minimise screens.

Concepts for using the book

There are several simple concepts with which the reader should become familiar, and these are:

1. Menu selection will be in bold type, e.g. **Draw**.
2. A menu sequence will be in bold type, e.g. **Draw-Circle-3 Points**.
3. User keyboard entry will also be highlighted in bold type, e.g.
   a) co-ordinate entry: **125,36; @100,50; @200<45**
   b) command entry: **LINE; MOVE; ERASE**
   c) response to a prompt: **15**.
4. Icon selection will be in bold type, e.g. select the **LINE** icon from the Draw toolbar.
5. The AutoCAD 2005 prompt will be in italics and the prompt text in typewriter face, e.g.
   a) *prompt*   Specify first point
   b) *prompt*   Specify second point of displacement.
6. The symbol <R> or <RETURN> will be used to signify pressing the RETURN or ENTER key. Pressing the mouse right-button will also give the <RETURN> effect – called right-click.
7. The term 'pick' is continually used with AutoCAD, and refers to the selection of a line, circle, text item, dimension, etc. The mouse left button is used to 'pick' an object – called left-click.
8. Keyboard entry can be LINE or line. Both are acceptable.

Saving drawings

All work should be saved for recall at some later time, and drawings can be saved:
a) to a formatted disk (floppy, zip, etc.)
b) in a named folder on the hard drive.

It is the user's preference as to which method is used, but for convenience purposes only I will assume that a named folder is being used. This folder is named BEGIN and when a drawing is being saved or opened, the terminology used will be:
a) save drawing as **BEGIN\WORKDRG**
b) open drawing **BEGIN\EXER1**.

Chapter 2

The AutoCAD 2005 graphics screen

In this chapter we will investigate the graphics screen and the user-interface. We will also discuss some of the basic AutoCAD terminology.

Starting AutoCAD 2005

1. AutoCAD 2005 is started:
   a) from the Windows 'Start screen' with a double left-click on the AutoCAD 2005 icon
   b) by selecting the windows taskbar sequence:
      Start-Programs-AutoCAD 2005-AutoCAD 2005 (or similar).

2. Both methods will briefly display the AutoCAD 2005 logo and then:
   either   a) the actual graphics screen
   or   b) the Startup dialogue box.

3. If the Startup dialogue box is displayed, then select Cancel at present. This will allow the user access to the graphics screen. We will discuss the Startup dialogue box later in this chapter.

The graphics screen

The AutoCAD 2005 graphics screen (Figure 2.1) generally displays the following, although your screen may differ in appearance to Figure 2.1. This is quite normal and the user should not be concerned about any visual differences at this stage.

1. The title bar
2. The 'Windows buttons'
3. The menu bar
4. The standard toolbar
5. Layer information
6. The Windows taskbar
7. The status bar
8. The command prompt window area
9. The co-ordinate system icon
10. The drawing area
11. The on-screen cursor and Grips and/or Pickfirst box
12. Scroll bars at right and bottom of drawing area
13. The Layout tabs
14. The Draw toolbar
15. The Modify toolbar
16. Tool palette
17. The Dimension toolbar.

Title bar

The title bar is positioned at the top of the screen and displays the AutoCAD 2005 icon, the AutoCAD Release version and the current drawing name.

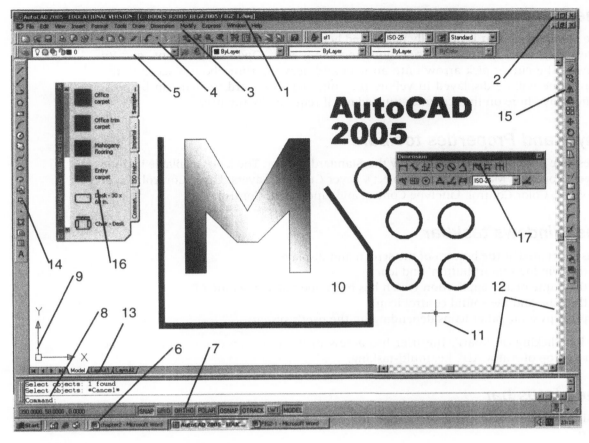

Figure 2.1 The AutoCAD 2005 graphics screen.

The Windows buttons

The Windows buttons are positioned at the right of the title bar, and are:
1. left button   minimise screen
2. centre button   maximise screen
3. right button   close current application.

The menu bar

1. The menu bar displays the default AutoCAD menu headings. By moving the mouse into the menu bar area, the cursor cross-hairs change to a **pick arrow** and with a left-click on any heading, the relevant 'pull-down' menu will be displayed. The full menu bar headings are:

   File Edit View Insert Format Tools Draw Dimension Modify (Express) Windows Help

2. Notes on menu bar
   a) Pull-down menu items with '. . .' after their name result in a screen dialogue box being displayed when the item is selected, i.e. left-clicked.
   b) Pull-down menu items with ▶ after their name result in a further menu being displayed when the item is selected. This is termed a cascade menu effect.
   c) Menu items in BOLD type are available for selection.
   d) Menu items in GREY type are not available for selection.
   e) Menu bar and pull-down menu items are selected (picked) with a mouse left-click.
   f) The (Express) menu bar selection will be displayed only if the express menu has been loaded.

The Standard toolbar

The Standard toolbar is normally positioned below the menu bar and allows the user access to several button icon selections including New, Open, Save, Print, etc. By moving the cursor pick arrow onto an icon and 'leaving it for about a second', the icon name will be displayed in yellow (default). The Standard toolbar can be positioned anywhere on the screen or 'turned off' if required by the user.

Layer and Properties toolbars

These are normally positioned below the Standard toolbar. The icons available include Layer Properties Manager, Make Object's Layer Current, Layers, Layer Control, Layer Previous, Color Control, Linetype Control and Lineweight Control.

The Windows taskbar

1. This is situated at the bottom of the screen and displays:
   a) the Windows 'Start button' and icon
   b) the name of any application which has been opened, e.g. AutoCAD
   c) the time and the sound control icon
   d) perhaps some other icons depending on the user's system.
2. By left-clicking on 'Start', the user has access to the other Programs which can be run 'on top of AutoCAD', i.e. multi-tasking.

The status bar

Positioned above the Windows taskbar, the status bar gives useful information to the user:

1. on-screen cursor X, Y and Z co-ordinates at the left
2. drawing aid buttons, e.g. SNAP, GRID, ORTHO, POLAR, OSNAP, OTRACK, LWT
3. MODEL/PAPER space toggle.

Command prompt window area

1. The command prompt area is where the user 'communicates' with AutoCAD 2005 to enter:
   a) a command, e.g. LINE, COPY, ARRAY
   b) co-ordinate data, e.g. 120,150; @15<30
   c) a specific value, e.g. a radius of 25.
2. The command prompt area is also used by AutoCAD to supply the user with information which could be:
   a) a prompt, e.g. `from point`
   b) a message, e.g. `object does not intersect an edge`.
3. The command area can be increased in size by 'dragging' the bottom edge of the drawing area upwards. I generally have a two- or three-line command area.

The co-ordinate system icon

This is the XY icon at the lower left corner of the drawing area. This icon gives information about the co-ordinate system in use. The default setting is the traditional Cartesian system with the origin (0,0) at the lower left corner of the drawing area. The co-ordinate icon will be discussed in detail later.

The drawing area

This is the user's drawing sheet and can be of any size as required. In general we will use A3-size paper, but will also investigate very large and very small drawing paper sizes.

The cursor cross-hairs

Used to indicate the on-screen position, and movement of the pointing device will result in the co-ordinates in the status bar changing. The 'size' of the on-screen cursor can be increased or decreased to suit user preference and will be discussed later.

The Grips/Pickfirst box

This is the small box which is normally 'attached' to the cursor cross-hairs. It is used to select objects for modifying and will be discussed in detail in a later chapter.

Scroll bars

Positioned at the right and bottom of the drawing area and are used to scroll the drawing area. They are very useful for large-size drawings and can be 'turned-off' if they are not required.

Layout tabs

Allows the user to 'toggle' between model and paper space for drawing layouts. The layout tabs will be discussed in a later chapter.

Toolbars

By default, Release 2005 displays the Draw and Modify toolbars although users may have them positioned differently from that shown in Figure 2.1. Other toolbars may also be displayed. Toolbars will be discussed later in this chapter.

Tool palette

Release 2005 displays All Palettes by default. This can be cancelled or repositioned by the user at any time.

Terminology

AutoCAD 2005 terminology is basically the same as previous releases, and the following gives a brief description of the items commonly encountered by new users to AutoCAD.

Menu

A menu is a list of options from which the user selects (picks) the one required for a particular task. Picking a menu item is achieved by moving the mouse over the required item and left-clicking. There are different types of menus, e.g. pull-down, cascade, screen, toolbar button icon.

Command

1. A command is an AutoCAD function used to perform some task. This may be to draw a line, rotate a shape or modify an item of text. Commands can be activated by:
    a) selection from a menu
    b) selecting the appropriate icon from a toolbar button

c) entering the command from the keyboard at the command line
d) entering the command abbreviation
e) using the Alt key as previously described.

2   Only the first three options will be used in the book.

Objects

Everything drawn in AutoCAD 2005 is termed an **object** or **entity**, e.g. lines, circles, text, dimensions, hatching, etc. are all objects. The user 'picks' the appropriate entity/object with a mouse left-click when prompted.

Default setting

All AutoCAD releases have certain values and settings which have been 'preset' and are essential for certain operations. These default settings are displayed with <> brackets, but can be altered by the user as and when required. For example:

1   From the menu bar select **Draw-Polygon** and:
    prompt   `_polygon Enter number of sides<4>`
    respond **press the ESC key** to cancel the command.

2   *Notes*
    a) <4> is the default value for the number of sides
    b) _polygon is the active command.

3   At the command line enter **LTSCALE <R>** and:
    prompt   `Enter new linetype scale factor<1.0000>` (or other value)
    enter   **0.5 <R>**.

4   *Notes*
    a) <1.0000> is the LTSCALE default value on my system
    b) we have altered the LTSCALE value to 0.5.

The escape (Esc) key

This is used to cancel any command at any time. It is very useful, especially when the user is 'lost in a command'. Pressing the Esc key will cancel any command and return the command prompt line.

Icon

An icon is a menu item in the form of a picture contained on a button within a named toolbar. Icons will be used extensively throughout the book.

Cascade menu

A cascade menu is obtained when an item in a pull-down menu with ▶ after its name is selected.

1   From the menu bar select the sequence **Draw-Circle** and the cascade effect as shown in Figure 2.2 will be displayed.

2   Cancel the cascade effect by:
    a) moving the pick arrow to any part of the screen and left-clicking
    b) pressing the Esc key – cancels the 'last' cascade menu, so two escapes are required.

Figure 2.2 Pull-down and cascade menu effect.

Dialogue boxes

A dialogue box is always displayed when an item with '. . .' after its name is selected.

1 Select the menu bar sequence **Format-Units** and:
prompt   Drawing Units dialogue box as Figure 2.3
respond   **select Cancel** to 'remove' the dialogue box from the screen.

2 Dialogue boxes allow the user to:
   a) alter parameter values
   b) toggle an aid ON/OFF
   c) select an option from a list.

3 Most dialogue boxes display the options OK, Cancel and Help which are used as follows:
   OK   accept the values in the current dialogue box
   Cancel   cancel the dialogue box without any alterations
   Help   gives further information in Windows format. The Windows effect can be cancelled with **File-Exit** or using the Windows Close button from the title bar (the right-most button).

Toolbars

1 Toolbars are aids for the user. They allow the Release 2005 commands to be displayed on the screen in button icon form. The required command is activated by picking (left-click) the appropriate button. The icon command is displayed as a **tooltip** in yellow (the default colour) by moving the pick arrow onto an icon and leaving it for a second.

2 There are 29 toolbars available for selection, and those normally displayed by default when AutoCAD 2005 is started are the Standard, Layers, Properties, Modify and Draw toolbars.

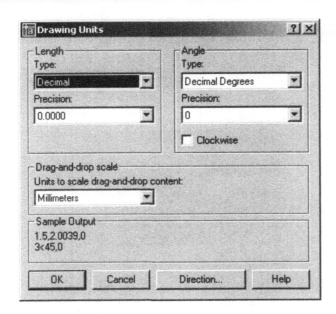

Figure 2.3 The Drawing Units dialogue box.

3 Toolbars can be:
   a) displayed and positioned anywhere in the drawing area
   b) customised to the user preference.

4 To activate a toolbar, select from the menu bar **View-Toolbars** and:
   prompt   Customize dialogue box
   with   five tabs – Commands, Toolbars, Properties, Keyboard and Tool Palettes
   respond   a) make the Toolbars tab active with a left pick to display a list of available toolbars and note that the toolbar names with a tick are active
               b) pick the box at Object Snap to display a tick
               c) pick Close.

5 The Object Snap toolbar will be displayed in the drawing area.

6 When toolbars are positioned in the drawing area as the newly activated Object Snap toolbar and the Dimension toolbar in Figure 2.1, they are called **FLOATING** toolbars.

7 Toolbars can be:
   a) Moved to a suitable position on the screen by the user. This is achieved by moving the pick arrow into the blue title area of the toolbar and holding down the mouse left button. Move the toolbar to the required position on the screen and release the left button.
   b) Altered in shape by 'dragging' the toolbar edges sideways or downwards.
   c) Cancelled at any time by picking the 'Cancel box' at the right of the toolbar title bar.

8 It is the user's preference as to what toolbars are displayed at any one time. In general I always display the Draw, Modify and Object Snap toolbars and activate others as and when required.

9 Toolbars can be **DOCKED** at the edges of the drawing area by moving them to the required screen edge. The toolbar will be automatically docked when the edge is reached. Figure 2.1 displays a floating and several docked toolbars:
   a) Docked   Standard, Layers and Properties at the top of the screen; Draw and Modify at either side of the screen. These toolbars 'were set' by default
   b) Floating   Dimension. This toolbar was 'activated' by me.

10 *Note*: Toolbars **do not have to be used** – they are an aid to the user. All commands are available from the menu bar, but it is recommended that toolbars are used, as they greatly increase draughting productivity. When used, it is the user's preference as to whether they are floating or docked.

Tool palette

1 A tool palette is an efficient method of organising, sharing, and placing blocks and hatches.

2 Tool palettes can be customised by the user.

3 By selecting **Auto-Hide** from the title bar, the tool palette can be minimised/maximised.

4 Tool palettes can be
   a) cancelled by selecting the Close (topmost) button from the title bar
   b) activated from the menu bar with **Tools-Tool Palettes Window**
   c) positioned by the user.

5 Like toolbars, the tool palettes do not need to be used. It is user preference.

6 We will discuss palettes in detail in a later chapter.

Fly-out menu

When a button icon is selected an AutoCAD command is activated. If the icon has a ◢ at the lower right corner of the icon box, and the left button of the mouse is held down, a **FLY-OUT** menu is obtained, allowing the user access to other icons.

1 Move the cursor pick arrow onto the Insert Block icon of the Draw toolbar

2 Hold down the left button and a fly-out menu is displayed allowing the user access to another four icons

3 Move the cursor to a clear area of the graphics screen and release the left button.

Wizard

Wizard allows the user access to various parameters necessary to start a drawing session, e.g. units, paper size, etc. There are several Wizard options available to the user and we will investigate how to use a Wizard in later chapters.

Template

A template allows the user access to different drawing standards with different sized paper, each template having a border and title box. AutoCAD 2005 supports the several drawing standards including ANSI, DIN, Gb, ISO, JIS and Metric. The use of templates will be investigated later in the book.

Toggle

This is the term used when a drawing aid is turned ON/OFF and usually refers to:

1 pressing a key

2 activating a parameter in a dialogue box, i.e. a tick/cross signifying ON, no tick/cross signifying OFF.

Function keys

Several of the keyboard function keys can be used as aids while drawing, these keys being:

F1	accesses the AutoCAD 2005 Help menu
F2	flips between the graphics screen and the AutoCAD Text window
F3	toggles the object snap on/off
F4	toggles the tablet on/off (if attached)
F5	toggles the isoplane top/right/left – for isometric drawings
F6	co-ordinates on/off toggle
F7	grid on/off toggle
F8	ortho on/off toggle
F9	snap on/off toggle
F10	polar tracking on/off toggle
F11	toggles object snap tracking off
F12	not used.

Help menu

AutoCAD 2005 has an 'on-line' help menu which can be activated at any time by selecting from the menu bar **Help-Help** or pressing the F1 function key. The Help dialogue box will be displayed as two distinct sections:
1. Left   with five tab selections – Contents, Index, Search, Favourites, Ask Me
2. Right   details about the topic.

File types

When a drawing has been completed it should be saved for future recall and all drawings are called *files*. AutoCAD 2005 supports different file formats, including:

.dwg	AutoCAD drawing
.dws	AutoCAD Drawing Standard
.dwt	AutoCAD Template Drawing template file
.dxf	AutoCAD Data Exchange Format.

Saved drawing names

Drawing names should be as simple as possible. While operating systems support file names which contain spaces and full stops(.), I would not recommend this practice. The following are typical drawing file names which I would recommend to be used:

EX1; EXER-1; EXERC_1; MYEX-1; DRG1, etc.

When drawings have to be saved during the exercises in the book, I will give the actual names to be used.

Customising the graphics screen

The graphics screen can be adapted or 'customised' to user requirements, e.g. the screen colour can be set, scroll bars can be turned off if required, a screen menu can be displayed, etc. There are several 'settings' which we will now investigate, but the user should decide for themselves whether they want to customise their graphics screen to my settings. This is **now your personal decision**.

From the menu bar select **Tools-Options** and:
prompt   Options dialogue box with nine tab selections
respond   **by picking the named tab and alter as described**.

A Display tab
   a) *Window elements*
      1. Display scroll bars in drawing area: active, i.e. tick
      2. Display screen menu: not active, i.e. blank
      3. Colors: pick and set Model tab background to white or black then Apply & Close (note 1).
   b) *Layout elements*
      1. Display Layout and Model tabs: active
      2. Display printable area: active
      3. Display paper background and paper shadow: both active
      4. Show Page Setup Manager for new layouts: active
      5. Create viewport in new layouts: active.
   c) *Cross-hair size*
      1. The default is 5
      2. Set as required (note 2).
   d) *Display resolution*: leave as given.
   e) *Display performance*: leave as given.
   f) *Reference Edit fading intensity*: leave as given.

   Notes
   1. Allows the user to set a background screen colour to their preference
   2. Sets on-screen cursor size. 100 gives a full-screen cursor.

B Open and Save tab
   a) *File Save*
      1. Save as: AutoCAD 2004 Drawing (*.dwg).
   b) *File Safety Precautions*
      1. Automatic save: active
      2. Minutes between saves: set as required, e.g. 10
      3. Create backup copy with each save: active.
   c) *File Open*
      1. Number of recently used files to list: set between 0 and 9
      2. Display full path in title: active.
   d) Leave rest as given.

C System tab
   a) *General Options*
      1. Display OLE properties dialogue: active
      2. Start up: show Startup dialogue box.

D Drafting tab
   a) *AutoSnap Settings*
      1. Marker: active
      2. Magnet: active
      3. Display AutoSnap tooltip: active
      4. Display AutoSnap aperture box: not active
      5. AutoSnap marker color: accept default or set as required.
   b) *Autosnap Marker Size*: accept default.
   c) *Object Snap Options*: ignore hatch objects active.
   d) *AutoTrack Settings*
      1. Display polar tracking vector: active
      2. Display full-screen tracking vector: active
      3. Display AutoTrack tooltip: active.

    e) *Alignment Point Acquisition*: automatic active
    f) *Aperture Size*: accept default.

E Selection tab
    a) *Pickbox size*: accept default or set as required.
    b) *Selection modes*
        1. Noun/verb selection: active
        2. Implied windowing: active
        3. Object grouping: active
        4. Rest: leave as given.
    c) *Grip size*: accept default.
    d) *Grips*
        1. Accept three default colours
        2. Enable grips: active
        3. Enable grips with blocks: not active
        4. Enable grip tips: active.

F User Preferences tab
    a) *Windows Standard Behaviour*
        1. Windows standard accelerator keys: active
        2. Shortcut menus in drawing area: active
        3. Pick (left-click) on right-click Customisation and:
            a) Default mode: Shortcut Menu active
            b) Edit mode: Shortcut Menu active
            c) Command mode: Shortcut Menu (command options) active
            d) pick Apply & Close.
    b) *Drag-and-Drop scale*
        1. Source content units: Millimeters
        2. Target Drawing Units: Millimeters.
    c) *Priority for Co-ordinate Data Entry*: Keyboard entry except scripts active.
    d) *Associative Dimensioning*: Make new dimensions associative active.
    e) *Fields*: Display background of fields active.

G Other tabs: leave at present.

H Now pick Apply then OK from the Options dialogue box to accept the changes and to return to the drawing screen.

These modifications will stay current until altered by the user.

Other modifications

There are two other alterations which we will be discussing before leaving this chapter and starting to draw some AutoCAD objects. These modifications can be considered as 'customising the system' to user requirements.

1 The co-ordinate icon at the lower left of the drawing area can be customised to be displayed in 2D or 3D, so from the menu bar select **View-Display-UCS Icon** and:
    a) On and Origin both active (tick).
    b) pick Properties and:
        prompt   UCS Icon dialogue box
        respond 1. UCS icon style: 2D
                    2. UCS icon size: 12
                    3. UCS icon color: black or pick to suit

     4. Layout tab icon color: blue or pick to suit
     5. pick OK
   and   the traditional AutoCAD 2D icon with X, Y and Z axes will be displayed.
2. The small box attached to the cursor cross-hairs is an aid to the user, but can cause confusion to new AutoCAD users. We will use these aids in later chapters, but at present 'will turn them off'. This can be achieved with the following keyboard entries:

Enter	*Prompt*	*Enter*
GRIPS <R>	Enter new value for GRIPS	0 <R>
PICKFIRST <R>	Enter new value for PICKFIRST	0 <R>

Finally

We have spent some time discussing the graphics screen and terminology in this rather long chapter and are now ready to start drawing, but before this, select from the menu bar **File-Exit** and pick **No** to Save changes if the AutoCAD message dialogue box is displayed – more on this later. We have thus customised our drawing screen and quit AutoCAD.

Chapter **3**

Drawing, erasing and the selection set

In this chapter we will investigate how lines and circles can be drawn and then erased. When several line and circle objects have been created by different methods, we will investigate the selection set which is a very powerful aid when modifying a drawing.

Starting a new drawing with Wizard

1. Start AutoCAD and:
   prompt   Startup dialogue box
   with   four selections:
                 Open a Drawing; Start from Scratch; Use a Template; Use a Wizard
   respond   *a*) pick Use a Wizard icon (right-most icon)
                 b) pick QuickSetup – Figure 3.1
                 c) pick OK
   prompt   QuickSetup (Units) dialogue box
   respond   *a*) Select Decimal Units – Figure 3.2
                 b) pick Next>
   prompt   QuickSetup (Area) dialogue box
   respond   *a*) enter Width: 420
                 b) enter Length: 297 – Figure 3.3
                 c) pick Finish.

2. The AutoCAD 2005 drawing screen will be displayed and should display the Standard and Properties toolbars at the top of the screen, and the Modify and Draw toolbars.

3. *Notes*
   a) The toolbars which are displayed will depend on how the last user 'left the system'. If you do not have the Draw and Modify toolbars displayed then:
       1. select from the menu bar **View-Toolbars**
       2. activate the Draw and Modify toolbars with a tick
       3. close the Toolbars dialogue box
       4. position the toolbars to suit, i.e. floating or docked.
   b) After selecting options from the Startup dialogue box, the New Features Workshop screen may be displayed. The user should read the options, decide on which should remain active and then close the dialogue box.

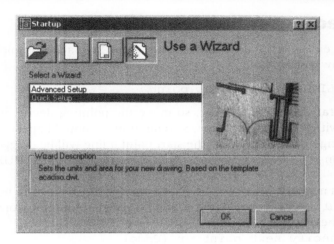

Figure 3.1 The Startup (Use a Wizard) dialogue box.

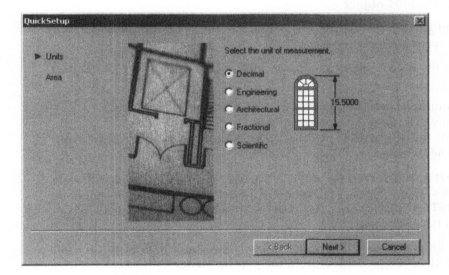

Figure 3.2 The QuickSetup (Units) dialogue box.

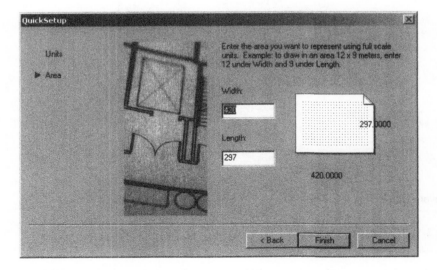

Figure 3.3 The QuickSetup (Area) dialogue box.

Drawing line objects

1. A line requires a start and end point. To draw a line with AutoCAD, activate (pick) the LINE icon from the Draw toolbar and:

   the command prompt displays: `LINE Specify first point`.

2. You now have to pick a **start** point for the line, so move the pointing device and pick (left-click) any point within the drawing area. Several 'things' should happen:
   a) as you move the pointing device away from the start point a line will be dragged from this point to the on-screen cursor position. This drag effect is termed **RUBBERBAND**
   b) as the pointing device is moved, a small coloured box **may** be displayed with text similar to: *Polar: 80.00<0*. If it does, do not panic and if it does not do not worry. We will discuss this in the next chapter
   c) the prompt becomes: `Specify next point or [Undo]`.

3. Move the pointing device to any other point on the screen and left-click. **This is your first AutoCAD 2005 object**.

4. The line command is still active with the rubberband effect and the prompt line is still asking you to specify the next point.

5. Continue moving the mouse about the screen and pick points to give a series of 'joined lines'.

6. Finish the LINE command with a right-click on the mouse and:
   a) a pop-up menu will be displayed as Figure 3.4(a)
   b) pick Enter from this dialogue to end the LINE command and the 'blank' command line will be returned.

7. From the menu bar select **Draw-Line** and the 'Specify first point' prompt will again be displayed at the command prompt. Draw some more lines then end the command by pressing the RETURN/ENTER key. The LINE command will be 'stopped', but no pop-up menu will have appeared.

8. At the command line enter **LINE <R>** and draw a few more lines. End the command with a right-click and pick **Enter** from the pop-up menu.

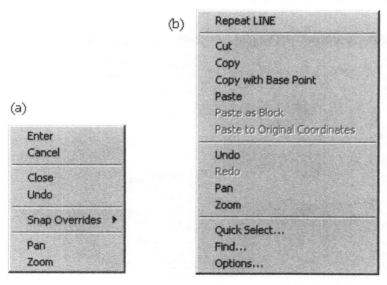

Figure 3.4 The pop-up menus to (a) end the LINE command and (b) repeat the LINE command.

9 Right-click on the mouse to display a pop-up menu as Figure 3.4(a) and pick **Repeat LINE**. Draw some more lines then end with a right-click and pick Enter.

10 *Notes*
   a) The different ways of activating the LINE command:
      1. with the LINE icon from the Draw toolbar
      2. from the menu bar with **Draw-Line**
      3. by entering LINE at the command line
      4. with a right-click of the mouse (if LINE was the last command).
   b) The two options to 'exit/stop' a command:
      1. with a right-click of the mouse – pop-up menu displayed
      2. by pressing the RETURN/ENTER key – no pop-up menu.
   c) Cancelling a command with a mouse right-click, **may** display the pop-up similar to Figure 3.4(a).
   d) When a command has been finished, a mouse right-click will display a pop-up menu as Figure 3.4(b) with the **LAST COMMAND** available for selection, e.g. Repeat LINE.
   e) The pop-up menu displayed with the mouse right-click is called a **shortcut menu**. It is a useful aid to the CAD user.

Drawing circle objects

1 All circles have a centre point and a radius. To draw a circle with AutoCAD, activate the CIRCLE icon from the Draw toolbar and:
   prompt   `_circle Specify center point for circle or [3P/2P/Ttr (tan tan radius)]`
   respond **pick any point on the screen as the circle centre**
   prompt   `Specify radius of circle or [Diameter]`
   respond **drag out the circle and pick any point for radius**.

2 From the menu bar select **Draw-Circle-Center, Radius** and pick a centre point and drag out a radius.

3 At the command prompt enter **CIRCLE <R>** and create another circle anywhere on the screen.

4 Using the icons, menu bar or keyboard entry, draw some more lines and circles until you are satisfied that you can activate and end the two commands.

5 Figure 3.5(a) displays some AutoCAD line and circle objects.

Erasing objects

Now that we have drawn some lines and circles, we will investigate how they can be erased, which seems rather silly? The erase command will be used to demonstrate different options available to us when it is required to modify a drawing. The actual erase command can be activated by one of three methods:
a) picking the ERASE icon from the Modify toolbar
b) with the menu bar sequence **Modify-Erase**
c) entering **ERASE <R>** at the command line.

1 Before continuing with the exercise, select from the menu bar the sequence **Tools-Options** and:
   prompt   `Options dialogue box`

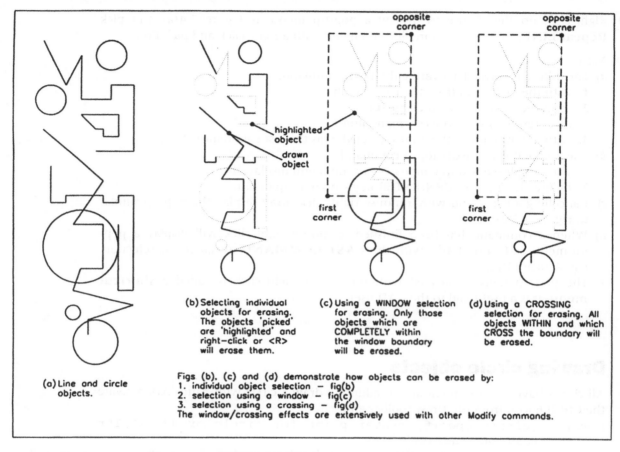

Figure 3.5 Line and circle objects (drawing and erasing).

    respond   pick the Selection tab and ensure:
- a) Noun/verb selection: not active
- b) Use shift to add to selection: not active
- c) Press and drag: not active
- d) Implied windowing: active
- e) Object grouping: active
- f) Pickbox size: set to suit (about 1/4 distance from left)
- g) pick Apply then OK when complete.

2 Now continue with the erase exercise.

3 Ensure you still have several lines and circles on the screen. Figure 3.5(a) is meant as a guide only.

4 From the menu bar select **Modify-Erase** and:
    prompt   `Select objects`
    and   cursor cross-hairs replaced by a 'pickbox' which moves as you move the mouse
    respond   **position the pickbox over any line and left-click**
    and   the following will happen:
- a) the selected line will 'change appearance', i.e. be 'highlighted'
- b) the prompt displays `Select objects: 1 found`
- c) then: Select objects

    respond   Continue picking lines and circles to be erased (about six) and each selected object will be highlighted.

5 When you have selected enough objects, right-click the mouse.
6 The selected objects will be erased, and the Command prompt will be returned blank.
7 Figure 3.5(b) demonstrates the individual object selection erase effect.

Oops

1 Suppose that you had erased the wrong objects.
2 Before you **DO ANYTHING ELSE**, enter **OOPS <R>** at the command line.
3 The erased objects will be returned to the screen.
4 Consider this in comparison to a traditional draughtsperson who has rubbed out several lines/circles – they would have to redraw each one.
5 OOPS must be used **IMMEDIATELY** after the last erase command and **must be entered from the keyboard**.

Erasing with a window/crossing effect

Individual selection of objects is satisfactory if only a few objects have to be modified (remember that we have only used the erase command so far). When a large number of objects require to be modified, the individual selection method is very tedious, and AutoCAD overcomes this by allowing the user to position a 'window' over an area of the screen which will select several objects 'at the one pick'.

To demonstrate the window effect, ensure you have several objects (about 20) on the screen and refer to Figure 3.5(c).

1 Select the ERASE icon from the Modify toolbar and:
   prompt   Select objects
   enter   **W <R>** (at the command line) – the window option
   prompt   Specify first corner
   respond **position the cursor at a suitable point and left-click**
   prompt   Specify opposite corner
   respond **move the cursor to drag out a window (rectangle) and left-click**
   prompt   ??? found and certain objects highlighted
   then   Select objects – i.e. any more objects to be erased?
   respond **right-click or <R>**.

2 The highlighted objects will be erased.
3 At the command line enter **OOPS <R>** to restore the erased objects.
4 From the menu bar select **Modify-Erase** and:
   prompt   Select objects
   enter   **C <R>** (at the command line) – the crossing option
   prompt   Specify first corner
   respond **pick any point on the screen**
   prompt   Specify opposite corner
   respond **drag out a window and pick the other corner**
   prompt   ??? found and highlighted objects
   respond **right-click**.

5 The objects highlighted will be erased – Figure 3.5(d).

Notes on window/crossing

1. The window/crossing concept of selecting a large number of objects will be used extensively with the modify commands, e.g. erase, copy, move, scale, rotate, etc. The objects which are selected when **W** or **C** is entered at the command line are as follows:
   (W) for window   all objects *completely within* the window boundary are selected
   (C) for crossing   all objects *completely within and also which cross* the window boundary are selected.

2. The window/crossing option **IS ENTERED FROM THE KEYBOARD**, i.e. W or C.

3. Figure 3.5 demonstrates the single object selection method as well as the window and crossing methods for erasing objects.

4. *Automatic window/crossing*
   In the example used to demonstrate the window and crossing effect, we entered a W or a C at the command line. AutoCAD allows the user to activate this window/crossing effect automatically by picking the two points of the 'window' in a specific direction. Figure 3.6 demonstrates this with:
   a) the window effect by picking the first point anywhere and the second point either upwards or downwards to the right
   b) the crossing effect by picking the first point anywhere and the second point either upwards or downwards to the left.

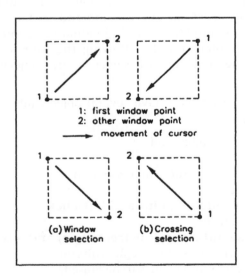

Figure 3.6   Automatic window/crossing selection.

The selection set

Window and crossing are only two options contained within the selection set, the most common selection options being:

Crossing, Crossing Polygon, Fence, Last, Previous, Window and Window Polygon.

During the various exercises in the book, we will use all of these options but will only consider three at present, so:

1. Erase all objects from the screen – window option.

2. Refer to Figure 3.7(a) and draw some new lines and circles – the actual layout is not important, but try and draw some objects 'inside' others.

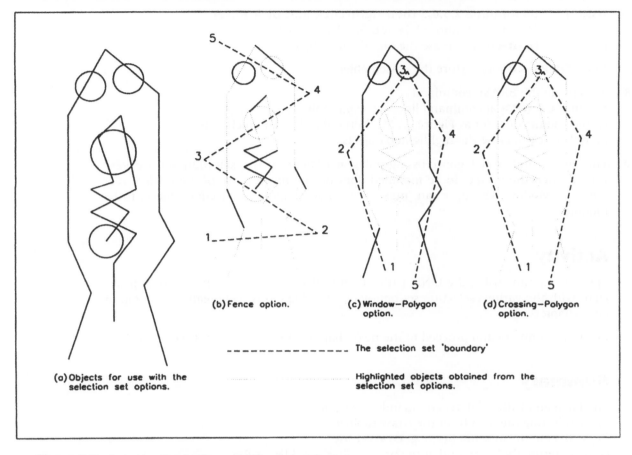

Figure 3.7 Investigating further selection set options.

3   Refer to Figure 3.7(b), select the ERASE icon from the Modify toolbar and:
    prompt   Select objects
    enter   **F <R>** – the fence option
    prompt   First fence point
    respond   **pick a point (pt 1)**
    prompt   Specify endpoint of line or [Undo]
    respond   **pick a suitable point (pt 2)**
    prompt   Specify endpoint of line or [Undo]
    respond   **pick point 3, then points 4 and 5 then right-click**
    prompt   Shortcut menu
    respond   **pick Enter**
    prompt   ??? found and certain objects highlighted
    respond   **right-click or <R>**.

4   The highlighted objects will be erased.

5   Enter **OOPS <R>** to restore these erased objects.

6   Menu bar with **Modify-Erase** and referring to Figure 3.7(c):
    prompt   Select objects
    enter   **WP <R>** – the window-polygon option
    prompt   First polygon point
    respond   **pick a point (pt 1)**
    prompt   Specify endpoint of line or [Undo]

respond	**pick points 2,3,4,5 then right-click and pick Enter**
prompt	`??? found and objects highlighted`
respond	**right-click** to erase the highlighted objects.

7. Enter **OOPS <R>** to restore the erased objects.

8. *a)* activate the ERASE command
   b) enter **CP <R>** at command line – crossing-polygon option
   c) pick points in order as Figure 3.7(d) then right-click and pick Enter
   d) right-click to erase the highlighted objects.

9. The fence/window-polygon/crossing-polygon options of the selection set are very useful when the 'shape' to be modified does not permit the use of the normal rectangular window. The user can 'make their own shape' for selecting objects to be modified.

Activity

Spend some time using the LINE, CIRCLE and ERASE commands and become proficient with the various selection set options for erasing – this will greatly assist you in later chapters.

Read the summary and proceed to the next chapter. Do not exit AutoCAD if possible.

Summary

1. The LINE and CIRCLE draw commands can be activated:
   a) by selecting the icon from the Draw toolbar
   b) with a menu bar sequence, e.g. **Draw-Line**
   c) by entering the command at the prompt line, e.g. LINE <R>.

2. The ERASE command can be activated:
   a) with the ERASE icon from the Modify toolbar
   b) from the menu bar with **Modify-Erase**
   c) by entering ERASE <R> at the command line.

3. All modify commands (e.g. ERASE) allow access to the selection set.

4. The selection set has several options including Window, Crossing, Fence, Window-Polygon and Crossing-Polygon.

5. The appropriate selection set option can be activated from the command line by entering the letters: W, C, F, WP, CP.

6. The term WINDOW refers to all objects completely contained within the window boundary.

7. CROSSING includes all objects which cross the window boundary and are also completely within the window.

8. It is possible to 'automatically' create the window effect without entering letters.

9. OOPS is a very useful command which 'restores' objects erased with the last erase command.

Chapter 4
The 2D drawing aids

Now that we know how to draw and erase lines and circles, we will investigate the aids which are available to the user. AutoCAD 2005 has several drawing aids which include:

Grid
: allows the user to place a series of imaginary dots over the drawing area. The grid spacing can be altered by the user at any time while the drawing is being constructed. As the grid is imaginary, it does **not** appear on the final plot.

Snap
: allows the user to set the on-screen cursor to a predetermined point on the screen, this usually being one of the grid points. The snap spacing can also be altered at any time by the user. When the snap and grid are set to the same value, the term **grid lock** is often used.

Ortho
: an aid which allows only horizontal and vertical movement of the on-screen cursor.

Polar tracking
: allows objects to be drawn at specific angles along an alignment path. The user can alter the 'polar angle' at any time.

Object snap
: the user can set a snap relative to a predetermined geometry. This drawing aid will be covered in detail in a later chapter.

Getting ready

1 Still have some line and circle objects from Chapter 3 on the screen?

2 Menu bar with **File-Close** and:
   prompt   AutoCAD Message dialogue box with Save changes options
   respond pick No – more on this in the next chapter.

3 Begin a new drawing with the menu bar sequence **File-New** and:
   prompt   Create New Drawing dialogue box
   respond a) pick Use a Wizard
              b) pick QuickSetup
              c) pick OK
   prompt   QuickSetup (Units) dialogue box
   respond pick Decimal then Next>
   prompt   QuickSetup (Area) dialogue box
   respond a) set Width: 420 and Length: 297
              b) pick Finish.

4 A blank drawing screen will be displayed.

5 Menu bar with **Draw-Rectangle** and:
   prompt Specify first corner point and enter: **0,0 <R>**
   prompt Specify other corner point and enter: **420,297 <R>**.

6 Menu bar with **View-Zoom-All** and the rectangle shape will 'fill the screen'. This rectangle will be 'our drawing paper'.

Grid and snap setting

The grid and snap spacing can be set by different methods and we will investigate setting these aids from the command line and from a dialogue box.

1 At the command line enter **GRID <R>** and:
   prompt Specify grid spacing (X) or [On/OFF/Snap/Aspect]<10.000>
   enter   **20 <R>**.

2 At the command line enter **SNAP <R>** and:
   prompt Specify snap spacing or...
   enter   **20 <R>**.

3 Refer to Figure 4.1 and use the LINE command to draw the letter H using the grid and snap settings of 20.

4 Using keyboard entry, change the grid and snap spacing to 15.

5 Use the LINE command and draw the letter E.

6 From the menu bar select **Tools-Drafting Settings** and:
   prompt   Drafting Settings dialogue box
   respond activate the Snap and Grid tab
   and   a) Snap on with X and Y spacing 15
              b) Grid on with X and Y spacing 15
              c) These values are from our previous step 4 entries

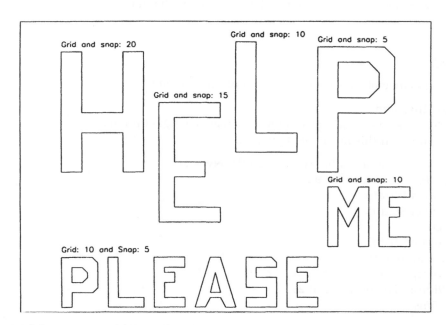

Figure 4.1 Using the GRID and SNAP drafting aids.

respond 1. alter the Snap X spacing to 10 by:
   a) click to right of last digit
   b) backspace until all digits are removed
   c) enter 10
   d) left-click at Snap Y spacing – alters to 10
2. alter the Grid X spacing by:
   a) position pick arrow to left of first digit
   b) hold down left button and drag over all digits – they will be highlighted
   c) enter 10
   d) left-click at Grid Y spacing – alters to 10 (Figure 4.2)
3. pick OK.

7 Use the LINE command to draw the letter L.

8 Use the Drafting Settings dialogue box to set both the grid and snap spacing to 5 and draw the letter P.

9 *Note*: The Drafting Settings dialogue box allows the user access to the following drawing aids:
   a) the grid and snap settings
   b) polar tracking
   c) object snap settings.

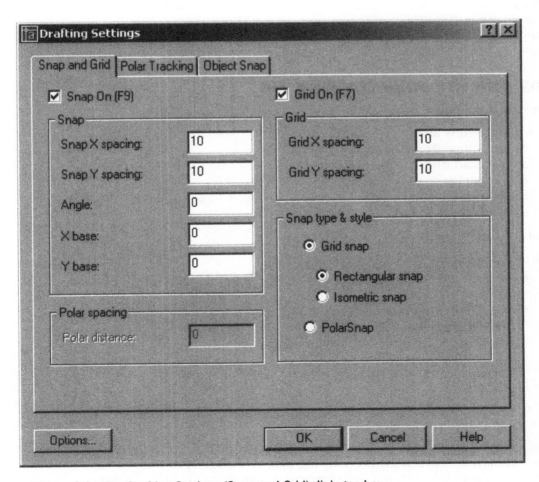

Figure 4.2 The Drafting Settings (Snap and Grid) dialogue box.

Toggling the grid/snap/ortho

1   The drawing aids can be toggled ON/OFF with:
    a) the function keys, i.e. F7 grid; F8 ortho; F9 snap
    b) the Drafting Settings dialogue where a tick in the box signifies that the aid is on, and a blank box means the aid is off
    c) the status bar with a right-click on Snap, Grid, Ortho and selecting On/Off as appropriate from the pop-up menu.

2   My preference is to set the grid and snap spacing values from the dialogue box or command line, then use the function keys to toggle the aids on/off as required.

3   Take care if the ortho drawing aid is on. Ortho only allows horizontal and vertical movement and lines may not appear as expected. I tend to work with ortho off.

4   The Drafting Settings dialogue box can be activated:
    a) from the menu bar with **Tools-Drafting Settings**
    b) with a right-click on Snap or Grid from the status bar and then picking Settings.

5   *Task*
    Refer to Figure 4.1 and:
    a) with the grid and snap set to 10, draw the letters M and E
    b) with the grid set to 10 and the snap set to 5, complete PLEASE to your own design specification
    c) when complete, do not erase any of the objects.

Drawing with the polar tracking aid

1   The screen should still display HELP ME PLEASE?

2   Menu bar with **File-New** and:
    prompt   Create New Drawing dialogue box
    respond   a) pick Start from Scratch icon (second left)
                b) pick Metric
                c) pick OK
                d) blank drawing screen returned.

3   Set the grid and snap on with settings of 20.

4   Right-click on POLAR in the status bar, pick Settings and:
    prompt   Drafting Settings dialogue box with Polar Tracking tab active
    respond   a) ensure Polar Tracking On (F10)
                b) scroll at Increment angle and pick 30
                c) ensure Track using all polar angle settings is active
                d) ensure Absolute is active
                e) dialogue box as Figure 4.3
                f) pick OK.

5   Activate the LINE command and pick a suitable grid/snap start point towards the top of the screen.

6   Move the cursor horizontally to the right and observe the polar tracking information displayed. Move until the tracking data is **Polar: 100.0000 < 0°** as Figure 4.4(a) then left-click. This is a line **segment** drawn using the polar tracking drawing aid.

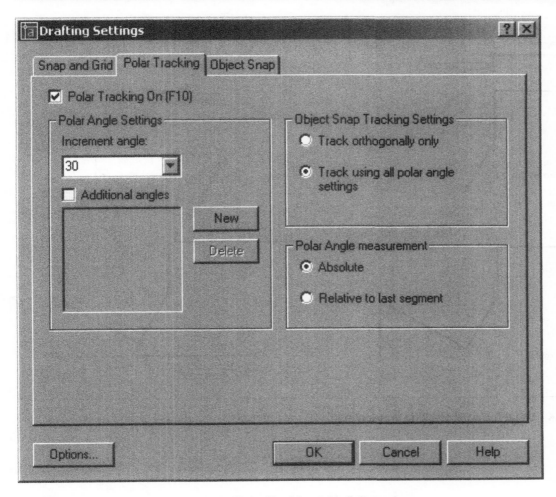

Figure 4.3 The Drafting Settings (Polar Tracking tab) dialogue box.

7   Now move the cursor vertically downwards until 40.0000 < 270° is displayed as Figure 4.4(b) then left-click.

8   Move the cursor downwards and to the right until a 300 degree angle is displayed as Figure 4.4(c) and enter 50 from the keyboard. The entered value of 50 is the length of the line segment.

9   Move upwards to right until a 30 degree angle is displayed in the polar tracking tip box as Figure 4.4(d) and enter 80 from the keyboard.

10 Complete the polar tracking line segments with:
    a) an angle of 270 and a keyboard entry of 50 as Figure 4.4(e). You will probably have to toggle (F9) the snap off for this part of the exercise
    b) a line segment length of 150 at an angle of 180 – Figure 4.4(f). Snap off again
    c) end the line command with right-click/enter.

11 Note that the polar tracking aid displays information of the format **100.0000 < 90°**. This is: **a line length and an angle**.

12 When this exercise is complete, proceed to the next chapter but try not to exit AutoCAD.

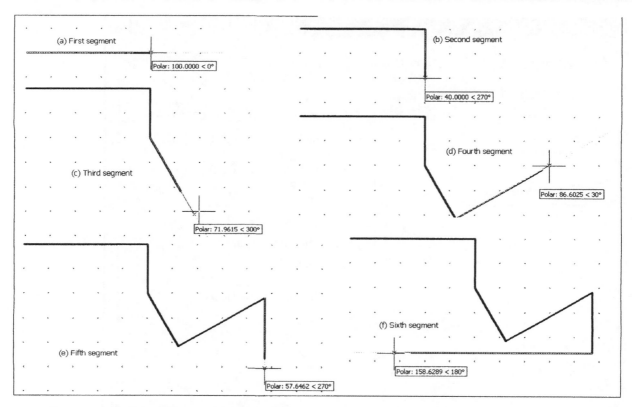

Figure 4.4 Using the polar tracking drawing aid.

Chapter 5

Saving and opening drawings

AutoCAD 2005 allows multiple drawings to be opened during a drawing session. It is thus essential that *all* users know how to save and open a drawing, and how to exit AutoCAD correctly.

In this and all the following chapters, all drawing work will be saved to the named folder **BEGIN**.

Saving a drawing and exiting AutoCAD

1 If the previous chapter work has been followed correctly, the user has two drawings opened:
   a) the line segments created using polar tracking
   b) the HELP ME PLEASE drawing created using the grid and snap drawing aids.

2 Menu bar with **File-Exit** and:
   prompt   AutoCAD message dialogue box – similar to Figure 5.1.

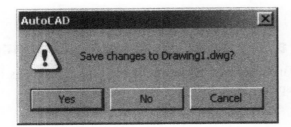

Figure 5.1 The AutoCAD message dialogue box.

3 This dialogue box is informing the user that since starting the current drawing session, changes have been made and that these drawing changes have not yet been saved. The user has to respond to one of the three options which are:
   Yes   picking this option will save a drawing with the name displayed, i.e. Drawing1.dwg or similar
   No   selecting this option means that the alterations made will not be saved
   Cancel   returns the user to the drawing screen.

4 At present, pick Cancel as we want to investigate how to save a drawing.

5   Select from the menu bar **File-Save As** and:
    prompt   Save Drawing As dialogue box
    respond   a) Scroll at Save in by picking the arrow at right
                b) pick (left-click) the C: drive to display folder names
                c) double left-click on the Begin folder
                d) File name: alter to DRG2
                e) Files of type: scroll and select AutoCAD 2004 Drawing (*.dwg)
                f) pick Save.

6   The screen drawing will be saved to the named folder, but will still be displayed on the screen.

7   Menu bar with **File-Close** and the line segments drawing will disappear from the screen and the HELP ME PLEASE drawing will be displayed.

8   Menu bar with **File-Save As** and using the Save Drawing As dialogue box:
    a) ensure the BEGIN folder is current
    b) alter File name to MYFIRST
    c) ensure File type is AutoCAD 2004 Drawing (*.dwg)
    d) pick Save.

9   Now menu bar with **File-Exit** to exit AutoCAD.

10 *Notes*
    a) When multiple drawings have been opened in AutoCAD 2005, the user is prompted to save changes to each drawing before AutoCAD can be exited.
    b) The Save Drawing As dialogue box displays other options, e.g. History, My Documents, Favorites, etc. These are typical Windows terms.
    c) The Save Drawing As dialogue box has icon selections for:
       1. back to previous folder
       2. up one level
       3. search the web
       4. create new folder.

Opening, modifying and saving an existing drawing

While AutoCAD is used to create drawings, it also allows existing drawings to be displayed and modified to user specifications. To demonstrate this:

1   Start AutoCAD and:
    prompt   Startup dialogue box
    respond   **pick the Open a Drawing tab**
    prompt   Startup Open a Drawing dialogue box
    respond   **pick Browse**
    prompt   Select File dialogue box
    respond   a) scroll at Look in
                b) pick (left-click) the C: drive
                c) double left-click the BEGIN folder
    and   all saved drawings displayed
    respond   a) pick MYFIRST
                b) preview displayed – Figure 5.2
                c) pick Open.

2   The HELP ME PLEASE drawing will be displayed.

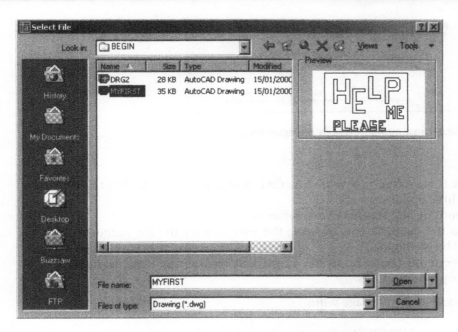

Figure 5.2 The Select File dialogue box with preview of MYFIRST drawing.

3   Set the grid to 10 and snap to 5 then:
   a) erase the ME letters
   b) draw lines around the letters H, E, L and P
   c) draw lines around the word PLEASE – Figure 5.3.

4   Menu bar with **File-Save As** and:
   prompt   Save Drawing As dialogue box
   with   MYFIRST.dwg as the File name
   respond   **pick Save**
   prompt   Save Drawing As message dialogue box – Figure 5.4
   with   C:\BEGIN\MYFIRST.dwg already exists (or a similar C: path name)
              Do you want to replace it?
   respond   Do nothing at present.

Figure 5.3 The modified MYFIRST drawing.

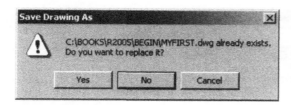

Figure 5.4 The Save Drawing As message dialogue box.

5 This dialogue box is very common with AutoCAD and it is important that the user understands the three options:
   Cancel   does nothing and returns the dialogue box
   No   returns the dialogue box allowing the user to alter the file name which should be highlighted
   Yes   will overwrite the existing file name and replace the original drawing with any modifications.

6 At this stage, respond to the message with:
   a) pick No
   b) alter the file name to MYFIRST1
   c) pick Save.

7 What have we achieved?
   a) we opened drawing MYFIRST from the C:\BEGIN folder
   b) we altered the drawing layout
   c) we saved the alterations as MYFIRST1
   d) the original MYFIRST drawing is still available and has not been modified.

8 Menu bar with **File-Open** and:
   prompt   Select File dialogue box
   respond *a*) pick DRG2 and note the preview
              b) pick Open.

9 The screen will display the line segments drawn with polar tracking.

10 We now have two opened drawings – DRG2 and MYFIRST1 (or is it MYFIRST?).

Closing files

We will use the two opened drawings to demonstrate how AutoCAD should be exited when several drawings have been opened in the one drawing session.

1 Erase the line segments with a window selection – easy?

2 Menu bar with **File-Close** and:
   prompt   AutoCAD Message dialogue box
   with   Save changes to C:\BEGIN\DRG2.dwg message
   respond **pick No** – can you reason out why we pick No?

3 The screen will display the MYFIRST1 (HELP PLEASE) modified drawing.

4 Menu bar with **File-Close** and a blank screen will be returned with a short menu bar display: File, View, Window, Help.

5 Select File from the menu bar to display a pull-down menu similar to Figure 5.5 with the last ?? opened drawings listed. The number of drawings listed will depend on the value you set in the User Preferences tab of the Options dialogue box in Chapter 2.

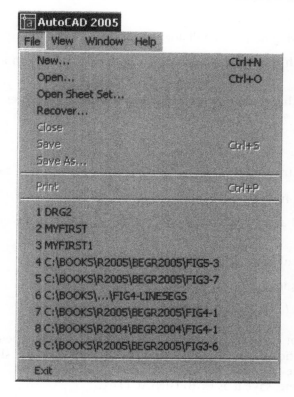

Figure 5.5 The File menu bar pull-down.

6 Respond to the pull-down menu by picking MYFIRST1 to display the modified HELP PLEASE drawing.

7 *a)* menu bar with File and pick MYFIRST
   b) menu bar with File and pick DRG2.

8 We have now opened three drawings with DRG2 displayed.

9 Menu bar with:
   a) **File-Close** to close DRG2 and display MYFIRST
   b) **File-Close** to close MYFIRST and display MYFIRST1
   c) **File-Close** to close MYFIRST1 and display a blank screen
   d) **File-Exit** to exit AutoCAD.

10 All drawings having been saved correctly and AutoCAD has been exited properly.

Save and Save As

The menu bar selection of File allows the user to pick either Save or Save As. New AutoCAD users should be aware of the difference between these two options.

Save

1 Will save the current drawing with the same name with which the drawing was opened.

2 No dialogue box will be displayed.

3 The original drawing will be automatically overwritten if alterations have been made to it.

Save As

1 Allows the user to enter a drawing name via a dialogue box.

2 If a drawing already exists with the entered name, a message is displayed in a dialogue box.

It is strongly recommended that the SAVE AS selection is used at all times.

Assignment

You are now in the position to try a drawing (well two drawings) for yourself, so:

1 Start AutoCAD and select Start from **Scratch-Metric-OK**.

2 Refer to activity drawing 1 (all activity drawings are grouped together at the end of the book).

3 Set a grid and snap spacing to suit, e.g. 10 and/or 5.

4 Menu bar with **Draw-Rectangle** and create a rectangle from 0,0 to 420,297.

5 Using only the LINE command (and perhaps ERASE if you make a mistake):
 a) draw the given shapes
 b) the size and position are not really important at this stage, the objective being to give you a chance to practice drawing using the drawing aids. All the shapes should 'fit into' the rectangle
 c) when the drawing is complete, save it as C:\BEGIN\ACT1-A.

6 Erase all the shapes within the rectangle then using only the CIRCLE command (no need for ERASE?):
 a) draw the given designs
 b) again all the shapes should 'fit into' the rectangle
 c) save the completed drawing as C:\BEGIN\ACT1-B.

Summary

1 The recommended procedure for saving a drawing is:
 a) menu bar with **File-Save As**
 b) select your named folder
 c) enter drawing name in File name box
 d) pick Save.

2 The procedure to open a drawing from within AutoCAD is:
 a) menu bar with **File-Open**
 b) select named folder
 c) pick drawing name from list
 d) preview obtained
 e) pick Open.

3 The procedure to open a drawing from start is:
 a) start AutoCAD
 b) pick Open a Drawing tab from the Startup dialogue box
 c) either   1. pick drawing name if displayed then OK
    or   2. pick Browse then:
                a) scroll at Look in
                b) select folder name
                c) pick the drawing name
                d) note preview then pick Open.

4   The recommended procedure to end an AutoCAD session is:
    a) complete the drawing
    b) save the drawing to your named folder
    c) menu bar with **File-Close** to close all opened drawings
    d) menu bar with **File-Exit** to quit AutoCAD.

5   The save and open command can be activated:
    a) by menu bar selection – recommended method
    b) by keyboard entry
    c) from icon selection in the Standard toolbar.

6   The menu bar method for saving a drawing is recommended as it allows the user the facility to enter the drawing name. This may be different from the opened drawing name.

7   The Save icon from the Standard toolbar and entering SAVE at the command line is a quick save option, and does not allow a different file name to be entered. These methods save the drawing with the opened name, and therefore overwrite the original drawing. This may not be what you want?

Chapter 6

Standard sheet 1

Traditionally one of the first things that a draughtsperson does when starting a new drawing is to get the correct size sheet of drawing paper. This sheet will probably have borders, a company logo and other details already printed on it. The drawing is then completed to 'fit into' the pre-printed layout material. A CAD drawing is no different from this, with the exception that the user does not 'get a sheet of paper'. Companies who use AutoCAD will want their drawings to conform to their standards in terms of the title box, text size, linetypes being used, the style of the dimensions, etc. Parameters which govern these factors can be set every time a drawing is started, but this is tedious and against CAD philosophy. It is desirable to have all standard requirements set automatically, and this is achieved by making a drawing called a standard sheet, prototype drawing or template drawing – you may have other names for it. Standard sheets can be 'customised' to suit all sizes of paper, e.g. A0, A1, etc. as well as any other size required by the customer. These standard sheets will contain the company's settings, and the individual draughtsman can add their own personal settings as required. It is this standard sheet which is the CAD operator's 'sheet of paper'.

We will create an A3 standard sheet, save it and use it for all future drawing work. At this stage, the standard sheet will not have many 'settings', but we will continue to refine it and add to it as we progress through the book.

The A3 standard sheet will be created using the Advanced Wizard so:

1 Start AutoCAD and:
   prompt   Startup dialogue box
   respond a) pick Use a Wizard
              b) pick Advanced Setup
              c) pick OK
   prompt   Advanced Setup dialogue box
   respond to each dialogue box with the following selections:
              a) Units: Decimal with 0.00 precision then Next>
              b) Angle: Decimal Degrees with 0.0 precision then Next>
              c) Angle Measure: East for 0 degrees then Next>
              d) Angle Direction: Counter-Clockwise then Next>
              e) Area: Width of 420, Length of 297 then Finish.

2 A blank drawing screen will be returned.

3 Menu bar with **Tools-Drafting Settings** and from the Drafting Settings dialogue box select:
   a) Snap and Grid tab with:
      1. Snap on with X and Y spacing set to 5
      2. Grid on with X and Y spacing set to 10
      3. Rectangular snap style active
   b) Polar Tracking tab with:
      1. Polar Tracking off

      c) Object Snap tab with:
         1. Object Snap off
         2. Object Snap Tracking off
         3. All snap modes off
      d) pick OK.

4 *Note*: The object snap drawing aid will be considered in a later chapter.

5 Menu bar with **Format-Drawing Limits** and:
   prompt   Specify lower left corner and enter: 0,0 <R>
   prompt   Specify upper right corner and enter: 420,297 <R>.

6 Menu bar with **View-Zoom-All** and the grid will 'fill the screen'.

7 At the command line enter:
   a) **GRIPS <R>** and set to 0
   b) **PICKFIRST <R>** and set to 0.

8 Display toolbars to suit. I would suggest:
   a) Standard and Properties docked at the top as default
   b) Draw and Modify docked to suit
   c) Other toolbars will be displayed as required.

9 Menu bar with **Draw-Rectangle** and:
   prompt   Specify first corner point
   enter   0,0 <R>
   prompt   Specify other corner point
   enter   420,297 <R>.

10 This rectangle will represent our 'drawing area'.

11 Menu bar with **File-Save As** and:
    a) scroll and pick your named folder, e.g. C:\BEGIN
    b) enter the file name as A3SHEET
    c) pick Save.

Notes

1 This completes our standard sheet at this stage. We have created an A3-size sheet of paper which has the units and screen layout set to our requirements. The status bar displays the co-ordinates to two decimal places with both the Snap and Grid on. The A3SHEET standard sheet has been saved to the C:\BEGIN folder as a drawing (.dwg) file.

2 Although we have activated several toolbars in our standard sheet, the user should be aware that these may not always be displayed when your standard sheet drawing is opened. AutoCAD displays the screen toolbars which were active when the system was 'shut down'. If other CAD operators have used 'your machine', then the toolbar display may not be as you left it. If you are the only user on the machine, then there should not be a problem. Anyway you should know how to display toolbars.

3 We will discuss AutoCAD's template files in more detail in a later chapter.

4 Do not confuse my standard sheet idea with the CAD Standards in AutoCAD 2005. The standard sheet idea is a reference name only.

5 You can now exit AutoCAD or continue to the next chapter.

Chapter 7

Line creation and co-ordinate entry

The line and circle objects so far created were drawn at random on the screen without any attempt being made to specify position or size. To draw objects accurately, co-ordinate input is required and AutoCAD 2005 allows different 'types' of co-ordinate entry including:
a) absolute, i.e. from an origin point
b) relative (or incremental), i.e. from the last point referenced.

In this chapter we will use our A3SHEET standard sheet to create several squares by different entry methods. The completed drawing will then be saved for future work.

Getting started

1 If continuing from Chapter 6, proceed to step 4.

2 *a*) If AutoCAD is active, then close any existing drawing then menu bar with **File-Open** and:
   prompt   Select File dialogue box
   respond 1. scroll at Look in
              2. pick C: drive
              3. double left-click on BEGIN
              4. pick A3SHEET and note Preview
              5. pick Open.
   b) If AutoCAD is not active, then start AutoCAD and:
   prompt   Startup dialogue box
   respond 1. select Open a Drawing icon
              2. pick Browse (see note)
              3. Select File dialogue box displayed
              4. scroll at Look in and pick C: drive
              5. double left-click on BEGIN
              6. pick A3SHEET
              7. pick Open.
   Note: You may have A3SHEET listed at this stage. If so, open it.

3 The A3SHEET standard sheet will be displayed, i.e. a black border with the grid and snap spacing set as required (10 and 5).

4 Display the Draw and Modify toolbars and position to suit and decide if you want to use polar tracking. Ensure that the object snap modes are off – they should be.

5 Refer to Figure 7.1.

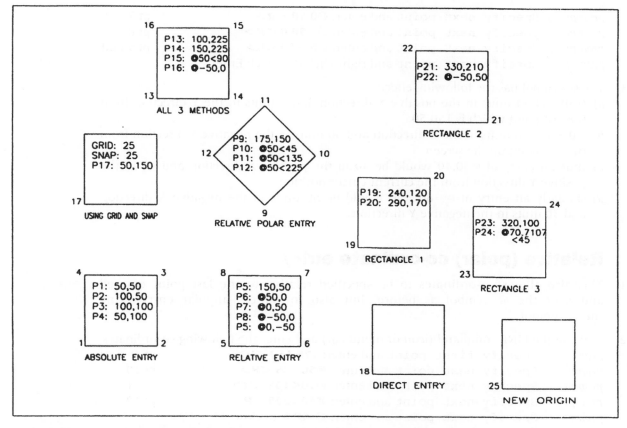

Figure 7.1 Line creation using different entry methods.

Absolute co-ordinate entry

1 This is the traditional XY Cartesian system, where the origin point is (0,0) at the lower left corner of the drawing area. This origin point can be 'moved' by the user as you will discover later in this chapter. When drawing a line, the co-ordinates of the start and end points of the line must be known.

2 Select the LINE icon from the Draw toolbar and: *Figure 7.1 ref*
   prompt Specify first point and enter: **50,50 <R>**   pt 1 start
   prompt Specify next point and enter: **100,50 <R>** pt 2
   prompt Specify next point and enter: **100,100 <R>** pt 3
   prompt Specify next point and enter: **50,100 <R>** pt 4
   prompt Specify next point and enter: **50,50 <R>**   pt 1 end
   prompt Specify next point and right-click
   then   pick Enter to end the line command.

Relative (absolute) co-ordinate entry

1 Relative co-ordinates are from the last point referenced and the **@ symbol** is used for the incremental entry.

2 From the menu bar select **Draw-Line** and enter the following XY co-ordinate pairs, remembering <R> after each entry.
   prompt Specify first point and enter: **150,50 <R>**   pt 5 start
   prompt Specify next point and enter: **@50,0 <R>**   pt 6

prompt	Specify next point and enter: **@0,50 <R>**	pt 7
prompt	Specify next point and enter: **@−50,0 <R>**	pt 8
prompt	Specify next point and enter: **@0,−50 <R>**	pt 5 end
prompt	Specify next point and right-click then pick Enter.	

3. The @ symbol has the following effect:
   a) @50,0 is 50 units in the positive X direction and 0 units in the Y direction from the last point which is 150,50
   b) @0,−50 is 0 units in the X direction and 50 units in the negative Y direction from the last point on the screen
   c) thus an entry of @30,40 would be 30 in the positive X direction and 40 in the positive Y direction from the current cursor position
   d) similarly an entry of @−80,−50 would be 80 units in the negative X direction and 50 units in the negative Y direction.

Relative (polar) co-ordinate entry

1. This also allows co-ordinates to be specified relative to the last point entered and uses the @ symbol as before, but also introduces angular entry using the < symbol.

2. Activate the LINE command (icon or menu bar) and enter the following co-ordinates:

prompt	Specify first point and enter: **175,150 <R>**	pt 9 start
prompt	Specify next point and enter: **@50<45 <R>**	pt 10
prompt	Specify next point and enter: **@50<135 <R>**	pt 11
prompt	Specify next point and enter: **@50<225 <R>**	pt 12
prompt	Specify next point and enter: **C <R>** to close the square and end line command.	

3. Notes
   a) The relative polar entries can be read as:
      1. @50<45 is 50 units at an angle of 45 degrees from the last point referenced which is 175,150
      2. @50<225 is 50 units at an angle of 225 degrees from the current cursor position.
   b) The entry **C <R>** is the **CLOSE** option and:
      1. closes the square, i.e. a line is drawn from the current screen position (point 12) to the start point 9
      2. ends the sequence, i.e. <R> not needed
      3. the close option works for any straight line shape.
   c) There is NO comma (,) with polar entries. This is a common mistake with new AutoCAD users, i.e.:
      1. @50<45 is correct
      2. @50,<45 is wrong and gives the command line error: **Point or option keyword required**.

Using all three entry methods

1. The different co-ordinate entry methods can be 'mixed and matched' when drawing a series of line segments.

2. Activate the LINE command then enter the following:

prompt	Specify first point and enter: **100,225 <R>**	pt 13 start
prompt	Specify next point and enter: **150,225 <R>** absolute entry	pt 14
prompt	Specify next point and enter: **@50<90 <R>** relative polar entry	pt 15

prompt   `Specify next point and enter:`
               @−50,0 <R> relative absolute entry   pt 16

prompt   `Specify next point and enter:` **C <R>**
               to close square and end command.

Grid and snap method

The grid and snap drawing aids can be set to any value suitable for current drawing requirements, so:

1. Set the grid and snap spacing to 25.

2. With the LINE command, draw a 50-unit square – the start point being at 50,150 which is pt 17 in Figure 7.1.

3. When the 50-unit square has been drawn, reset the grid and snap to original values, i.e. 10.

Direct distance entry

1. This method uses the position of the on-screen cursor and is very suitable when polar tracking is active, but remember that our polar tracking is off.

2. Activate the LINE command and:
   prompt   `Specify first point`
   enter   **250,30 <R>** – this is pt 18 in Figure 7.1
   prompt   `Specify next point`
   respond   move cursor horizontally to right and enter: **50 <R>**
   prompt   `Specify next point`
   respond   move cursor vertically upwards and enter: **50 <R>**
   prompt   `Specify next point`
   respond   move cursor horizontally to left and enter: **50 <R>**
   prompt   `Specify next point`
   enter   **C <R>**.

Rectangles

1. Rectangular shapes (in our case, squares) can be created from co-ordinate input by specifying two points on a diagonal of the rectangle, and the command can be used with absolute or relative input.

2. From the menu bar select **Draw-Rectangle** and:
   prompt   `Specify first corner point`
   enter   **240,120 <R>**   pt 19
   prompt   `Specify other corner point`
   enter   **290,170 <R>**   pt 20

3. Select the rectangle icon from the Draw toolbar and:
   prompt   `Specify first corner point`
   enter   **330,210 <R>**   pt 21
   prompt   `Specify other corner point`
   enter   **@−50,50 <R>**   pt 22

4. At the command line enter RECTANG <R> and:
   prompt   `Specify first corner point and enter:` **320,100 <R>**   pt 23
   prompt   `Specify other corner point and enter:` **@70.7107<45 <R>**   pt 24
   Question: Why the 70.7107 length and 45 angle entries?

Moving the origin

All the squares created so far have been with the origin point (0,0) at the lower left corner of 'our drawing paper'. If you move the cursor onto this position (snap on) and observe the status bar, it will display 0.00, 0.00, 0.00. These are the x, y and z co-ordinates of this point. At present we are drawing in 2D and thus the third co-ordinate will always be 0.00. The origin can be moved to any point on the screen and we will reset the origin and draw a 50-unit square from this new origin position, so:

1. Menu bar with **View-Display-UCS Icon** and ensure that both On and Origin are active (tick at name).

2. Menu bar with **Tools-New UCS-Origin** and:
   prompt   Specify new origin point
   enter   **340,20 <R>**.

3. The UCS icon should move to this position. If it does not, then repeat step 1.

4. Move the cursor (snap on) to the + in the icon and observe the status bar.

5. The co-ordinates read 0.00, 0.00, 0.00, i.e. we have reset the origin point which is point 25 in Figure 7.1.

6. Now draw a square of side 50 from this new origin using any of the methods previously described.

7. At the command line enter **UCS <R>** and:
   prompt   Enter an option [New/Move...
   enter   **P <R>** – the previous option.

8. The UCS icon will be 'returned' to its original origin position at the lower left corner of our drawing paper.

Saving the squares

1. The drawing screen should now display ten squares positioned as Figure 7.1 but without the text. This drawing must be saved as it will be used in other chapters.

2. From the menu bar select **File-Save As** and:
   prompt   Save Drawing As dialogue box
   respond   a) scroll at Save in and pick C: drive
              b) double left-click on BEGIN folder
              c) enter file name as **DEMODRG**
              d) pick Save.

3. The complete file name path for the saved drawing is **C:\BEGIN\DEMODRG**.

Conventions

When using co-ordinate input, the user must know the positive and negative directions for both linear and angular inputs. The two conventions are as follows:

1. *Co-ordinate axes*
   The XY axes convention used by AutoCAD is shown in Figure 7.2(a) and displays four points with their co-ordinate values. When using the normal XY co-ordinate system:
   a) a positive X direction is to the right, and a positive Y direction is upwards
   b) a negative X direction is to the left, and a negative Y direction is downwards.

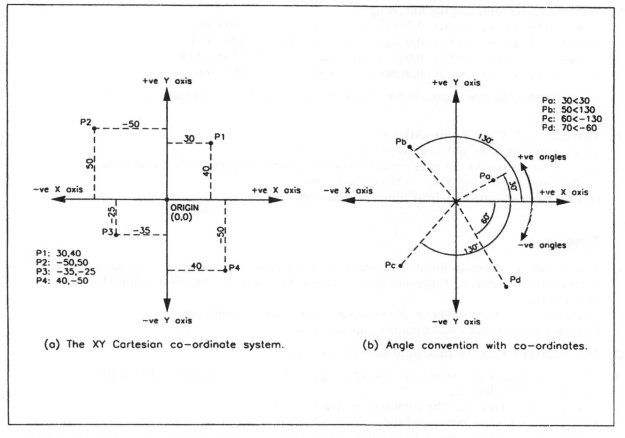

Figure 7.2 Co-ordinate and angle convention.

2 *Angles*
  When angles are being used:
  a) positive angles are anti-clockwise
  b) negative angles are clockwise
  c) Figure 7.2(b) displays the angle convention with four points with their polar co-ordinate values.

Task

Before leaving this exercise, try the following:

1 Make sure you have saved the squares as C:\BEGIN\DEMODRG.

2 Erase a square created with a **Draw-Rectangle** sequence. It is erased with a 'single pick' as it is a polyline (more on polylines in a later chapter).

3 Now erase all the squares from the screen.

4 Draw a line sequence using the following entries:
  a) Specify first point: **30,40**
  b) Specify next point: **−100,100**
  c) Specify next point: **−150,−200**
  d) Specify next point: **80,−100**
  e) Specify next point: **C <R>**.

5   Draw four lines, entering the following:
    a) Specify first point: **0,0**; Specify next point: **100<30**
    b) Specify first point: **0,0**; Specify next point: **200<150**
    c) Specify first point: **0,0**; Specify next point: **250<−130**
    d) Specify first point: **0,0**; Specify next point: **90<−60**.

6   When these eight line segments have been drawn, not all are completely visible on the screen.

7   Menu bar with **View-Zoom-All** and:
    a) all eight lines are visible, i.e. we can draw 'off the screen'?
    b) move the cursor to the intersection of the four polar lines, and with the snap on, the status bar displays 0,0,0 as the co-ordinates
    c) now **View-Zoom-Previous** from the menu bar.

Notes

1   When using co-ordinate input with the LINE command, it is very easy to make a mistake with the entries. If the line 'does not appear to go in the direction it should', then either:
    a) enter U <R> from the keyboard to 'undo' the last line segment drawn, or
    b) right-click and pick undo from the pop-up menu.

2   This 'undo effect' can be used until all segments are erased.

3   The @ symbol is very useful if you want to 'get to the last point referenced on the screen'. Try the following:
    a) draw a line and cancel the command with a <RETURN>
    b) re-activate the line command and enter @ <R>
    c) the cursor 'snaps to' the endpoint of the drawn line.

4   After a command has been cancelled, a <RETURN> keyboard press will always activate the last command.

5   A right-click on the mouse will activate a 'pop-up' dialogue box, allowing the last command to be activated.

Assignment

This activity only uses the LINE command (and ERASE?), but requires co-ordinate entry (and some 'sums') for you to complete the drawing.

1   Close all existing drawings then open your A3SHEET standard sheet.

2   Refer to Activity 2 and draw the three shapes using co-ordinate input. Any entry method can be used but I would recommend you to:
    a) position the start points with absolute entry
    b) use relative entry as much as possible.

3   When the drawing is complete, save it as C:\BEGIN\ACT2.

4   Read the summary then proceed to the next chapter.

Summary

1   Co-ordinate entry can be ABSOLUTE or RELATIVE.

2   ABSOLUTE entry is from an origin – the point (0,0). Positive directions are UP and to the RIGHT, negative directions are DOWN and to the LEFT. The entry format is X, Y, e.g. 30,40.

3   RELATIVE entry refers to the co-ordinates of the last point entered and uses the @ symbol. The entry format is:
    a) relative absolute: @X, Y, e.g. @50,60
    b) relative polar: @X<A, e.g. @100<50 and note – no comma.

4   An angle of −45 degrees is the same as an angle of +315 degrees.

5   The following polar entries are the same:
    a) @−50<30
    b) @50<210
    c) @50<−150
    Try them if you are not convinced.

6   All entry methods can be used in a line sequence.

7   A line sequence is terminated with:
    a) the <RETURN> key
    b) a mouse right-click and pick Enter
    c) 'closing' the shape with a C <R>.

8   The rectangle command is useful, but it is a 'single object' and not four 'distinct lines'.

9   The user can activate the LINE command by:
    a) icon selection from the Draw toolbar
    b) menu bar selection with **Draw-Line**
    c) entering LINE <R> at the command line.

Chapter 8

Circle creation

In this chapter we will investigate how circles can be created by adding several to the squares created in the previous chapter, so:
a) Open your **C:\BEGIN\DEMODRG** to display the ten squares created in Chapter 7.
b) Refer to Figure 8.1 and ensure the Draw and Modify toolbars are displayed.

AutoCAD 2005 allows circles to be created by six different methods and the command can be activated by icon selection, menu bar selection or keyboard entry. When drawing circles, absolute co-ordinates are usually used to specify the circle centre, although the next chapter will introduce the user to the object snap modes. These object snap modes allow greater flexibility in selecting existing entities for reference.

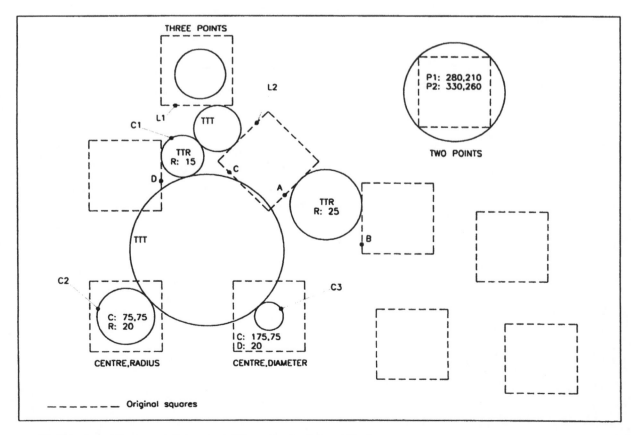

Figure 8.1 Circle creation using different selection methods.

Centre-radius

Select the CIRCLE icon from the Draw toolbar and:
prompt Specify center point for circle or [3P/2P/Ttr (tan tan radius)]:
enter   75,75 <R> – the circle centre point
prompt Specify radius of circle or [Diameter]
enter   20 <R> – the circle radius.

Centre-Diameter

From the menu bar select **Draw-Circle-Center,Diameter** and:
prompt Specify center point for circle or...
enter   175,75 <R>
prompt Specify diameter of circle
enter   20 <R>.

Two points on circle diameter

At the command line enter **CIRCLE <R>** and:
prompt Specify center point for circle or...
enter   2P <R> – the two point option
prompt Specify first end point on circle's diameter
enter   280,210 <R>
prompt Specify second end point on circle's diameter
enter   330,260 <R>.

Three points on circle circumference

Menu bar selection with **Draw-Circle-3 Points** and:
prompt   Specify first point on circle
respond **pick any point within the top left circle**
prompt   Specify second point on circle
respond **pick another point within the top left circle**
prompt   Specify third point on circle
respond **drag out the circle and pick a point**.

TTR: tangent-tangent-radius

1 Menu bar with **Draw-Circle-Tan,Tan,Radius** and:
   prompt   Specify point on object for first tangent of circle
   respond **move cursor over the line A and leave for a second**
   and   a) a small marker is displayed
              b) Deferred Tangent tooltip displayed
   respond **pick line A** – i.e. left-click on it
   prompt   Specify point on object for second tangent of circle
   respond **pick line B**
   prompt   Specify radius of circle
   enter   25 <R>
   and   a circle is drawn tangential to the two selected lines.

2   At the command line enter **CIRCLE <R>** and:
    prompt   Specify center point for circle or...
    enter   **TTR <R>** – the tan,tan,radius option
    prompt   First tangent point prompt and: **pick line C**
    prompt   Second tangent point prompt and: **pick line D**
    prompt   radius prompt and enter: **15 <R>**
    and   a circle is drawn tangential to the two selected lines, line C being assumed extended.

TTT – tangent-tangent-tangent

1   Menu bar with **Draw-Circle-Tan,Tan,Tan** and:
    prompt   Specify first point on circle and: **pick line L1**
    prompt   Specify second point on circle and: **pick line L2**
    prompt   Specify third point on circle and: **pick circle C1**.

2   Activate the **Draw-Circle-Tan,Tan,Tan** sequence and:
    prompt   Specify first point on circle and: **pick circle C1**
    prompt   Specify second point on circle and: **pick circle C2**
    prompt   Specify third point on circle and: **pick circle C3**.

3   Circles have been drawn tangentially to selected objects, these being:
    a) two lines and a circle
    b) three circles.

4   *Questions*
    a) How long would it take to draw a circle as a tangent to three other circles by conventional draughting methods, i.e. drawing board, T-square, set squares, etc.?
    b) Can a circle be drawn as a tangent to two circles and a line or to three inclined lines? Try these for yourself.

Saving the drawing

1   Assuming that the CIRCLE commands have been entered correctly, your drawing should resemble Figure 8.1 (without the text) and is ready to be saved for future work.

2   From the menu bar select **File-Save As** and:
    prompt   Save Drawing As dialogue box
    with   1. Begin folder name active
               2. File name: **DEMODRG**
    respond **pick Save**
    prompt   Drawing already exists message
    respond **pick Yes** – obvious?

Task

1   The two Tan,Tan,Tan circles have been created without anything being known about their radii.

2   From the menu bar select **Tools-Inquiry-List** and:
    prompt   Select objects
    respond **pick the smaller TTT circle**
    prompt   1 found and Select objects
    respond **right-click**
    prompt   AutoCAD Text window with information about the circle.

3   Note the information then cancel the text window by picking the right (X) button from the title bar.

4   Repeat the **Tools-Inquiry-List** sequence for the larger TTT circle.

5   The information for my two TTT circles is as follows:

	Smaller	*Larger*
Centre point	139.52, 208.53	131.88, 122.06
Radius	16.47	53.83
Circumference	103.51	338.21
Area	852.67	9102.42

6   Could you calculate these figures manually as easily as has been demonstrated?

Assignment

1   Open your A3SHEET standard sheet.

2   Refer to the Activity 3 drawing which can be completed with only the CIRCLE command.

3   The method of completing the drawings is at your discretion.

4   Remember that absolute co-ordinates are recommended for circle centres and that the TTR method is very useful.

5   You may require some 'sums' for certain circle centres, but the figures are relatively simple.

6   When the drawing is complete, save it as **C:\BEGIN\ACT3**.

Summary

1   Circles can be created by six methods, the user specifying:
    a) a centre point and radius
    b) a centre point and diameter
    c) two points on the circle diameter
    d) any three points on the circle circumference
    e) two tangent specification points and the circle radius
    f) three tangent specification points.

2   The TTR and TTT options can be used with lines, circles, arcs and other objects.

3   The centre point and radius can be specified by:
    a) co-ordinate entry
    b) picking a point on the screen
    c) referencing existing entities – next chapter.

Chapter **9**

Object snap

The lines and circles drawn so far have been created by co-ordinate input. While this is the basic method of creating objects, it is often desirable to 'reference' existing objects already displayed on the screen, e.g. we may want to:
a) draw a circle with its centre at the midpoint of an existing line
b) draw a line, from a circle centre perpendicular to another line.

These types of operations are achieved using the **object snap modes** – generally referred to as **OSNAP** – and are one of the most useful (and powerful) draughting aids.

Object snap modes are used transparently, i.e. whilst in a command, and can be activated:
a) from the Object Snap toolbar
b) by direct keyboard entry.

While the toolbar method is the quicker and easier to use, we will investigate both methods.

Getting ready

1 Open your **C:\BEGIN\DEMODRG** drawing of the squares and circles.
2 Erase the two TTT circles and the lower right square.
3 Display the Draw, Modify and Object Snap toolbars and position them to suit and refer to Figure 9.1.

Using object snap from the keyboard

Activate the LINE command and:
prompt	Specify first point
enter	**MID <R>**
prompt	of
respond	1. move cursor to line D1 and leave for few seconds
	2. coloured triangular marker at line midpoint
	3. Midpoint tooltip displayed in colour
now	**pick line D1** – i.e. left-click
and	line 'snaps to' the midpoint of D1
prompt	Specify next point
enter	**PERP <R>**
prompt	to
respond	**pick line D2** – note coloured Perpendicular marker
prompt	Specify next point
enter	**CEN <R>**

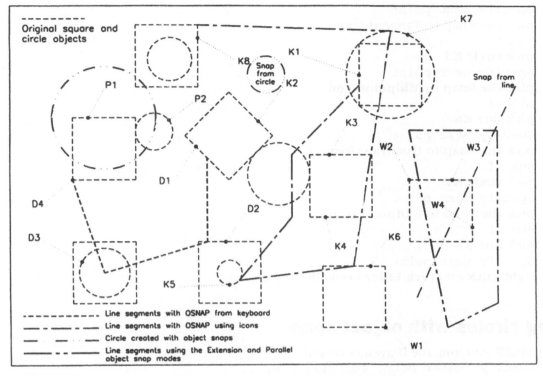

Figure 9.1 Using the object snap modes with C:\BEGIN\DEMODRG.

prompt of
respond **pick circle D3** – note coloured Center marker
prompt Specify next point
enter **INT <R>**
prompt of
respond **pick point D4** – note coloured Intersection marker
prompt Specify next point
respond **right-click and pick Enter** to end the line sequence.

Using object snap from the toolbar

Activate the LINE command and:
prompt Specify first point
respond **pick the Snap to Nearest icon**
prompt nea to
respond **pick any point on line K1**
prompt Specify next point
respond **pick the Snap to Apparent Intersection icon**
prompt appint of
respond **pick line K2**
prompt and
respond **pick line K3**
prompt Specify next point
respond **pick the Snap to Perpendicular icon**
prompt per to
respond **pick line K4**

prompt	Specify next point
respond	**pick the Snap to Tangent icon**
prompt	tan to
respond	**pick circle K5**
prompt	Specify next point
respond	**pick the Snap to Midpoint icon**
prompt	mid of
respond	**pick line K6**
prompt	Specify next point
respond	**pick the Snap to Quadrant icon**
prompt	qua of
respond	**pick circle K7**
prompt	Specify next point
respond	**pick the Snap to Endpoint icon**
prompt	end of
respond	**pick line K8**
prompt	Specify next point
respond	**right-click and pick Enter** to end the line sequence.

Drawing circles with object snap

Select the CIRCLE icon from the Draw toolbar and:

prompt	Specify center point for circle or...
respond	**pick the Snap to Midpoint icon**
prompt	mid of
respond	**pick line P1**
prompt	Specify radius of circle or...
respond	**pick the Snap to Center icon**
prompt	cen of
respond	**pick circle P2**.

Notes

1 Save your drawing at this stage as **C:\BEGIN\DEMODRG**.

2 The endpoint 'snapped to' depends on which part of the line is 'picked'. The coloured marker indicates which line endpoint.

3 A circle has four quadrants, these being at the 3,12,9,6 o'clock positions. The coloured marker indicates which quadrant will be snapped to.

The extension and parallel object snap modes

1 The object snap modes selected so far should have been self-explanatory to the user, i.e. endpoint will snap to the end of a line, center will snap to the centre of a circle, etc. The extension and parallel modes are used as follows:
   a) Extension   used with lines and arcs and gives a temporary extension line as the cursor is passed over the endpoint of an object.
   b) Parallel   used with straight line objects only and allows a vector to be drawn parallel to another object.

2 To demonstrate these two object snap modes, continue with the DEMODRG and turn on the grid and snap with a spacing to 10 for both.

Activate the LINE command and:
prompt   Specify first point
respond   **pick the Snap to Extension icon**
prompt   ext of
respond   move cursor over point W1 then drag to right
and   a) highlighted extension line dragged out
              b) information displayed about distance from endpoint as a tooltip
respond   a) move cursor until Extension: 50.00<0.0 displayed
               b) left-click
prompt   Specify next point
respond   **pick the Snap to Extension icon**
prompt   ext of
respond   move cursor over point W2 and drag vertically up
and   a) highlighted extension line dragged out
              b) information displayed as a tooltip
respond   a) move cursor until Extension: 40.00<90.0 displayed
               b) left-click
prompt   Specify next point
respond   **pick the Snap to Parallel icon**
prompt   par to
respond   a) move cursor over line W3, leave for a few seconds and note the display
               b) move cursor to right of last pick point and:
                   1. highlighted line
                   2. information about distance and angle displayed
               c) move cursor to right until 70.00<0.0 displayed
               d) left-click
prompt   Specify next point
respond   **pick the Snap to Parallel icon**
prompt   par to
respond   a) move cursor over line W4
               b) move cursor vertically below last pick point
               c) move cursor until 140.00<270.0 displayed
               d) left-click
prompt   Specify next point
enter   **C <R>** to close the shape and end the line command.

3   The line segments should be as displayed in Figure 9.1.

4   Save your layout as **C:\BEGIN\DEMODRG**, updating the existing DEMODRG.

Running object snap

1   Using the object snap icons from the toolbar will increase the speed of the draughting process, but it can still be 'tedious' to have to pick the icon every time an ENDpoint (for example) is required. It is possible to 'preset' the object snap mode to ENDpoint, MIDpoint, CENter, etc., and this is called a running object snap. Pre-setting the object snap does not preclude the user from selecting another mode, i.e. if you have set an ENDpoint running object snap, you can still pick the INTersection icon.

2   The running object snap can be set:
     a) from the menu bar with **Tools-Drafting Settings** and pick the Object Snap tab
     b) entering OSNAP <R> at the command line

c) picking the object snap settings icon from the Object Snap toolbar
d) with a right-click on OSNAP in the status bar and picking Settings.

3 Each method displays the Drafting Settings dialogue box with the Object Snap tab active.

4 *Tasks*
   a) Select **Tools-Drafting Settings** from the menu bar and:

   prompt   Drafting Settings dialogue box
   respond 1. ensure the Object Snap tab is active
            2. ensure Object Snap On (F3) is active (tick)
            3. ensure Object Snap Tracking on is not active (no tick)
            4. activate Endpoint, Midpoint and Nearest by picking the appropriate box – tick means active
            5. dialogue box as Figure 9.2
            6. pick OK.

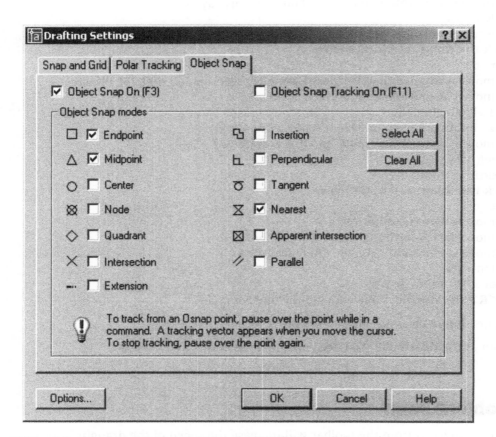

Figure 9.2 The Object Snap tab of the Drafting Settings dialogue box.

b) Now activate the LINE command and move the cursor cross-hairs onto any line and leave it.
c) A coloured marker will be displayed at the Nearest point of the line – it may be that you have the midpoint marker displayed.
d) Press the TAB key to cycle through the set running object snaps, i.e. the line should display the cross, square and triangular coloured markers for the Nearest, Endpoint and Midpoint object snap settings.
e) Cancel the line command with ESC.

AutoSnap and AutoTrack

1. The Object Snap tab of the Drafting Settings dialogue box allows the user to select **Options**. This selection will display the Drafting tab of the Options dialogue box. The user can control both the AutoSnap and the AutoTrack settings which can be simply defined as:
   a) AutoSnap   a visual aid for the user to see and use object snaps efficiently, i.e. marker and tooltip displayed.
   b) AutoTrack   an aid to the user to assist with drawing at specific angles, i.e. polar tracking.

2. The settings which can be altered with AutoSnap and AutoTrack are:

AutoSnap	AutoTrack
Marker	Display polar tracking vector
Magnet	Display full-screen tracking vector
Display tooltip	Display autotrack tooltip
Display aperture box	Alter alignment point acquisition
Alter marker colour	Alter aperture size
Alter marker size	

3. The various terms should be self-explanatory at this stage (?):
   a) Marker   is the geometric shape displayed at a snap point
   b) Magnet   locks the aperture box onto the snap point
   c) Tooltip   is a flag describing the name of the snap location.

4. The rest of the options should be apparent. It is normal to have the marker, magnet and tooltip active (ticked). The colour of the marker and the aperture box sizes are at the user's discretion, as is having object snap tracking 'on'.

Cancelling a running object snap

A running object snap can be left 'active' once it has been set, but this can cause problems if the user 'forgets' about it. The running snap can be cancelled:

1. Using the Object Snap settings dialogue box and selecting Clear All.

2. By entering **–OSNAP <R>** at the command line and:
   prompt   Enter list of object snap modes
   enter   **NONE <R>**.

3. *Notes*
   a) Selecting the Snap to None icon from the Object Snap toolbar will turn off the object snap running modes for the next point selected.
   b) Using the Object Snap dialogue box is the recommended way of activating and de-activating object snaps.

The Snap From object snap

This is a very useful object snap, allowing the user the reference points relative to existing objects.

1. Using the squares and circles which should still be displayed, erase the lines and circles created with the object snaps. This is to give 'some space'. Ensure the Object Snap toolbar is displayed.

2 Activate the circle command and:
   prompt   Specify center point for circle
   respond **pick Snap From icon** from the Object Snap toolbar
   prompt   from Base Point
   respond **pick Snap to Midpoint icon**
   prompt   mid of
   respond **pick line K8**
   prompt   <Offset>
   enter   **@50,−15 <R>**
   prompt   Specify radius of circle
   enter   **15 <R>**.

3 A circle is drawn with its centre 50 mm horizontally from the midpoint of the selected line.

4 With the LINE command active:
   prompt   Specify first point
   respond **pick Snap From icon**
   prompt   from Base point
   respond **pick Intersection icon**
   prompt   int of
   respond **pick point W1**
   prompt   <Offset>
   enter   **@25,25 <R>**
   prompt   Specify next point
   respond **pick Snap From icon**
   prompt   from Base point
   respond **pick Snap to Centre icon**
   prompt   cen of
   respond **pick circle K7**
   prompt   <Offset>
   enter   **@100<−5 <R>**
   prompt   Specify next point
   respond **right-click and Enter**.

5 A line is drawn between the specified points. The endpoints of this line have been 'offset' from the selected objects by the entered co-ordinate values.

6 Do not save these additions to your drawing layout.

Object snap tracking

Object snap tracking allows the user to 'acquire co-ordinate data' from the object snap modes which have been set. To demonstrate this drawing aid:

1 Erase all squares, circles, etc. from the screen to leave the basic rectangular sheet outline.

2 Refer to Figure 9.3 and draw a line from 50,50 to 100,150 and a circle, centre at 200,200 with radius 50 as Figure 9.3(a).

3 Right-click **OSNAP from the status bar**, pick **Settings** and:
   prompt   Drafting Settings dialogue box
   respond a) Object Snap tab active
            1. Object Snap On active
            2. Endpoint and Center object snap modes active
            3. Object Snap Tracking On active

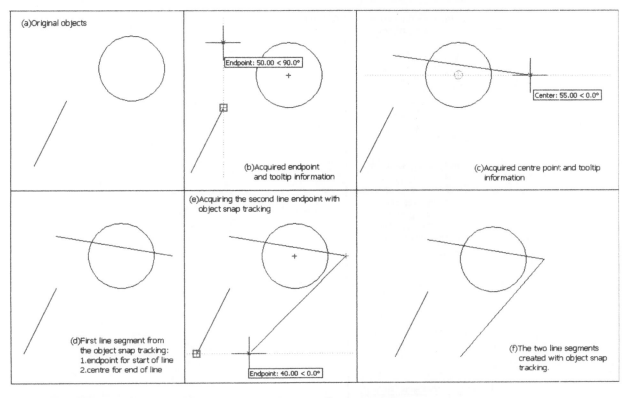

Figure 9.3 Using object snap tracking to draw two line segments.

       b) Polar Tracking tab
          1. Polar Tracking On active
          2. Incremental angle: 30
       c) pick OK.

4   Activate the LINE command and:
    a) move the cursor over the top end of the drawn line and the endpoint marker (square) will be displayed
    b) move the cursor vertically upwards to display object snap tracking information, similar to Figure 9.3(b)
    c) enter 80 <R> and the start point of the line will be obtained
    d) move the cursor to the centre of the circle and the centre marker (circle) will be displayed
    e) move the cursor horizontally to the right and object snap tracking information will be displayed similar to Figure 9.3(c)
    f) enter 78 <R> and a line segment is drawn – Figure 9.3(d)
    g) move cursor onto bottom end of first line to acquire the endpoint marker then move horizontally to the right to display object snap tracking data similar to Figure 9.3(e)
    h) enter 100 <R> then right-click/Enter to end the line command
    i) the second line segment is complete – Figure 9.3(f).

5   Task 1
    a) from the object snap settings, deactivate the Endpoint and Center snap modes and activate the Midpoint snap mode
    b) from the Polar Tracking Settings, check the incremental angle is set to 30
    c) refer to Figure 9.4

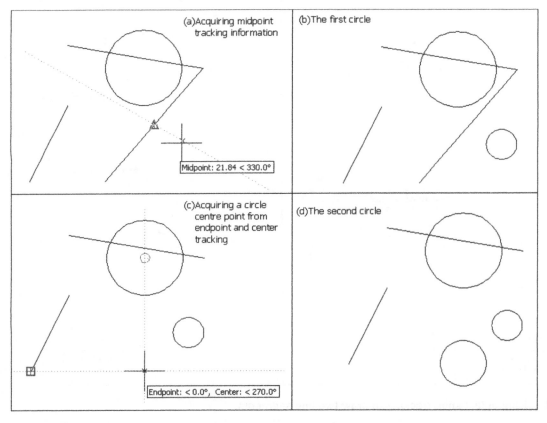

Figure 9.4 Acquiring circle centre points using object snap tracking.

    d) activate the circle command and acquire the midpoint of the second line drawn by object snap tracking
    e) move downwards to the right until object snap tracking data displays an angle of 330 degrees as Figure 9.4(a)
    f) enter 50 <R> for the circle centre point, then 20 <R> for circle radius
    g) circle drawn at selected point as Figure 9.4(b).

6 *Task 2*
    a) erase the last line segment drawn and set the polar tracking angle to 90
    b) turn off the Midpoint object snap mode, and activate the Endpoint and Center modes
    c) with the circle command:
        1. acquire the endpoint of lower end of line
        2. acquire the centre point of the circle
        3. move cursor vertically downwards until it is horizontally in line with the lower end of line
        4. the two acquired point object snap tracking data should be displayed similar to Figure 9.4(c)
        5. pick this point as the circle centre
        6. enter a radius value, e.g. 30 to position the circle as Figure 9.4(d)
        7. think of the benefits of this type of operation, i.e. acquiring centre point data without any co-ordinate input.

7 This completes the object snap tracking exercise. Do not save.

Assignment

1. Open your C:\BEGIN\A3SHEET standard sheet.
2. Refer to Activity 4 and draw the three components using lines and circles.
3. The object snap modes will require to be used and hints are given.
4. When complete, save as C:\BEGIN\ACT4.
5. The diameter of the inner circle is 62.38. Use the LIST command to check.
6. Read the summary then progress to the next chapter.

Summary

1. Object snap (OSNAP) is used to reference existing objects.
2. The object snap modes are invaluable aids to draughting and should be used whenever possible.
3. The user can 'pre-set' a running object snap.
4. Geometric markers will indicate the snap points on objects.
5. The Drafting Settings dialogue box allows the user to 'control' the geometric markers.
6. Object snap is an example of a transparent command, as it is activated when another command is being used.
7. The object snap modes can be set and cancelled using the dialogue box or toolbar.
8. Object snap tracking allows the user to 'acquire' data for selected points 'set' by the object snap modes.
9. *Drawing Aids*
   At this stage the user now has knowledge about the basic 2D draughting aids, these being: Grid; Snap; Ortho; Object snap modes; Polar tracking; Object snap tracking.

Chapter 10

Arc, donut and ellipse creation

These three drawing commands will be discussed in turn using the square and circle exercise. Each command can be activated from the toolbar, menu bar or by keyboard entry and both co-ordinate entry and referencing existing objects (OSNAP) will be demonstrated.

Getting started

1 Open your C:\BEGIN\DEMODRG to display the squares, circles and object snap lines, etc.
2 Erase the objects created during the object snap exercise.
3 Refer to Figure 10.1 and activate the Draw, Modify and Object Snap toolbars.

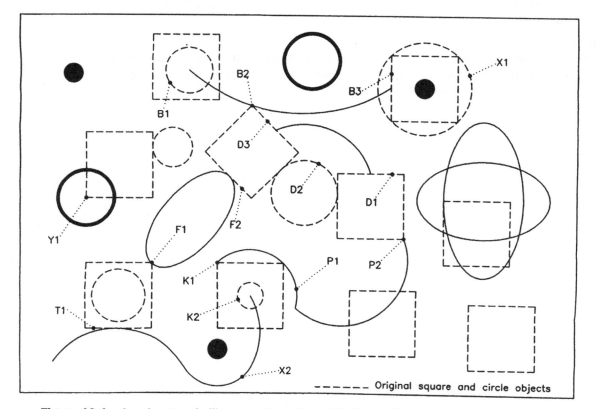

Figure 10.1 Arc, donut and ellipse creation with C:\BEGIN\DEMODRG.

Arcs

There are ten different arc creation methods. Arcs are normally drawn in an anti-clockwise direction with combinations of the arc start point, end point, centre point, radius, included angle, length of arc, etc. We will investigate four different arc creation methods as well as continuous arcs. You can try the others for yourself.

Start,Center,End

From the menu bar select **Draw-Arc-Start,Center,End** and:
prompt   `Specify start point of arc`
respond   **Snap to Midpoint icon and pick line D1**
prompt   `Specify center point of arc`
respond   **Snap to Center icon and pick circle D2**
prompt   `Specify end point of arc`
respond   **Snap to Midpoint icon and pick line D3**.

Start,Center,Angle

1 Menu bar with **Draw-Arc-Start,End,Angle** and:
   prompt   `Specify start point of arc`
   respond   **Snap to Intersection icon and pick point K1**
   prompt   `Specify center point of arc`
   respond   **Snap to Center icon and pick circle K2**
   prompt   `Specify included angle`
   enter   **−150 <R>**.

2 Note that negative angle entries draw arcs in a clockwise direction.

Start,End,Radius

Menu bar again with **Draw-Arc-Start,End,Radius** and:
prompt   `Start point and` **snap to Endpoint of arc P1**
prompt   `End point and` **snap to Intersection of point P2**
prompt   `Radius and` **enter 50 <R>**.

Three points (on arc circumference)

Activate the **3 Points arc** command and:
prompt   `Start point and` **snap to Center of circle B1**
prompt   `Second point and` **snap to Intersection of point B2**
prompt   `End point and` **snap to Midpoint of line B3**.

Continuous arcs

1 Activate the **3 Points arc** command again and:
   prompt   `Start point and enter` **25,25 <R>**
   prompt   `Second point and` **snap to Midpoint of line T1**
   prompt   `End point and enter` **@50,−30 <R>**.

2 Select from the menu bar **Draw-Arc-Continue** and:
   prompt   `End point – and cursor snaps to end point of the last arc drawn`
   enter   **@50,0 <R>**.

3 Repeat the **Arc-Continue** selection and:
   prompt   `End point`
   respond   **snap to Center of circle K2**.

Donut

A donut is a 'solid filled' circle or annulus (a washer shape), the user specifying the inside and outside diameters and then selecting the donut centre point.

1   Menu bar with **Draw-Donut** and:
    prompt   Specify inside diameter of donut and enter: **0 <R>**
    prompt   Specify outside diameter of donut and enter: **15 <R>**
    prompt   Specify center of donut and enter: **40,245 <R>**
    prompt   Specify center of donut
    respond   **Snap to Center of circle X1**
    prompt   Specify center of donut
    respond   **Snap to Center of arc X2**
    prompt   Specify center of donut and right-click.

2   Repeat the donut command and:
    prompt   Specify inside diameter of donut and enter: **40 <R>**
    prompt   Specify outside diameter of donut and enter: **45 <R>**
    prompt   Specify center of donut and enter: **220,255 <R>**
    prompt   Specify center of donut
    respond   **Snap to Intersection of point Y1**
    prompt   Specify center of donut and right-click.

3   *Note*: The donut command allows repetitive entries to be made by the user, while the circle command only allows one circle to be created per command – I do not know why this is!

Ellipse

Ellipses are created by the user specifying:
either   *a*) the ellipse centre and two axes endpoints
or   *b*) three points on the axes endpoints.

1   Select from the menu bar **Draw-Ellipse-Center** and:
    prompt   Specify center of ellipse and enter: **350,150 <R>**
    prompt   Specify endpoint of axis and enter: **400,150 <R>**
    prompt   Specify distance to other axis or [Rotation] and enter: **350,120 <R>**.

2   Select the ELLIPSE icon from the Draw toolbar and:
    prompt   Specify axis endpoint of ellipse or [Arc/Center]
    enter   **C <R>** – the center option
    prompt   Specify center of ellipse
    respond   Snap to Center icon and **pick the existing ellipse**
    prompt   Specify endpoint of axis and enter: **@30,0 <R>**
    prompt   Specify distance to other axis and enter: **@0,60 <R>**.

3   Menu bar with **Draw-Ellipse-Axis,End** and:
    prompt   Specify axis endpoint of ellipse
    respond   **Snap to Intersection of point F1**
    prompt   Specify other endpoint of axis
    respond   **Snap to Midpoint of line F2**
    prompt   Specify distance to other axis or [Rotation]
    enter   **R <R>** – the rotation option
    prompt   Specify rotation around major axis and enter: **60 <R>**.

Notes

1 At this stage your drawing should resemble Figure 10.1 but without the text.

2 Save the layout as **C:\BEGIN\DEMODRG** for future recall if required. Remember that this will 'over-write' the existing C:\BEGIN\DEMODRG file.

3 Arcs, donuts and ellipses have centre points and quadrants which can be 'snapped to' with the object snap modes.

4 It is also possible to use the tangent snap icon and draw tangent lines, etc. between these objects. Try this for yourself.

Solid fill

Donuts are generally displayed on the screen 'solid', i.e. 'filled in'. This solid fill effect is controlled by the FILL system variable and can be activated from the menu bar or the command line.

1 Do you still have the DEMODRG layout on the screen?

2 Menu bar with **Tools-Options** and:
prompt   Options dialogue box
respond   pick the Display tab
then   a) Apply solid fill OFF – i.e. no tick in box
           b) pick OK.

3 Menu bar with **View-Regen** and the donuts will be displayed without the fill effect.

4 At the command line enter **FILL <R>**
prompt   Enter mode [ON/OFF] <OFF>
enter   **ON <R>**.

5 At the command line enter **REGEN <R>** to 'refresh' the screen and display the donuts with the fill effect.

6 *Note*: This fill effect also applies to polylines which will be discussed in a later chapter.

Assignment

1 Open your A3SHEET and refer to Activity 5.

2 Draw the shapes using arcs, donuts and ellipses.

3 No sizes are given, so use your imagination.

4 Setting the grid to 10 and the snap to 5 will help.

5 Save as C:\BEGIN\ACT5 when complete.

Summary

Arcs, donuts and ellipses are Draw commands and can be created by co-ordinate entry or by referencing existing objects.

General

1. The three objects have a centre point and quadrants.
2. They can be 'snapped to' with the object snap modes.
3. Tangent lines can be drawn to and from them.

Arcs

1. Several different creation options.
2. Normally drawn in an anti-clockwise direction.
3. Very easy to draw in the wrong 'sense' due to the start and end points being selected wrongly.
4. Continuous arcs are possible.
5. A negative angle entry will draw the arc clockwise.

Donuts

1. Require the user to specify the inside and outside diameters.
2. Can be displayed filled of unfilled.
3. An inside radius of 0 will give a 'filled circle'.
4. Repetitive donuts can be created.

Ellipses

1. Two creation methods.
2. Partial ellipses (arcs) are possible.
3. The created ellipses are 'true', i.e. have a centre point.

Chapter 11
Layers and standard sheet 2

All the objects that have been drawn so far have had a continuous linetype and no attempt has been made to introduce centre or hidden lines, or even colour. Auto-CAD has a facility called LAYERS which allows the user to assign different linetypes and colours to named layers. For example, a layer may be for red continuous lines, another may be for green hidden lines and yet another for blue centre lines. Layers can also be used for specific drawing purposes, e.g. there may be a layer for dimensions, one for hatching, one for text, etc. Individual layers can be 'switched' on/off by the user to mask out drawing objects which are not required.

The concept of layers can be imagined as a series of transparent overlays, each having its own linetype, colour and use. The overlay used for dimensioning could be switched off without affecting the remaining layers.

Figure 11.1 demonstrates the layer concept with:
a) Five layers used to create a simple component with each 'part' of the component created on 'its own' layer.
b) The layers 'laid on top of each other'. The effect is that the user 'sees' one component.

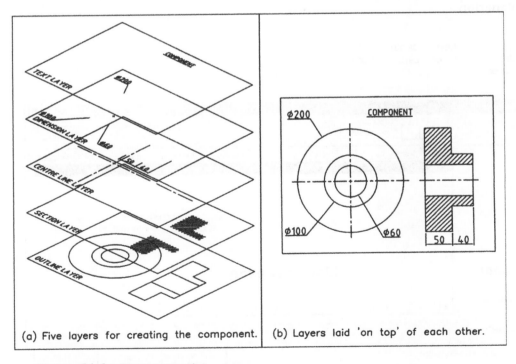

Figure 11.1 Layer concept.

The following points are worth noting when considering layers:
a) All objects are drawn on layers.
b) Layers should be used for each 'part' of a drawing, i.e. dimensions should not be on the same layer as centre lines (for example).
c) New layers must be 'created' by the user, using the Layer Properties Manager dialogue box.
d) Layers are one of the most important concept in AutoCAD.
e) Layers are essential for good and efficient draughting.

As layers are very important, and as the user must have a sound knowledge of how they are used, this chapter is rather long (and perhaps boring). I make no apology for this, as all CAD operators must be able to use layers correctly.

Note: I would recommend that this chapter be completed at 'one sitting'.

Getting started

Several different aspects of layers will be demonstrated in this chapter. Once these concepts have been discussed, we will modify our existing standard sheet, so:

1 Open your C:\BEGIN\A3SHEET standard sheet.

2 Draw a horizontal and vertical line each of length 200, and a circle of radius 75 anywhere on the screen.

The Layer Properties Manager dialogue box

1 From the menu bar select **Format-Layer** and:
   prompt   Layer Properties Manager dialogue box similar to Figure 11.2
   respond **Study the layout of the dialogue box and read the following explanation**

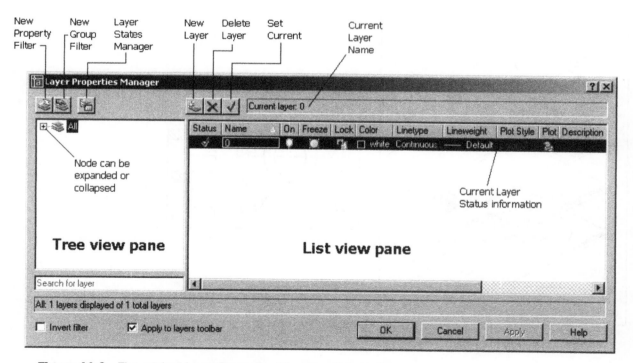

Figure 11.2 The original Layer Properties Manager dialogue box.

Note: Much of the terminology in the explanation which follows may be entirely new to the reader, but should become apparent as the chapter progresses.

2. The Layer Properties Manager dialogue box has two distinct 'sections', these being:
   a) *The tree view pane*
      1. displays a hierarchical list of layers and filters in the drawing
      2. the top node (**All**) displays all layers in the drawing
      3. filters are displayed in alphabetical order
      4. a node is expanded to display a nested filter list
      5. the All Used Layers filter is read-only.
   b) *The list view pane*
      1. displays layers and layer filters with their properties and descriptions
      2. if a layer filter is selected in the tree view, the list view displays only the layers in that layer filter
      3. the All filter in the tree view displays all layers and layer filters in the drawing
      4. when a layer property filter is selected and there are no layers that fit its definition, the list view is empty.

3. The icons displayed in the Layer Properties Manager dialogue box are:
   a) *New Property Filter icon*
      1. displays the Layer Filter Properties dialogue box
      2. the user can create a layer filter based on one or more properties of the layers.
   b) *New Group Filter icon*
      1. creates a layer filter created by the user.
   c) *Layer States Manager icon*
      1. displays the Layer States Manager
      2. the user can save the current property settings for layers in a named layer state and then restore those settings later.
   d) *New Layer icon*
      1. creates a new layer
      2. the list displays a layer named LAYER 1
      3. the name is selected so that the user can enter a new layer name immediately
      4. the new layer inherits the properties of the currently selected layer in the layer list.
   e) *Delete Layer icon*
      1. marks selected layers for deletion
      2. layers are deleted when the user clicks Apply or OK
      3. only unreferenced layers can be deleted
      4. referenced layers include layers 0 and DEFPOINTS, layers containing objects and the current layer.
   f) *Set Current icon*
      1. sets the selected layer as the current layer
      2. objects that you create are drawn on the current layer.

4. Also displayed in the Layer Properties Manager dialogue box are:
   a) *Current Layer name*
      1. displays the name of the current layer
      2. this should be **0** at present.
   b) *Current Layer Status* information with:
      1. Status: indicates the item type, e.g. layer or filter
      2. Name: displays the layer or filter names. At present this is 0
      3. On: turns highlighted layers on and off and yellow is on, blue is off
      4. Freeze: freezes or thaws a highlighted layer. Yellow is thawed, blue is frozen
      5. Lock: locks or unlocks a highlighted layer. The icon display should be obvious to the user
      6. Color: allows the user to alter the colour of a highlighted layer

7. Linetype: allows the user to alter the linetype of a highlighted layer
8. Lineweight: allows the user to alter the lineweight of a highlighted layer
9. Plot Style: allows the plot style of highlighted layers to be altered
10. Plot: determines whether highlighted layers are plotted
11. Description: the user can enter a text description for the highlighted layer.

5 *a)* at present, all objects have been created on layer 0
   b) this is the default layer and is 'supplied' with AutoCAD
   c) it is the current layer and is displayed in the Layers toolbar with the layer state icons at the top of the drawing area.

6 *Tasks*
   a) The layer 0 line should be highlighted. If it is not, move the pointing arrow onto any part of this line and left-click.
   b) Move the cursor along the highlighted area and pick the yellow On/Off icon and:
   prompt   AutoCAD layer message box – Figure 11.3
   respond   pick **No** from the message box
   and   icon now displayed in blue indicating off
   respond   pick **OK** from the Layer Properties Manager dialogue box
   and   the drawing screen will be returned and no objects are displayed. The lines and circle were drawn on layer 0 which has been turned off.

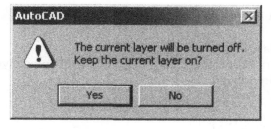

Figure 11.3 The AutoCAD layer warning message.

   c) Right-click the mouse button and pick **Repeat Layer** from the pop-up menu to re-activate the Layer Properties Manager dialogue box and:
      1. layer 0 highlighted in colour
      2. pick (left-click) the blue 'On' icon and it changes to yellow – i.e. layer is ON
      3. pick OK.
   d) The line and circle objects are again displayed as layer 0 is on.

Linetypes

AutoCAD allows the user to display objects with different linetypes, e.g. continuous, centre, hidden, dotted, etc. Until now, all objects have been displayed with a continuous linetype.

1 Activate the menu selection **Format-Layer** and:
   prompt   Layer Properties Manager dialogue box
   respond   **pick Continuous from layer 0 'line'**
   prompt   Select Linetype dialogue box
   with   Loaded linetypes:
             Linetype   *Appearance*   *Description*
             Continuous   _____   Solid line
   respond   **pick Load . . .**

prompt	Load or Reload Linetype dialogue box
with	a) Filename: acadiso.lin
	b) a list of all available linetypes in the acadiso.lin file
respond	a) scroll (at right) and pick CENTER2 – Figure 11.4
	b) hold down the Ctrl key
	c) scroll and pick HIDDEN
	d) pick OK from Load or Reload Linetypes dialogue box
prompt	Select Linetype dialogue box

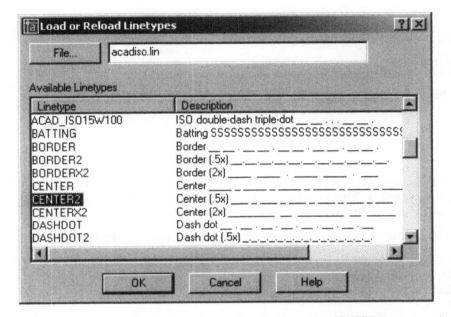

Figure 11.4 The Load or Reload Linetypes dialogue box with CENTER linetype selected.

with	Loaded Linetypes:
	Linetype *Appearance* *Description*
	CENTER2 ___ _ ___ _ ___ Center (.5x)___ _ ___ _ ___
	Continuous _____ Solid line
	HIDDEN ___ ___ ___ Hidden__ __ __ __
result	Figure 11.5 i.e. we have loaded the CENTER2 and HIDDEN linetypes from the acadiso.lin file into the Layer Properties Manager dialogue box and they can now be used in the current drawing
respond	a) pick CENTER2 and it becomes highlighted
	b) pick OK from Select Linetype dialogue box
prompt	Layer Properties Manager dialogue box
with	layer 0 having CENTER2 linetype
respond	pick OK from Layer Properties Manager dialogue box.

2 The drawing screen will display the three objects and the border with center linetype appearance. Remember that AutoCAD is an American package – hence center, and not centre.

3 *Note*: Although we have used the Layer Properties Manager dialogue box to 'load' the CENTER2 and HIDDEN linetypes, linetypes can also be loaded by selecting from the menu bar with **Format-Linetype** and the Linetype Manager dialogue box will be displayed. The required linetypes are loaded 'into the current drawing' in the same manner as described.

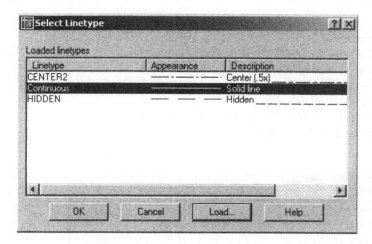

Figure 11.5 The Select Linetype dialogue box with the CENTER2 and HIDDEN linetypes loaded.

Colour

Individual objects can be displayed on the screen in different colours, but I prefer to use layers for the colour effect. This is achieved by assigning a specific colour to a named layer and will be discussed later in the chapter. For this exercise:

1 Activate the Layer Properties Manager dialogue box. Layer 0 should be highlighted.

2 Left-click on 'white' and:
   prompt   Select Color dialogue box
   respond   1. pick the red square from the standard color bar – Figure 11.6
              2. pick OK
   prompt   Layer Properties Manager dialogue box
   with   highlighted layer 0 line having color red – named and coloured square
   respond   pick OK.

3 Does the screen display the three objects with red centre lines, and the border as well?

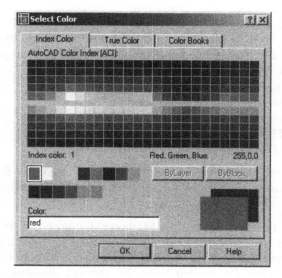

Figure 11.6 The Select Color dialogue box.

4 Note the toolbars below the Standard toolbar:
   a) the Layer toolbar displays layer 0 with a red square
   b) the Properties toolbar displays:
      1. Color Control: ByLayer is a red square
      2. Linetype Control: ByLayer is a center linetype.

5 *Task*
   By selecting the Layer Properties Manager icon from the Layers toolbar, use the Layer Properties Manager dialogue box to:
   a) set layer 0 linetype to Continuous
   b) set layer 0 colour to White/Black
   c) screen should display the original continuous linetype objects
   d) note the Layers and Properties toolbar.

6 *Notes on colour*
   a) The default AutoCAD colour for objects is dependent on your screen configuration, but is generally:
      either 1. white background with black lines
      or   2. black background with white lines.
   b) The white/black linetype can be confusing to new AutoCAD users.
   c) All colours in AutoCAD are numbered. There are 255 colours available, but the seven standard colours are:
      1: red, 2: yellow, 3: green, 4: cyan, 5: blue, 6: magenta, 7: black/white.
   d) The number or colour name can be used to select a colour. The numbers are associated with colour pen plotters.
   e) The complete 'colour palette' can be activated:
      1. from the Layer Program Manager dialogue box by picking the colour square/name under Color
      2. by selecting **Format-Color** from the menu bar.
   f) The Select Color dialogue box allows the user further colour options, these being:
      1. Index Color: default
      2. True Color
      3. Color Books.
   g) Generally only the seven standard colours will be used, but there may be the odd occasion when a colour from the default Index Color palette will be selected.

Creating new layers

Layers should be made to suit individual or company requirements, but for our purposes the layers which will be made for all future drawing work are:

Usage	Layer name	Layer colour	Layer linetype
General	0	white/black	continuous
Outlines	OUT	red	continuous
Centre lines	CL	green	center2
Hidden detail	HID	number 12	hidden
Dimensions	DIMS	magenta	continuous
Text	TEXT	blue	continuous
Hatching	SECT	number 74	continuous
Construction	CONS	to suit	continuous

These seven layers must be 'made' by us (remember that layer 0 is given to us). AutoCAD allows new layers to be created from the command line or by using the Layer Properties Manager dialogue box. We will use both methods, so:

1 Close the existing drawing (no to save changes) then open your A3SHEET standard sheet.

2   At the command line enter **–LAYER <R>** and:
    prompt   `Current layer: "0"`
               `Enter an option [?/Make/Set/New/ON/OFF/Color/Ltype/`
               `Lweight/Plot/Freeze...`
    enter   **N <R>** – the new option
    prompt   `Enter name list for new layer(s)`
    enter   **OUT, CL<R>** – outline and centre line layer names
    prompt   `Enter an option [?/Make/Set/New/ON/OFF/Color/Ltype...`
    enter   **C <R>** – the color option
    prompt   `New color [Truecolor/Colorbook]`
    enter   **1 <R>** – colour red
    prompt   `Enter name list of layer(s) for color 1 (red)`
    enter   **OUT <R>**
    prompt   `Enter an option [?/Make/Set/New/ON/OFF/Color/Ltype...`
    enter   **C <R>**
    prompt   `New color [Truecolor/Colorbook]`
    enter   **3 <R>** – colour green
    prompt   `Enter name list of layer(s) for color 3 (green)`
    enter   **CL <R>**
    prompt   `Enter an option [?/Make/Set/New/ON/OFF/Color/Ltype...`
    enter   **L <R>** – the linetype option
    prompt   `Enter loaded linetype name or [?]`
    enter   **CENTER2 <R>**
    prompt   `Enter name list of layer(s) for linetype "CENTER"`
    enter   **CL <R>**
    prompt   `Enter an option [?/Make/Set/New/ON/OFF/Color/Ltype...`
    enter   **<R>** – to end the command line sequence.

3   We have now created two new layers (OUT and CL), but still have five to create. We will use the dialogue box for these layers, so menu bar with **Format-Layer** and:
    prompt   Layer Properties Manager dialogue box
    with   three listed layers:

Name	*Color*	*Linetype*
0	White	Continuous: the AutoCAD default
CL	Green	CENTER2
OUT	Red	Continuous

    respond **pick the New Layer icon**
    and   Layer 1 added to layer list with the same properties as layer 0 – i.e. white colour and continuous linetype
    respond **pick New another four times** until there are eight listed layers in total:
               a) the original layer 0
               b) OUT and CL created from the command line
               c) the added new layers, named Layer 1–Layer 5.

4   *Naming the new layers*
    a) move pick arrow onto the **Layer 1 name** and pick it and:
        1. it becomes highlighted
        2. a box effect is placed around the Layer 1 name
    b) press the **F2 key**
    c) enter HID from the command line
    d) repeat steps 1–3 and rename layer 2–layer 5 as follows:
        layer 2: DIMS   layer 3: TEXT   layer 4: SECT   layer 5: CONS

5 *Assigning linetypes to the new layers*
   a) pick the layer HID
   b) move pick arrow to **Linetype-Continuous** and pick it
      prompt   Select Linetype dialogue box
      with   CENTER2 and Continuous linetype names
      respond   **pick Load**
      prompt   Load or Reload Linetypes dialogue box
      respond   1. scroll and pick HIDDEN
                    2. pick OK
      prompt   Select Linetype dialogue box
      with   HIDDEN added to the linetype names
      respond   1. pick HIDDEN
                    2. pick OK
      prompt   Layer Properties Manager dialogue box
      with   layer HID with HIDDEN linetype
   c) the remaining layers all have continuous linetype.

6 *Assigning colours to the new layers*
   a) highlight layer HID if not still highlighted
   b) move pick arrow to **Color-White** and pick it
      prompt   Select Color dialogue box
      respond   1. at Color: white, enter 12
                    2. pick OK
      prompt   Layer Properties Manager dialogue box
      with   layer HID with color number 12
   c) pick the other named layer lines and set the following colours:
      DIMS: magenta   TEXT: blue   SECT: number 74   CONS: colour to suit

7 *Adding a layer description*
   a) select the layer 0 line
   b) move the arrow and pick under description to display a box effect
   c) press **F2** and enter **GENERAL** from the keyboard
   d) pick the other layer lines and enter your own layer description.

8 At this stage the Layer Properties Manager dialogue box should resemble Figure 11.7.

9 Study the information added, then pick OK. This will save the created layers with linetypes and colours.

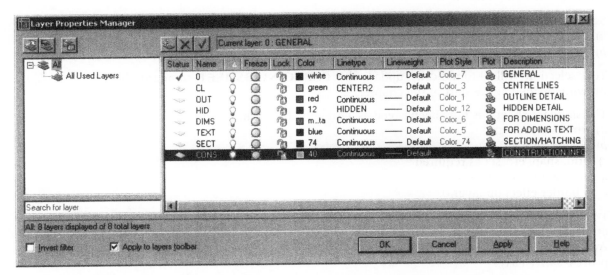

Figure 11.7 The Layer Properties Manager dialogue box with the new layer information.

The current layer

The current layer is the one on which all objects are drawn. The current layer name appears in the Layers toolbar which is generally docked below the Standard toolbar at the top of the screen. The current layer is also named in the Layer Properties Manager dialogue box when it is activated and is displayed with a green status tick. The current layer is 'set' by the user.

1 Activate the Layer Properties Manager dialogue box and:
   prompt   Layer Properties Manager dialogue box
   with   the layers created earlier displayed in numeric then alphabetical order, i.e. 0, CL, CONS . . . TEXT
   respond 1. pick layer line OUT – becomes highlighted
              2. pick the Set Current icon and the green status tick is placed at the OUT line layer
              3. pick OK.

2 The drawing screen will be returned, and the Layers toolbar will display OUT with a red box, and ByLayer will also display a red box.

3 The current layer can be set:
   a) from the Layer Properties Manager dialogue box as described in step 1
   b) from the Layers toolbar by scrolling at the right of the layer name and selecting the name of the layer to be current.

4 I generally start a drawing with layer OUT current, but this is a personal preference. Other users may want to start with layer CL or 0 as the current layer, but it does not matter as long as the objects are eventually 'placed' on their correct layers.

5 Having created layers it is now possible to draw objects with different colours and linetypes, simply by altering the current layer. All future work should be completed with layers used correctly, i.e. if text is to be added to a drawing, then the TEXT layer should be current.

Saving the layers to the standard sheet

1 Ensure that layer OUT is current.

2 Erase the two lines and circle.

3 Select one of the following:
   a) the Save icon from the Standard toolbar
   b) menu bar with **File-Save**.

4 This selection will automatically update the **C:\BEGIN\A3SHEET** standard sheet drawing opened at the start of the chapter.

5 The standard sheet has now been saved as a drawing file with:
   a) units set to metric
   b) sheet size A3
   c) grid, snap, etc. set as required
   d) several new layers
   e) a border effect on layer 0.

6 With the layers having been saved to the A3SHEET standard sheet, the layer creation process does not need to be undertaken every time a drawing is started. Additional layers can be added to the standard sheet at any time – is not the process fairly easy?

Layer states

Layers can have different 'states', for example, they could be:
a) ON or OFF
b) THAWED or FROZEN
c) LOCKED or UNLOCKED.

The layer states are displayed both in the Layer Control box of the Properties toolbar as well as in the Layer Properties Manager dialogue box itself. In both, the layer states are displayed in icon form. These icon states are:
a) yellow – ON, THAWED
b) bluish grey – OFF, FROZEN
c) lock and unlock should be obvious?

The following exercise will investigate layer states:

1 Your A3SHEET standard sheet should be displayed with the black border and layer OUT current.

2 Using the eight layers (our created seven and layer 0):
   a) make each layer current in turn
   b) draw a 50-radius circle on each layer anywhere within the border
   c) make layer 0 current
   d) toggle the grid off to display eight coloured circles.

3 The green circle will be displayed with center linetype and the hidden linetype circle displayed with colour number 12.

4 Activate the Layer Properties Manager dialogue box and:
   a) pick the yellow On icon of the CL layer line and:
      1. the line becomes highlighted
      2. the icon changes to blue
   b) pick OK and the drawing screen is returned with no green circle displayed
   c) the CL layer has been turned off.

5 Layer Properties Manager dialogue box and:
   a) pick the yellow Freeze icon of DIMS layer line and:
      1. line becomes highlighted
      2. icon changes to blue
   b) pick OK and drawing screen is returned with no magenta circle displayed
   c) the DIMS layer has been frozen.

6 Pick the scroll arrow from the Layers toolbar and:
   a) note CL and DIMS display information!
   b) pick Lock/Unlock icon on HID layer – note icon appearance
   c) left-click to side of pull-down menu
   d) hidden linetype circle still displayed although the HID layer has been locked.

7 From the Layers toolbar pull-down menu:
   a) pick the Freeze and Lock icons for the SECT layer
   b) left-click to side
   c) no dark green circle, as layer SECT was frozen.

8 Using the pull-down layer menu effect:
   a) turn off, freeze and lock the TEXT layer
   b) no blue circle as layer TEXT was turned off and frozen.

9 Activate the Layer Properties Manager dialogue box and:
   a) note the icon display for all the layers
   b) make layer OUT current

c) pick Freeze icon for layer OUT
   d) Warning message – **Cannot freeze the current layer**
   e) pick OK
   f) turn layer OUT off
   g) Warning message:
      The current layer will be turned off
      Keep the current layer on?
   h) pick **No** from this message dialogue box
   i) pick OK from Layer Properties Manager dialogue box
   j) no red circle displayed as layer OUT has been turned off although it is the current layer.

10 The screen should now display:
   a) a black circle – drawn on layer 0
   b) a hidden linetype circle (colour 12) – on frozen layer HID
   c) a circle on layer CONS coloured to your own selection
   d) the black border – drawn on layer 0.

11 **Thus objects will not be displayed if their layer is OFF or FROZEN.**

12 Make layer 0 current.

13 a) erase the hidden linetype circle – you cannot and the prompt line displays: **1 was on a locked layer**
   b) cancel the Erase command.

14 Using the CEN object snap, draw a line from the centre of the hidden linetype circle to the centre of the black circle. You can reference the hidden linetype circle, although it is on a locked layer.

15 a) make layer OUT current
   b) Warning message: **The current layer is turned off**
   c) pick OK
   d) draw a line from 50,50 to 200,200
   e) no line is displayed and the reason should be obvious.

16 Erase the coloured circle on layer CONS then activate the Layer Properties dialogue box.

17 Pick the **New Property Filter icon** and:
   prompt   Layer Filter Properties dialogue box
   respond   1. alter filter name to **TRY1**
               2. pick the box at Status, scroll and select the layer in use (bluish) icon
               3. pick the box at Name and * displayed – wildcard for all layer names
               4. pick the box at On, scroll and select the On icon (dialogue box as Figure 11.8)
               5. pick OK
   prompt   Layer Properties Manager dialogue box
   with   TRY1 added to the tree hierarchy side and only the four filtered layers displayed.

18 Using the tree side of the Layer Properties Manager dialogue box:
   a) TRY1 current – 4 layers displayed
   b) pick All Used Layers – 7 layers displayed with no CONS layer
   c) pick All – 8 layers displayed
   d) collapse the tree hierarchy by picking the (−) at All and 8 layers displayed
   e) expand the tree side by picking the (+) to display the other options
   f) pick the TRY1 filter to display 4 layers

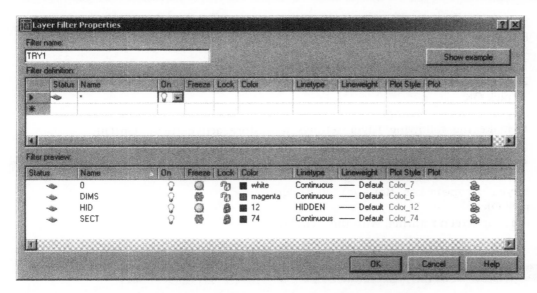

Figure 11.8 The Layer Filter Properties dialogue box for filter TRY1.

    g) pick on the **Search for layer** line and:
        prompt   `* displayed`
        respond   pick to right of * and backspace
        and   * disappears and no layers displayed
        respond   1. enter OUT and no layer displayed – why?
                       2. pick to right of OUT and backspace to remove the name
                       3. enter HID and hidden line details displayed
                       4. pick All from tree side then OK from dialogue box
        prompt   drawing screen with two circles and a line.

19 Using the Layer Properties Manager dialogue box:
    a) ensure all layers are: ON, THAWED and UNLOCKED
        1. On   – 3 layers were turned off
        2. Thaw – 3 layers were frozen
        3. Unlock – 3 layers were locked
    b) pick OK
    c) 7 circles and 2 lines displayed
    d) remember that the CONS layer circle was erased.

20 The layer states can be activated from:
    a) the Layer Properties Manager dialogue box
    b) the Layers toolbar pull-down menu.

21 This rather long exercise is now complete.

Saving layer states

Layer states can be saved so that the user can restore the original layer 'settings' at any time. To demonstrate this:

1   Erase the two lines to leave the circles and the A3 border.

2   At the command line enter LAYER <R> and:
    prompt   `Layer Properties Manager dialogue box`
    respond   **pick the Layer States Manager icon** from the tree side

prompt	Layer States Manager dialogue box
respond	pick New
prompt	New layer State to Save dialogue box
respond	1. New layer state name: enter **A3SHEET** 2. Description: enter My layer states for an A3-size paper 3. pick OK
prompt	Layer States Manager dialogue box
respond	1. make the following Layer settings to restore active (tick): On/Off; Frozen/Thawed; Locked/Unlocked; Color; Linetype; Lineweight 2. dialogue box as Figure 11.9 3. pick Close
prompt	Layer Properties Manager dialogue box
respond	pick OK
and	drawing screen returned with the 7 circles.

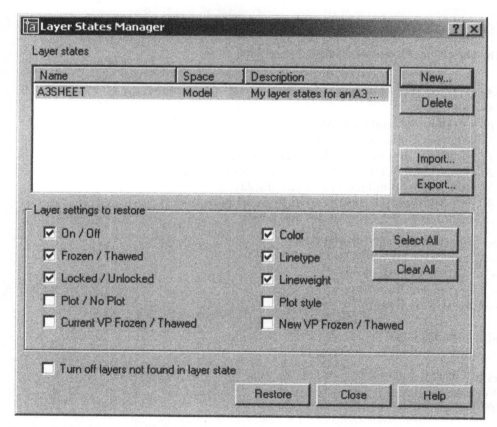

Figure 11.9 The Layer States Manager dialogue box for A3SHEET.

3 Scroll at layer information from the Layers toolbar and:
   a) turn off and lock all layers
   b) message about current layer displayed – pick OK
   c) screen display is blank – obviously?

4 Pick the **Layer Properties Manager icon** from the Layers toolbar and:
   prompt   Layer Properties Manager dialogue box
   with   all layers with off and locked icons
   respond **pick the Layer States Manager icon**
   prompt   Layer States Manager dialogue box

respond	1. pick A3SHEET (probably highlighted)
	2. pick Restore
prompt	Layer Properties Manager dialogue box
with	all layers with on and unlocked icons
respond	pick OK.

5   The original screen objects will be displayed.

6   It may be useful to have the saved A3SHEET layer states added to the A3SHEET standard sheet. I will let you decide for yourself if you think it would be useful.

The Layers toolbar

The Layers toolbar is generally displayed at all times and is usually docked below the Standard toolbar at the top of the drawing screen under the title bar. This toolbar gives information about layers and their states, and Figure 11.10 displays the toolbar details for our current drawing.

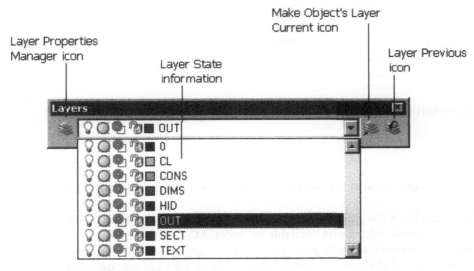

Figure 11.10   The Layers toolbar description.

The various 'parts' of the toolbar are:

1   *Layer Properties Manager icon*
    Selecting this icon will display the Layer Properties Manager dialogue box. It is a quicker alternative to the menu bar sequence **Format-Layer**.

2   *Make Object's Layer Current icon*
    Select this icon and:

prompt	Select object whose layer will become current
respond	**pick the blue circle**
prompt	TEXT is now the current layer
and	blue square and TEXT displayed in Layers toolbar.

3   *Layer Previous icon*
    Select this icon and:

prompt	Restored previous layer status
and	Layer OUT will again be current.

4   *Pull-down layer information*
    Allows the user to quickly set a new current layer, or activate one of the layer states, e.g. off, freeze, lock, etc.

The Properties toolbar

The Properties toolbar is also generally displayed at all times and is docked beside the Layers toolbar. Figure 11.11 displays the Properties toolbar for our A3SHEET drawing at this stage. The three main options of the Properties toolbar are:

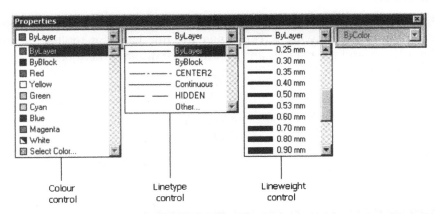

Figure 11.11   The Properties toolbar description.

1   *Color Control*
    a) By scrolling at the colour control arrow, the user can select one of the standard colours.
    b) By picking *Select Color* from the pull-down menu, the Select Color dialogue box is displayed allowing access to all colours available in AutoCAD.
    c) Selecting a colour by this method will allow objects to be drawn with that particular colour, irrespective of the current layer colour. This is **not recommended** but may be useful on certain occasions – i.e. coloured objects should (ideally) be created on their own layer.
    d) It is recommended that **Color Control displays ByLayer**.

2   *Linetype Control*
    a) Similar to colour control, the scroll arrow displays all linetypes loaded in the current drawing.
    b) By picking *Other* the Linetype Manager dialogue box will be displayed, allowing the user to load any other linetype from a named file – usually **acadiso.lin**.
    c) By selecting a linetype from the pull-down menu, the user can create objects with this linetype, independent of the current layer linetype. **This is not recommended**.
    d) Layers should be used to display a certain linetype, but it may be useful to have different linetypes displayed on the one layer occasionally.
    e) It is recommended that **Linetype Control displays ByLayer**.

3   *Lineweight Control*
    a) Lineweight allows objects to be displayed (and plotted) with different thicknesses from 0 mm to 2.11 mm.
    b) The lineweight control scroll arrow allows the user to select the required thickness.

c) We will be investigating this topic in greater detail in a later chapter, and will not discuss at this stage.
d) **Lineweight Control should display ByLayer**.

Renaming and deleting layers

Unwanted or wrongly named layers can easily be deleted or renamed in AutoCAD 2005. To investigate how this is achieved:

1. The A3SHEET drawing should still be displayed with 7 coloured circles (the circle on layer CONS was erased).
2. Erase any lines.
3. Make layer OUT current.
4. Activate the Layer Properties Manager dialogue box by picking the Layer icon from the Layers toolbar and:
   a) pick the New Layer twice to add two new layers to the list – Layer 1 and Layer 2
   b) rename Layer 1 as NEW1 and Layer 2 as NEW2
   c) pick **All Used Layers** from the tree side and CONS, NEW1 and NEW2 will not be listed
   d) pick **All** from the tree side to redisplay all layers
   e) pick OK from the Layer Properties Manager dialogue box.
5. Making each new layer current in turn, draw a circle anywhere on the screen.
6. Make layer OUT current.
7. *a)* Activate the Layer Properties Manager dialogue box and:
   respond   pick NEW1 layer line then the Delete icon
   prompt   **AutoCAD message** – The selected layer was not deleted.
   b) Pick OK from this message box (Figure 11.12) then pick OK from the Layer Properties Manager dialogue box.

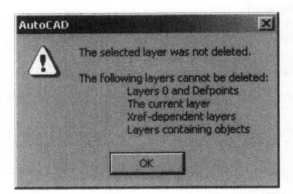

Figure 11.12 The AutoCAD layer warning message.

8. *a)* erase the two added circles
   b) activate the Layer Properties Manager dialogue box
   c) pick NEW1 then Delete icon
   d) pick NEW2 then Delete icon
   e) an X is placed at Status for the NEW1 and NEW2 layers
   f) pick OK.
9. Re-activate the Layer Properties Manager dialogue box and layers NEW1 and NEW2 will not be listed.

10 This completes this chapter, so:
   a) do not exit AutoCAD
   b) attempt the assignment
   c) read the summary and the note which follows.

Assignment

1 Ensure that your A3SHEET has been saved with the named layers from this chapter.

2 Complete Activity 6 which requires you to:
   a) open your A3SHEET drawing
   b) complete the assignment
   c) save the completed work as C:\BEGIN\ACT6.

Summary

1 Layers are one of the most important concept in AutoCAD. **Perhaps even the most important?**

2 Layers allow objects to be created with different colours and linetypes.

3 Layers are created using the Layer Properties Manager dialogue box – often called layer control.

4 There are over 250 colours available, but the seven standard colours should be sufficient for most user requirements.

5 Linetypes are loaded by the user as required.

6 Layers saved to a standard sheet need only be created once.

7 New layers can easily be added as and when required.

8 The layer states are:
   ON   all objects are displayed and can be modified
   OFF   objects are not displayed
   FREEZE   similar to OFF but the screen regenerates faster
   THAW   undoes a frozen layer
   LOCK   objects are displayed but cannot be modified
   UNLOCK   undoes a locked layer.

9 Layer states are displayed and activated in icon form from the Layer Properties Manager dialogue box or using Layer Control from the Layers toolbar.

10 Care must be taken when modifying a drawing with layers which are turned off or frozen. More on this later.

11 Layers can be renamed at any time.

12 Unused layers can be deleted at any time.

13 Layer states can be saved and restored at any time.

Notes

1 By now you should have the confidence and ability to create line and circle objects by various methods, e.g. co-ordinate entry, referencing existing objects with OSNAP, etc.

2 We are now ready to create a working drawing which will be used to introduce several new concepts.

Chapter 12

User exercise 1

1. Open your A3SHEET standard sheet.
2. Refer to Figure 12.1 and:
   a) draw full size the component given
   b) use layer OUT
   c) a start point is given – **Use it as it is important for future work**
   d) *do not attempt to add the dimensions*
   e) use absolute co-ordinates for the (50,50) start point and relative co-ordinates for the outline
   f) use absolute co-ordinates for the circle centres – some 'sums' are required, but these should not give you any problems.

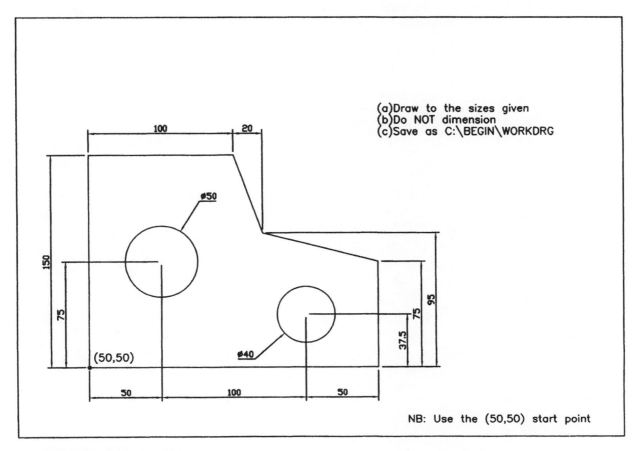

Figure 12.1 User exercise 1.

3 When the drawing is complete, menu bar with **File-Save As** and:
   prompt   Save Drawing As dialogue box
   with   File name: A3SHEET
   respond 1. ensure C:\BEGIN is current folder to Save in
              2. alter file name to: WORKDRG
              3. pick Save.

4 We have now opened our A3SHEET standard sheet, completed a drawing exercise and saved this drawing with a different name to that which was opened.

5 This is (at present) the method which will be used to complete all new drawing exercises.

6 Now continue to the next chapter.

7 *Note*: It should be apparent that the circles have been drawn without centre lines. This is deliberate, as it will allow us to investigate other CAD draughting techniques in a later chapter.

Chapter **13**

Fillet and chamfer

In this chapter we will investigate how the fillet and chamfer commands can be used to modify an existing drawing. Both commands can be activated by icon, menu bar selection or keyboard entry.

1 Open the C:\BEGIN\WORKDRG created in the previous chapter or simply continue from the previous chapter.
2 Ensure layer OUT is current and display the Draw and Modify toolbars.
3 Refer to Figure 13.1.

Fillet

A fillet is a radius added to existing line/arc/circle objects. The fillet radius must be specified before the objects to be filleted can be selected.

1 Select the FILLET icon from the Modify toolbar and:
   prompt   Current settings: MODE=TRIM, Radius=??
              Select first object or [Polyline/Radius/Trim/mUltiple]
   enter   **R <R>** – the radius option
   prompt   Specify fillet radius<??>
   enter   **15 <R>**

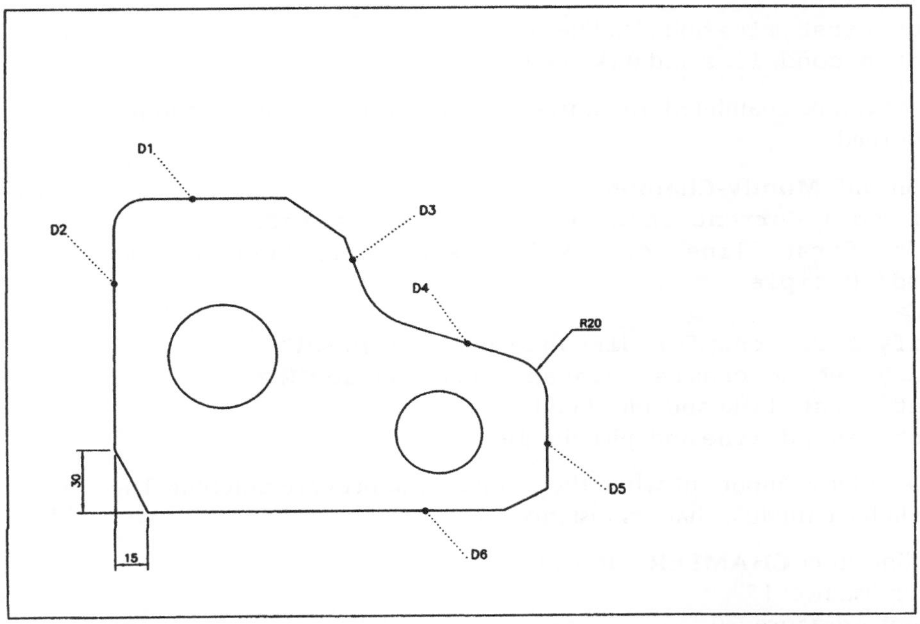

Figure 13.1 C:\BEGIN\WORKDRG after using the fillet and chamfer commands.

prompt   Select first object or [Polyline/Radius/Trim/mUltiple]
respond   **pick line D1**
prompt   Specify second object
respond   **pick line D2**.

2   The corner selected will be filleted with a radius of 15, and the two 'unwanted line portions' will be erased and the command line will be returned.

3   From the menu bar select **Modify-Fillet** and:
prompt   Current settings: MODE=TRIM, Radius=15
           Select first object or [Polyline/Radius/Trim/mUltiple]
enter   **R <R>** – the radius option
prompt   Specify fillet radius<15.00>
enter   **30 <R>**
prompt   Select first object and pick line D3
prompt   Select second object and pick line D4.

4   At the command line enter **FILLET <R>** and:
     a) set the fillet radius to 20
     b) fillet the corner indicated.

Chamfer

A chamfer is a straight 'cut corner' added to existing line objects. The chamfer distances must be 'set' prior to selecting the object to be chamfered.

1   Select the CHAMFER icon from the Modify toolbar and:
prompt   (TRIM mode) Current chamfer Dist1=?? Dist2=??
           Select first line or[Polyline/Distance/Angle/Trim/Method/mUltiple]
enter   **D <R>** – the Distance option
prompt   Specify first chamfer distance<??>
enter   **20 <R>**
prompt   Specify second chamfer distance<25.00>
enter   **20 <R>**
prompt   Select first line and pick line D1
prompt   Select second line and pick line D3.

2   The selected corner will be chamfered, the unwanted line portions removed and the command line returned.

3   Menu bar selection with **Modify-Chamfer** and:
prompt   (TRIM mode) Current chamfer Dist1=25, Dist2=25
           Select first line or [Polyline/Distance/Angle/Trim/Method/mUltiple]
enter   **D <R>**
prompt   Specify first chamfer distance and enter: **10 <R>**
prompt   Specify second chamfer distance and enter: **20 <R>**
prompt   Select first line and: pick line D5
prompt   Select second line and: pick line D6.

4   Note that the pick order is important when the chamfer distances are different. The first line picked will have the first chamfer distance set.

5   At the command line enter **CHAMFER <R>** and:
     a) set first chamfer distance: 15
     b) set second chamfer distance: 30
     c) chamfer the corner indicated.

Saving the modified work drawing

When the three fillets and three chamfers have been added to the component, select from the menu bar **File-Save**. This will automatically update the existing **C:\BEGIN\WORKDRG** drawing file.

Error messages

The fillet and chamfer commands are generally used without any problems, but the following error messages may be displayed at the command prompt:

1. Radius is too large.
2. Distance is too large.
3. Chamfer requires 2 lines (not arc segments).
4. No valid fillet with radius ??
5. Lines are parallel – this is a chamfer error.
6. Cannot fillet an entity with itself.

These error messages should be self-evident to the user.

Fillet and chamfer options

Although simple to use, the fillet and chamfer commands have several options. It is in your own interest to attempt the following exercises, so:

1. Erase the filleted/chamfered component from the screen but ensure that C:\BEGIN\WORKDRG has been saved.
2. Refer to Figure 13.2.
3. Draw three–four inclined lines then use the fillet and chamfer commands with the radius (R) and distance (D) values given. The effect of using the fillet/chamfer commands with inclined lines is displayed:
   a) fillet effect – Figure 13.2(a)
   b) chamfer effect – Figure 13.2(b).
4. Both the fillet and chamfer commands have a Polyline option and this option will be discussed when we have investigated the polyline command in a later chapter.
5. The two commands have a TRIM option and this effect is displayed in Figure 13.2(c). The option is obtained by entering T <R> at the prompt line after activating the fillet/chamfer command and setting the radius/distance value. The response is:
   prompt   Enter trim mode option [Trim/No Trim]<Trim>
   enter   a) **T <R>**: corner removed. This is the default
             b) **N <R>**: corners not removed.
6. Note that if the no trim (N) option is used, this effect will always be obtained until the user 'resets' the trim (T) option.
7. Both commands have a multiple option which allows the command to be used repetitively with the same radius/distances. The user enters U <R> at the prompt line when the command has been activated. This option is only valid for the current command.

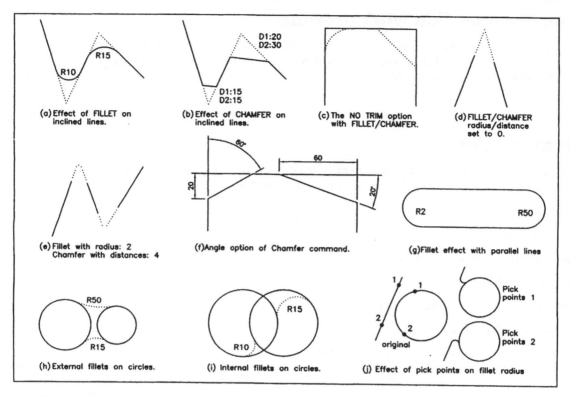

Figure 13.2 The fillet and chamfer options.

8 The two commands can be used to extend two inclined lines to a point as demonstrated in Figure 13.2(d) with:
   either   a) the fillet radius set to 0
   or   b) both chamfer distances set to 0.

9 Chamfer with one distance set to 20 and the other set to 0. Think about this effect!

10 Interesting effects can be created with inclined lines if the fillet radius or the chamfer distances are 'small'. The lines are extended if required, and the fillet/chamfer effect 'added at the ends'. This effect is displayed in Figure 13.2(e).

11 a) Chamfer has an angle option and when the command is selected:
   prompt   Select first line or [Polyline/Distance/Angle/Trim/
            Method/mUltiple]
   enter   A <R> – the angle option
   prompt   Specify chamfer length on the first line and enter:
            20<R>
   prompt   Specify chamfer angle from the first line and enter:
            60<R>.
   b) The required lines can now be chamfered as Figure 13.2(f) which displays:
      1. a length of 20 and angle of 60
      2. a length of 60 and angle of 20.
   c) The user has to be careful when selecting the first line as this is used for the distance value entered.

12 The fillet command can be used with parallel lines, the effect being to add 'an arc' to the ends selected. This arc is added independently of the set fillet radius as Figure 13.2(g) demonstrates. The fillet radii was set to 2 and 50, but the added 'arc' is the same at both ends. In my Figure 13.2(g) example, the actual arc radius is 15. Any idea why?

13 Two circles can be filleted but are not 'trimmed' as lines. The fillet effect on circles is displayed:
   a) externally as Figure 13.2(h)
   b) internally as Figure 13.2(i).

14 Two circles cannot be chamfered. If any circle is selected:
   prompt   Chamfer requires 2 lines (not arc segments).

15 Lines, circles and arcs can be filleted, but the position of the pick points is important as displayed in Figure 13.2(j).

16 Chamfer has a Method option and when M is entered:
   prompt   Enter trim method [Distance/Angle]
   and   a) entering D sets the two distance method (default)
               b) entering A sets the length and angle method.

Summary

1 FILLET and CHAMFER are Modify commands, activated from the menu bar, by icon selection or by keyboard entry.

2 Both commands require the radius/distances to be set before they can be used.

3 When values are entered, they become the defaults until altered by the user.

4 Lines, arcs and circles can be filleted.

5 Only lines can be chamfered.

6 A fillet/chamfer value of 0 is useful for extending two inclined lines to meet at a point.

7 Both commands have several useful options.

Chapter **14**

The offset, extend, trim and change commands

In this chapter we will investigate OFFSET, EXTEND and TRIM – three of the most commonly used draughting commands. We will also investigate how an object's properties can be altered with the CHANGE command, and finally we will discuss LTSCALE, a system variable.

To demonstrate these new commands:
a) Open C:\BEGIN\WORKDRG – easy by now?
b) Ensure layer OUT is current and display the Draw and Modify toolbars.

Offset

The offset command allows the user to draw objects parallel to other selected objects. Lines, circles and arcs can all be offset. The user specifies:
a) an offset distance
b) the side to offset the selected object.

1   Refer to Figure 14.1.

2   Pick the **OFFSET icon** from the Modify toolbar and:
    prompt   Specify offset distance or [Through]
    enter   **50 <R>** – the offset distance
    prompt   Select object to offset or <exit>
    respond   **pick line D1**
    prompt   Specify point on side to offset
    respond   **pick any point to right of line D1 as indicated**
    and   line D1 will be offset by 50 units to right
    prompt   Select object to offset – i.e. any more 50-offsets
    respond   **pick line D2**
    prompt   Specify a point on side to offset
    respond   **pick any point to left of line D2 as indicated**
    and   line D2 will be offset 50 units to left
    prompt   Select object to offset – i.e. any more 50-offsets
    respond   **right-click** to end command.

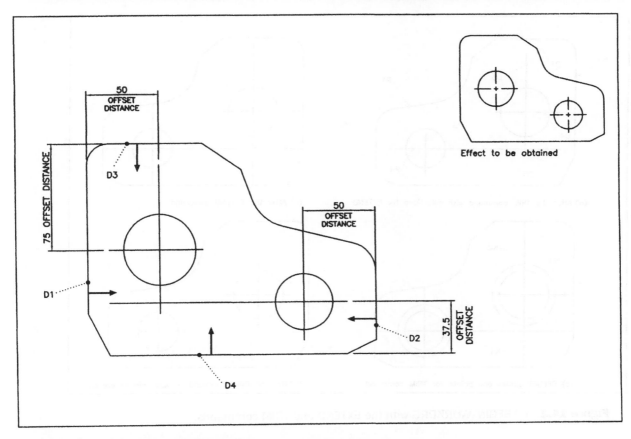

Figure 14.1 C:\BEGIN\WORKDRG after the OFFSET command.

3   Menu bar with **Modify-Offset** and:
    prompt   Specify offset distance or [Through]<50.00>
    enter   **75 <R>**
    prompt   Select object to offset and **pick line D3**
    prompt   Specify a point on side to offset and pick as indicated.

4   At the command line enter **OFFSET <R>** and:
    a) set an offset distance of 37.5
    b) offset line D4 as indicated.

5   We have now created lines through the two circle centres and later in the chapter we will investigate how these lines can be modified to be 'real centre lines'.

6   Continue to the next part of the exercise.

Extend

This command will extend an object 'to a boundary edge', the user specifying:
a) the actual boundary – an object
b) the object which has to be extended.

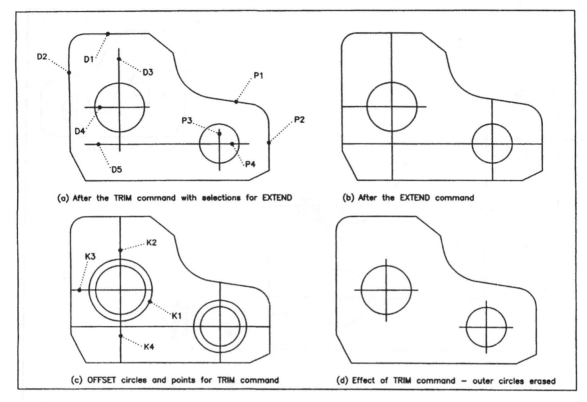

Figure 14.2 C:\BEGIN\WORKDRG with the EXTEND and TRIM commands.

1 Refer to Figure 14.2(a) and with SNAP OFF, select the EXTEND icon from the Modify toolbar and:
   prompt   Current settings: Projection=UCS, Edge=Extend
              Select boundary edges...
              Select objects
   respond **pick line D1**
   prompt   1 found
              Select objects – i.e. any more boundary edges
   respond **pick line D2**
   prompt   1 found, 2 total
              Select objects
   respond **right-click** to end boundary edge selection
   prompt   Select object to extend or shift-select to trim or [Project/Edge/Undo]
   respond **pick lines D3, D4 and D5 then right-click-Enter**.

2 The three lines will be extended to the selected boundary edges.

3 From the menu bar select **Modify-Extend** and:
   prompt   Select objects – i.e. the boundary edges
   respond **pick lines P1 and P2 then right-click**
   prompt   Select object to extend
   respond **pick lines P3 and P4 then right-click and pick Enter**.

4 At the command line enter **EXTEND <R>** and extend the two vertical 'centre lines' to the lower horizontal outline.

5 When complete, the drawing should resemble Figure 14.2(b).

Trim

Allows the user to trim an object 'at a cutting edge', the user specifying:
a) the cutting edge – an object
b) the object to be trimmed.

1 Refer to Figure 14.2(c) and OFFSET the two circles for a distance of 5 'outwards' – easy?

2 Extend the top horizontal 'circle centre line' to the offset circle – should be obvious why?

3 Select the TRIM icon from the Modify toolbar and:
    prompt   `Current settings: Projection=UCS, Edge=Extend`
                    `Select cutting edges...`
                    `Select objects`
    respond **pick circle K1 then right-click**
    prompt   `Select object to trim or shift-select to extend or [Project/Edge/Undo]`
    respond **pick lines K2, K3 and K4 then right-click-Enter**.

4 From menu bar select **Modify-Trim** and:
    prompt   `Select objects` – i.e. the cutting edge
    respond **pick the other offset circle then right-click**
    prompt   `Select object to trim`
    respond **pick the four circle 'centre lines' then right-click**.

5 Now we have 'neat lines' through the circle centres as Figure 14.2(d).

6 At this stage select the Save icon from the Standard toolbar to automatically update C:\BEGIN\WORKDRG. We will recall it shortly.

Additional exercises

Offset, extend and trim are powerful commands and can be used very easily. To demonstrate additional use for the commands, try the examples which follow:

1 Erase all objects from the screen – have you saved WORKDRG?

2 Refer to Figure 14.3 and attempt the exercises which follow.

3 *Offset for circle centre point*
Using OFFSET to obtain a circle centre point is one of the most common uses for the command.
Figure 14.3(a) demonstrates offsets of 18.5 horizontally and 27.8 vertically to position the circle centre point.
Question: What about inclined line offsets?

4 *Offset through option*
   a) This is a very useful option, as it allows an object to be offset through a specified point. When the command is activated:
        prompt   `Specify offset distance or [Through]`
        enter   **T <R>** – the through option
        prompt   `Select object to offset or <exit>`
        respond **pick the required object** – e.g. a line
        prompt   `Specify through point`
        respond **Snap to Center icon and pick the circle**
        prompt   `Specify object to offset` – i.e. any more through offsets
        respond **<RETURN>** to end command.
   b) The line will be offset through the circle centre – Figure 14.3(b).

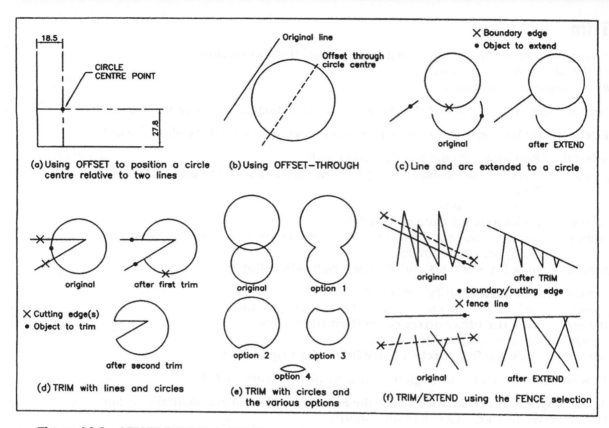

Figure 14.3 OFFSET, EXTEND and TRIM exercises.

5 *Extending lines and arcs*
   Lines and arcs can be extended to other objects, including circles – Figure 14.3(c).

6 *Trim lines and circles*
   Lines, circles and arcs can be trimmed to 'each other' and Figure 14.3(d) demonstrates trimming lines with a circle.

7 *Trimming circles*
   Circles can be trimmed to each other, but the selected objects to be trimmed can give different effects. Can you obtain the various options in Figure 14.3(e)?

8 *Trim/Extend with a fence selection*
   When several objects have to be trimmed or extended, the fence selection option can be used – Figure 14.3(f). The effect is achieved by:
   a) activating the command
   b) selecting the boundary (extend) or cutting edge (trim)
   c) entering F <R> – the fence option
   d) draw the fence line then right-click.

9 *Trim/Extend toggle effect*
   When the trim/extend command is activated and an object selected for the cutting edge/boundary, the prompt line will display:

   ```
   Select object to trim (extend) or shift-select to extend (trim)
   ```

   By using the shift key the user can 'toggle' between the two commands. This is very useful. Try it with a few lines.

   This completes the additional exercises which do not need to be saved.

Changing the offset centre lines

The WORKDRG drawing was saved with 'centre lines' obtained using the offset, extend and trim commands. These lines pass through the two circle centres, but they are continuous lines and not centre lines. We will modify these lines to be centre lines using the CHANGE command. Changing the properties of an object will be fully investigated in a later chapter, but for now:

1 Re-open C:\BEGIN\WORKDRG saved earlier in this chapter.

2 At the command line enter **CHANGE <R>** and:
   prompt   Select objects
   respond   **pick the four offset centre lines then right-click**
   prompt   Specify change point or [Properties]
   enter   **P <R>** – the properties option
   prompt   Enter property to change [Color/Elev/LAyer/LType/ltScale/LWeight/Thickness]
   enter   **LA <R>** – the layer option
   prompt   Enter new layer name<OUT>
   enter   **CL <R>**
   prompt   Enter property to change – i.e. any more changes?
   respond   **right-click-Enter** to end command.

3 The four selected lines will be displayed as green centre lines, as they were changed to the CL layer. This layer was made with centre linetype and colour green. Confirm with **Format-Layer** if you are not convinced!

4 Although the changed lines are centre lines, their 'appearance' may not be ideal and an additional command is required to 'optimise' the centre line effect. This command is LTSCALE.

LTSCALE

LTSCALE is a system variable used to 'alter the appearance' of non-continuous lines on the screen. It has a default value of 1.0 and this value is altered by the user to 'optimise' centre lines, hidden lines, etc. To demonstrate its use:

1 At the command line enter **LTSCALE <R>** and:
   prompt   Enter new linetype scale factor<1.0000>
   enter   **0.6 <R>**.

2 The four centre lines may be 'better defined' for the user?

3 The value entered for LTSCALE depends on the type of lines being used in a drawing, and can be further refined – more on this in a later chapter.

4 Try other LTSCALE values and optimise the centre lines to your needs, then save the drawing as C:\BEGIN\WORKDRG.

5 *Notes*
   a) LTSCALE must be entered from the command line. There is no icon or menu bar sequence to activate the command.
   b) The LTSCALE system variable is GLOBAL. This means that when its value is altered, all linetypes (centre, hidden, etc.) will automatically be altered to the new value. In a later chapter we will discover how this can be 'overcome'.
   c) The actual LTSCALE value entered varies greatly and some users may find that the 0.6 value is totally unsuitable and that a larger value (e.g. 12, 15, 25) is more suited to display the centre lines with a reasonable definition. This difference in

the LTSCALE value is dependant on other factors which will not be discussed at this stage. Ensure that you have an LTSCALE value entered which displays the green centre lines to your satisfaction.

Question

In our offset exercise we obtained four circle centre lines using the offset command. These lines were then changed to the CL layer to display them as 'real centre lines'.

The question I am repeatedly asked by new AutoCAD users is: 'Why not use offset with the CL layer current?' This question is reasonable so to investigate it:

1   C:\BEGIN\WORKDRG should still be displayed with the four 'changed' centre lines.

2   Make layer CL current.

3   Set an offset distance of 30 and offset:
    a) any red perimeter line
    b) any green centre line.

4   The effect of the offset command is:
    a) the offset red line is a red continuous line
    b) the offset green centre line is a green centre line.

5   The offset command will therefore offset an object 'as it was drawn' and is independent of the current layer.

Match properties

This is a very useful 'tool' to the user, as it does exactly what it says – it matches properties.

1   Erase the two offset lines, still with layer CL current.

2   Select the Match Properties icon from the Standard toolbar and:
    prompt   Select source object
    respond   **pick any green centre line**
    prompt   Select destination object(s) or [Settings]
    respond   **pick the two circles then right-click-Enter**.

3   The circles will now be displayed as green centre lines.

4   Menu bar with **Modify-Match Properties** and:
    a) pick any red line as source
    b) pick the two circles as the destination objects
    c) right-click-enter.

5   The circles will now be displayed as red outlines.

6   *Settings*
    The Match Properties command has a settings option which allows the user to determine which properties can be matched. To use this option:
    a) activate the Match Properties command
    b) pick any object as the source
    c) at the destination prompt, enter **S <R>** to display the Property Settings dialogue box – Figure 14.4

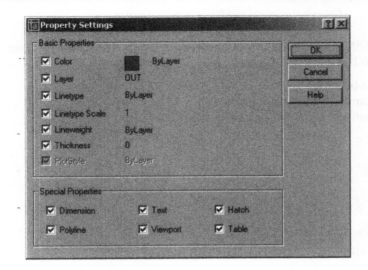

Figure 14.4 The Property Settings dialogue box.

d) generally the Basic and the Special Properties as displayed in Figure 14.4 are active at all times, but the user can alter these if they do not want a specific property matched

e) cancel the dialogue box then ESC to end the command.

This completes the exercises in this chapter. Ensure you have saved the WORKDRG with the four green centre lines.

Assignments

Three assignments for you to practice your skills with the OFFSET, TRIM and EXTEND commands as well as FILLET and CHAMFER. The assignments should not give you any problems. The procedure is:

1 Open your A3SHEET standard drawing paper.

2 Complete the drawing using layers correctly.

3 Save the completed drawing as C:\BEGIN\ACT??

4 Do NOT attempt to add dimensions.

Activity 7

Three interesting shapes to complete. One requires a bit of thought to complete the original shape, the other two require concentration to obtain the final shape. The commands to use are at your decision.

Activity 8

Two shapes to complete using the commands so far investigated. Make use of offset and trim as much as possible. No dimensions.

Activity 9

Three 'logo' type shapes. The CH is created from ellipses and lines. The circular shape uses fillets and the B shape should give you no trouble. Note that with these drawings I positioned the centre lines first.

Summary

1. OFFSET, TRIM and EXTEND are modify commands and can be activated by icon selection, from the menu bar or by keyboard entry.

2. Lines, arcs and circles can be offset by:
   a) entering an offset distance and selecting the object
   b) selecting an object to offset through a specific point.

3. Extend requires:
   a) a boundary edge
   b) objects to be extended.

4. Trim requires:
   a) a cutting edge
   b) objects to be trimmed.

5. Lines and arcs can be extended to other lines, arcs and circles.

6. Lines, circles and arcs can be trimmed.

Chapter 15
User exercise 2

With this exercise we will create another working drawing using the commands discussed in previous chapters. This drawing will be saved for future reference.

1 Open your **C:\BEGIN\A3SHEET** standard sheet with layer OUT current and display the Draw, Modify and Object Snap toolbars.

2 Refer to Figure 15.1 and complete the drawing using:
   a) the basic shape from six lines: USE THE 140,100 START POINT
   b) three offset lines
   c) four extended lines
   d) trim to give the final shape.

3 When complete save as **C:\BEGIN\USEREX**.

4 Do not exit AutoCAD.

So far, so good. At this stage of our learning process, we have completed the following drawings:

1 A demonstration drawing DEMODRG, used in the creation of lines, circles and arcs, and for investigating object snap. This drawing has 'served its purpose' and will not be used again.

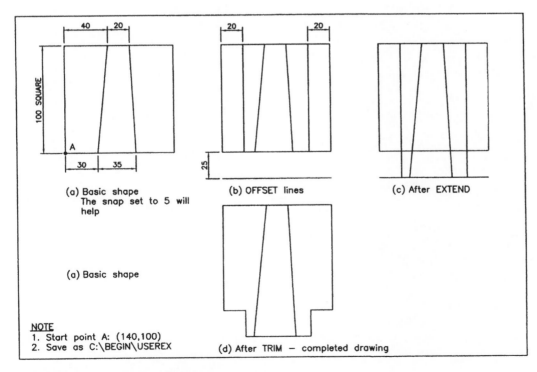

Figure 15.1 C:\BEGIN\USEREX construction.

2   A standard sheet with layers and other settings, saved as A3SHEET. This standard sheet will have further modifications and be used for all new drawing work.

3   A working drawing WORKDRG which was used to demonstrate several new topics. This drawing will be used to investigate other new topics.

4   A new working drawing named USEREX which will also be used to demonstrate additional new topics.

5   Nine activity exercises, ACT1 to ACT9, which you have been completing and saving as the book progresses. Or have you?

Now proceed to the next chapter in which we will discuss how text can be added to a drawing.

Chapter 16

Text

Text should be added to a drawing whenever possible. This text could simply be a title and date, but could also be a parts list, a company title block, notes on costing, etc. AutoCAD 2005 allows the user to enter:
a) short text entries, i.e. a single line or a few lines
b) larger text entries, i.e. several lines
c) via a table.

In this chapter we will consider the short-entry type text, and leave the other two types of text (multiline and table entry) to a later chapter, so:
a) USEREX still on the screen – it should be. If it is not, then open the appropriate drawing file.
b) Make layer Text (blue) current and refer to Figure 16.1.

One line of text

1 Menu bar with **Draw-Text-Single Line Text** and:
  prompt Specify start point of text or [Justify/Style]
  enter   **15,55 <R>**

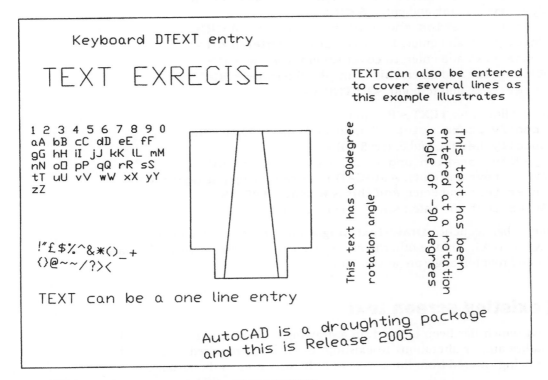

Figure 16.1 Adding text using C:\BEGIN\USEREX.

	prompt	Specify height<??> and enter: **8 \<R\>**
	prompt	Specify rotation angle of text<0.0> and enter: **0 \<R\>**
	prompt	Enter text
	enter	**TEXT can be a one line entry \<R\>**
	prompt	Enter text – i.e. any more lines of text
	respond	press the **\<RETURN\>** key to end the command.

2. The entered text item will be displayed at the entered point and must be ended with a return key press.

3. Repeat the **Draw-Text-Single Line Text** sequence with:
   a) start point: 20,240 \<R\>
   b) height: 15 \<R\>
   c) rotation: 0 \<R\>
   d) text: TEXT EXRECISE \<R\>\<R\> – two returns (**NB: this spelling is deliberate!!!**).

4. At the command line enter **DTEXT \<R\>** and enter:
   a) start point: 45,275
   b) height: 7.5
   c) rotation: 0
   d) text: Keyboard DTEXT entry \<R\>\<R\>.

5. *Notes*
   a) the text is displayed on the screen as you enter it from the keyboard
   b) this is referred to as *dynamic* text.

Several lines of text

1. Select from the menu bar **Draw-Text-Single Line Text** and:

	prompt	Specify start point of text and enter: **275,245 \<R\>**
	prompt	Specify height and enter: **6 \<R\>**
	prompt	Specify rotation angle of text and enter: **0 \<R\>**
	prompt	Enter text and enter: **TEXT can also be entered \<R\>**
	prompt	Enter text and enter: **to cover several lines as \<R\>**
	prompt	Enter text and enter: **this example illustrates \<R\>**
	prompt	Enter text and respond: **\<RETURN\>**.

2. At the command line enter **TEXT \<R\>** and:

	prompt	Specify start point of text and enter: **150,20 \<R\>**
	prompt	Specify height and enter: **8 \<R\>**
	prompt	Specify rotation angle of text and enter: **5 \<R\>**
	prompt	Enter text and enter: **AutoCAD is a draughting package \<R\>**
	prompt	Enter text and enter: **and this is Release 2005 \<R\>**
	prompt	Enter text and enter: **\<R\>** – to end command.

3. Using the menu bar sequence **Draw-Text-Single Line Text** or the command line entries DTEXT or TEXT, add the other items of text shown in Figure 16.1. The start point, height and rotation angle are at your discretion.

Editing existing screen text

Hopefully text which has been entered on a drawing is correct, but there may be spelling mistakes and/or alterations to existing text may be required. Text can be edited as it is being entered from the keyboard if the user notices the mistake.

Screen text which needs to be edited requires a command.

1   From the menu bar select **Modify-Object-Text-Edit** and:
    prompt   Select an annotation object or [Undo]
    respond **pick the TEXT EXRECISE item**
    prompt   Edit Text dialogue box – Figure 16.2 with the text phrase highlighted
    either   a) retype the highlighted phrase correctly then OK
    or   b) 1. left-click at right of the text item
                  2. backspace to remove error
                  3. retype correctly
                  4. pick OK
    or   c) 1. move cursor to TEXT EXR|ECISE and left-click
                  2. backspace to give TEXT EX|ECISE
                  3. move cursor to TEXT EXE|CISE and left-click
                  4. enter R to give TEXT EXER|CISE
                  5. pick OK
    prompt   Select an annotation object – i.e. any more selections
    respond **right-click and Enter**.

2   The text item will now be displayed correctly.

3   *Note*: You could always erase the text item and enter it correctly.

4   AutoCAD has a built-in spellchecker which 'uses' a dictionary to check the spelling. This dictionary can be changed to suit different languages and must be 'loaded' before it can be used. We will use a British English dictionary so from the menu bar select **Tools-Spelling** and:
    prompt   Select object
    respond **pick AutoCAD is a draughting package then right-click**
    prompt   Check Spelling dialogue box
    with   a named current dictionary
    respond **pick Change Dictionaries**
    prompt   Change Dictionaries dialogue box
    respond a) scroll at Main dictionary
               b) pick British English (ise)
               c) pick Apply & Close
    prompt   Check Spelling dialogue box
    with   a) Current dictionary: British English (ise)
               b) Current word: draughting
               c) Suggestions: draughtiness – Figure 16.3(a)
    respond **pick Ignore**
    prompt   AutoCAD Message box – Figure 16.3(b)
    respond **pick OK** – spell check is complete.

5   Save the screen layout if required, **but not as** C:\BEGIN\USEREX.

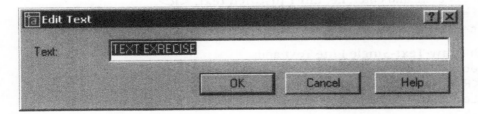

Figure 16.2   The Edit Text dialogue box.

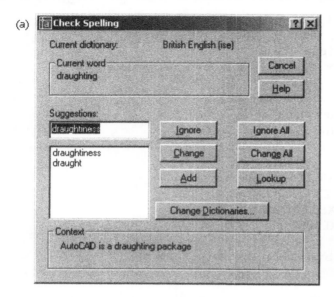
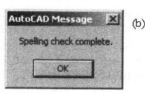

Figure 16.3 The Check Spelling dialogue box and the AutoCAD Message.

Text justification

Text items added to a drawing can be 'justified' (i.e. positioned) in different ways, and AutoCAD 2005 has several justification positions, these being:
a) six basic: left, align, fit, centre, middle, right
b) nine additional: TL, TC, TR, ML, MC, MR, BL, BC, BR.

1   Open your A3SHEET standard sheet and refer to Figure 16.4.

2   With layer OUT current, draw the following objects:
    a) a 100-sided square, the lower left point at (50,50)
    b) a circle of radius 50, centred at (270,150)
    c) five lines: 1. start point at 220,20 and end point at 320,20
                    2. start point at 220,45 and end point at 320,45
                    3. start point at 120,190 and end point at 200,190
                    4. start point at 225,190 and end point at 275,250
                    5. start point at 305,250 and end point at 355,100.

3   Make layer TEXT (blue) current and display the Draw, Modify and Object Snap toolbars.

4   At the command line enter **TEXT <R>** and:
    prompt   Specify start point of text or [Justify/Style]
    enter   **25,240 <R>**
    prompt   Specify height and enter: **6 <R>**
    prompt   Specify rotation angle of text and enter: **0 <R>**
    prompt   Enter text and enter: **This is NORMAL <R>**
    prompt   Enter text and enter: **i.e. LEFT justified text. <R>**
    prompt   Enter text and enter: **It is the default <R>**
    prompt   Enter text and enter: **<R>**.

5   Menu bar with **Draw-Text-Single Line Text** and:
    prompt   Specify start point of text or [Justify/Style]
    enter   **J <R>** – the justify option
    prompt   Enter an option [Align/Fit/Center/Middle/Right/TL/TC/TR/...

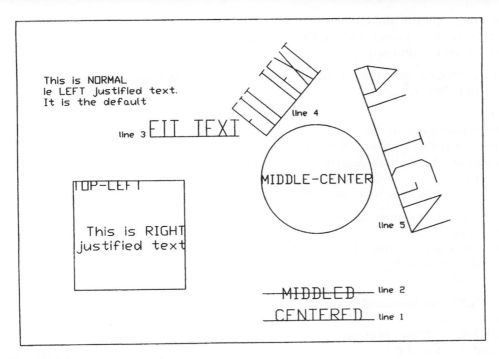

Figure 16.4 Text Justification exercise.

    enter   **R <R>** – the right justify option
    prompt   Specify right endpoint of text baseline
    respond **Snap to Midpoint icon and pick right vertical line of the square**
    prompt   Specify height and enter: **8 <R>**
    prompt   Specify rotation angle of text and enter: **0 <R>**
    prompt   Enter text and enter: **This is RIGHT <R>**
    prompt   Enter text and enter: **justified text <R><R>**.

6 Repeat the single line text command and:
    prompt   Specify start point of text or [Justify/Style]
    enter   **J <R>**
    prompt   Enter an option [Align/Fit...
    enter   **C <R>** – center option
    prompt   Specify center point of text
    respond **Snap to Midpoint icon and pick line 1**
    prompt   Specify height and enter: **10 <R>**
    prompt   Specify rotation angle of text and enter: **0 <R>**
    prompt   Enter text and enter: **CENTERED <R><R>**.

7 Enter TEXT <R> at the command line then:
   a) enter J <R> for justify
   b) enter M <R> for middle option
   c) Middle point: Snap to Midpoint icon and pick line 2
   d) Height: 10 and Rotation: 0
   e) Text: MIDDLED <R><R>.

8 Activate the single line text command with the Fit justify option and:
    prompt   Specify first endpoint of text baseline
    respond **Snap to Endpoint icon and pick left end of line 3**
    prompt   Specify second endpoint of text baseline
    respond **Snap to Endpoint icon and pick right end of line 3**

prompt   Specify height and enter: **15 \<R\>**
prompt   Enter text and enter: **FIT TEXT \<R\>\<R\>**
NB: This option has no rotation prompt.

9. Using the TEXT command with the Fit justify option, add the text item FIT TEXT with a height of 40 to line 4.

10. With TEXT again:
    a) select the Align justify option
    b) pick 'top' of line 5 as first endpoint of text baseline
    c) pick 'bottom' of line 5 as second endpoint of text baseline
    d) text item: ALIGN
    NB: This option has no height or rotation prompt.

11. Activate the single line text command and:
    a) justify with TL option
    b) Top/Left point: Intersection icon – pick top left of square
    c) Height of 8 and rotation of 0
    d) text item: TOP-LEFT.

12. Finally with the TEXT command and:
    a) justify with MC option
    b) Middle point: Snap to center of circle
    c) height of 8 and rotation of 0
    d) text item: MIDDLE-CENTER.

13. Your drawing should now resemble Figure 16.4 although the actual text appearance may differ from what is displayed due to the text style. The drawing can be saved, but we will not use it again.

14. The text justification options are easy to use. Simply enter the appropriate letter for the justification option at the command line. The entered letters are:

A: Align	F: Fit	C: Center
M: Middle	R: Right	TL: Top left etc.
MC: Middle center	BR: Bottom right	

Text style

When the single line text command is activated, one of the options available is *Style*. This option allows the user to select any one of several previously created text styles. This will be discussed in a later chapter which will investigate text styles and fonts.

Assignment

Attempt Activity 10 which has four simple components for you to complete. Text should be added, but not dimensions. The procedure is as always:

1. Open A3SHEET standard sheet.
2. Complete the drawings using layers correctly.
3. Save as C:\BEGIN\ACT10.
4. Read the summary then progress to the next chapter.

Summary

1. Text is a draw command which allows single/several lines of text to be entered from the keyboard.

2. The command is activated:
   a) from the keyboard with TEXT <R> or DTEXT <R>
   b) from the menu bar with **Draw-Text-Single Line Text**.

3. The entered text is dynamic, i.e. the user 'sees' the text on the screen as it is entered from the keyboard.

4. Text can be entered with varying height and rotation angle.

5. Text can be justified to user specifications, and there are several justification options.

6. Screen text can be modified with:
   a) menu bar **Modify-Object-Text-Edit**
   b) keyboard entry: DDEDIT.

7. Fitted and aligned text are similar, the user selecting both the start and end points of the text item, but:
   a) fitted: allows the user to enter a text height
   b) aligned: text is automatically 'adjusted' to suit the selected points.
   Neither has a rotation option.

8. Centered and middled text are similar, the user selecting the center/middle point, but:
   a) centered: is about the text baseline
   b) middled: is about the text middle point.

9. Text styles and multiple line (or paragraph) text can be set by the user. These two topics will be investigated in later chapters.

Chapter 17

Dimensioning

AutoCAD has both automatic and associative dimensioning, the terms meaning:

Automatic   when an object to be dimensioned is selected, the actual dimension text, arrows, extension lines, etc. are all added in one operation.

Associative   the arrows, extension lines, dimension text, etc. which 'make up' a dimension are treated as a single object.

AutoCAD allows different 'types' of dimensions to be added to drawings. These are displayed in Figure 17.1 and can be categorised as:

a) Linear: horizontal, vertical and aligned
b) Baseline and Continue
c) Ordinate: both X-datum and Y-datum
d) Angular
e) Radial: diameter and radius
f) Leader: taking the dimension text 'outside' the object.

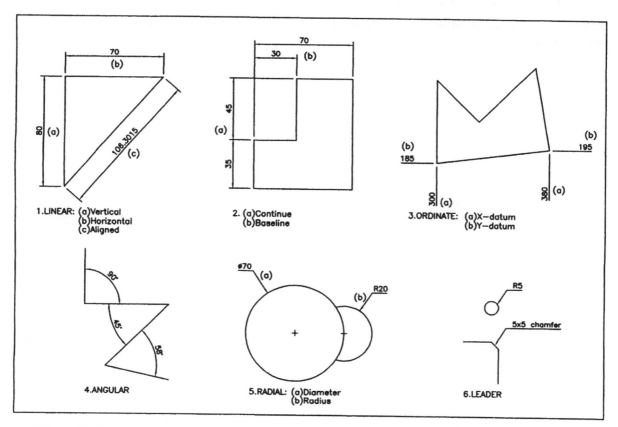

Figure 17.1   Dimension types.

Dimension exercise

To demonstrate how dimensions are added to a drawing, we will use one of our working drawings, so:

1. Open **C:\BEGIN\USEREX** and with layer OUT current add the following three circles:
   a) centre: 220,180, radius: 15
   b) centre: 220,115, radius: 3
   c) centre: 190,75, radius: 30
   d) trim the large circle to the line as Figure 17.2.

2. Make layer DIMS current and activate the Object Snap and Dimension toolbars.

3. At the command line enter **DIMSCALE <R>** and:
   prompt   Enter new value for DIMSCALE <??>
   enter   **2 <R>**.

4. *Notes*
   a) The true benefit of dimensioning drawings will not become apparent to the user until we have investigated Dimension Styles in the next chapter. The appearance of the your added dimensions in the exercise which follows may differ from those in Figure 17.2. This is to be expected and the user should not be concerned.
   b) The DIMSCALE entry in step 3 was to set an overall dimension scale of 2 to allow the added dimensions to be 'more readable'.

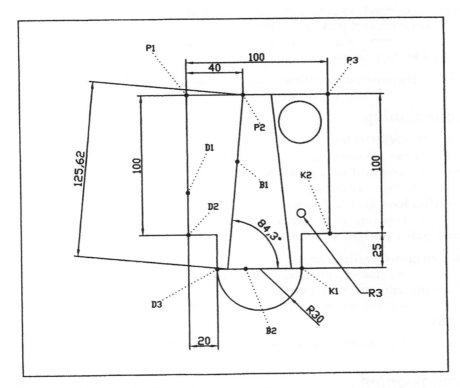

Figure 17.2 Adding dimensions to C:\BEGIN\USEREX.

Linear dimensioning

1. Select the **LINEAR DIMENSION icon** from the Dimension toolbar and:
   prompt   Specify first extension line origin or <select object>
   respond   **Endpoint icon and pick line D1**

prompt	Specify second extension line origin
respond	**Endpoint icon and pick the other end of line D1**
prompt	Specify dimension line location or [Mtext/Text...
respond	**pick any point to the left of line D1**.

2. From the menu bar select **Dimension-Linear** and:

prompt	Specify first extension line origin or <select object>
respond	**Intersection icon and pick point D2**
prompt	Specify second extension line origin
respond	**Intersection icon and pick point D3**
prompt	Specify dimension line location or...
respond	**pick any point below object**.

Baseline dimensioning

1. Select the **LINEAR DIMENSION** icon and:

prompt	Specify first extension line origin
respond	**Intersection icon and pick point P1**
prompt	Specify second extension line origin
respond	**Intersection icon and pick point P2**
prompt	Specify dimension line location
respond	**pick any point above the line**.

2. Select the **BASELINE DIMENSION icon** from the Dimension toolbar and:

prompt	Specify a second extension line origin
respond	**Intersection icon and pick point P3**
prompt	Specify a second extension line origin
respond	**press the ESC key** to end command.

3. The menu bar selection **Dimension-Baseline** could have been selected for step 2.

Continue dimensioning

1. Select the **LINEAR DIMENSION** icon and:

prompt	Specify first extension line origin
respond	**Intersection icon and pick point K1**
prompt	Specify second extension line origin
respond	**Intersection icon and pick point K2**
prompt	Specify dimension line location
respond	**pick any point to right of the line**.

2. Menu bar with **Dimension-Continue** and:

prompt	Specify a second extension line origin
respond	**Intersection icon and pick point P3**
prompt	Specify a second extension line origin
respond	**press ESC**.

3. The Continue-Dimension icon could have been selected for step 2.

Diameter dimensioning

Select the **DIAMETER DIMENSION icon** from the Dimension toolbar and:

prompt	Select arc or circle
respond	**pick the larger circle**
prompt	Dimension text=30
then	Specify dimension line location
respond	**drag out and pick a suitable point**.

Radius dimensioning

Select the **RADIUS DIMENSION icon** and:
prompt	Select arc or circle
respond	pick the arc
prompt	Dimension text=30
then	Specify dimension line location
respond	**drag out and pick a suitable point.**

Angular dimensioning

Select the **ANGULAR DIMENSION icon** and:
prompt	Select arc, circle, line or...
respond	pick line B1
prompt	Select second line
respond	**pick line B2**
prompt	Specify dimension arc line location
respond	**drag out and pick a point to suit.**

Aligned dimensioning

Select the **ALIGNED DIMENSION icon** and:
prompt	Specify first extension line origin
respond	**Endpoint icon and pick line B1**
prompt	Specify second extension line origin
respond	**Endpoint icon and pick other end of line B1**
prompt	Specify dimension line location
respond	**pick any point to suit.**

Leader dimensioning

Select the **QUICK LEADER DIMENSION icon** and:
prompt	Specify first leader point
respond	**Nearest icon and pick any point on smaller circle**
prompt	Specify next point
respond	**drag to a suitable point and pick**
prompt	Specify next point
respond	**right-click**
prompt	Specify text width<0>
respond	**right-click**
prompt	Enter first line of annotation text
enter	**R3 <R>**
prompt	Enter next line of annotation text
respond	**right-click to end leader dimension command.**

Notes

1. At this stage your drawing should resemble Figure 17.2. Save it as **C:\BEGIN\USEREX-A**.

2. As stated earlier, the dimensions in Figure 17.2 may differ in appearance from those which have been added to your drawing. This is because I have used the 'default AutoCAD dimension style' and made no attempt to alter it (although I did scale by 2). The object of the exercise was to investigate the dimensioning process.

3. The <RETURN> selection is useful if a single object is to be dimensioned.

4   Object snap is used extensively when dimensioning. This is one time when a running object snap (e.g. Endpoint) will assist, but remember to cancel the running object snap!

5   From the menu bar select **Format-Layer** to display the Layer Properties Manager dialogue box. Note the layer **Defpoints**. We did not create this layer. It is *automatically* made by AutoCAD any time a dimension is added to a drawing. This layer can be turned off or frozen but cannot be deleted. **It is best left untouched**.

Dimension terminology

All dimensions used with AutoCAD objects have a terminology associated with them. It is important that the user has an understanding of this terminology especially when creating dimension styles. It is in the users interest to know the various terms used and Figure 17.3 explains the basic terminology when dimensions are added to a drawing.

1   *Dimension and extension lines*
    These two features consist of:
    a) the dimension line
       - the actual line
       - the dimension text
       - arrowheads
       - extension lines
    b) the extension line
       - an origin offset from the object
       - an extension beyond the line
       - spacing (for baseline)

2   *Centre marking*
    Circular features can be displayed with: a mark, a line or nothing.

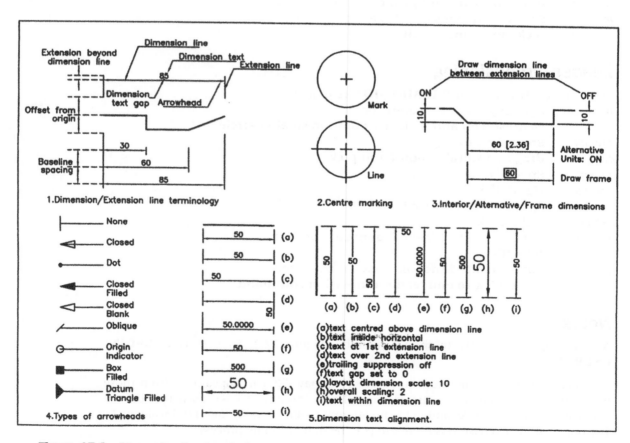

Figure 17.3   Dimension line terminology.

3 *Dimension text*
   It is possible for the dimension text to:
   a) have interior dimension line drawn or not drawn
   b) display alternative units, i.e. [imperial]
   c) draw a frame around the dimension text.

4 *Arrowheads*
   AutoCAD 2005 has several arrowheads for the dimension line and has the facility for user-defined arrowheads. A selection is displayed.

5 *Dimension text alignment*
   The dimension text can be aligned relative to the dimension line by altering certain dimension variables. A selection of dimension text positions is displayed.

Summary

1 AutoCAD 2005 has automatic and associative dimensions.

2 Dimensioning can be linear, radial, angular, ordinate or leader.

3 The diameter and degree symbols are automatically added when using radial or angular dimensions.

4 Object snap modes are useful when dimensioning.

5 A layer DEFPOINTS is created when dimensioning. The user has no control over this layer.

6 Dimensions should be added to a drawing using a Dimension Style.

   Now continue to the next exercise, still with the dimensioned USEREX displayed.

Chapter 18
Dimension styles 1

Dimension styles allow the user to set dimension variables to individual/company requirements. This permits various styles to be customised and saved for different clients.

To demonstrate how a dimension style is 'set and saved', we will create a new dimension style called A3DIM, use it with WORKDRG and then save it to our standard sheet.

Notes

a) The exercise which follows will display several new dialogue boxes and certain settings will be altered within these boxes. It is important for the user to become familiar with these Dimension Style dialogue boxes, as a good knowledge of their use is essential if different dimension styles have to be used.
b) The settings used in the exercise are my own, designed for our A3SHEET standard sheet.
c) You can alter the settings to your own values at this stage.

Getting started

1. Close the dimensioned USEREX drawing.
2. Open C:\BEGIN\WORKDRG to display the outline shape with two circles and centre lines.
3. Display the Draw, Modify, Dimension and Object Snap toolbars.

Creating the dimension style A3DIM

1. Either   *a*) menu bar with **Dimension-Style**
   or   *b*) Dimension Style icon from Dimension toolbar
   prompt   `Dimension Style Manager dialogue box`
   with   *a*) Current Dimstyle: ISO-25 or similar
           b) Several other named styles for selection?
           c) Preview of the current dimension style
           d) Description of current dimension style
   respond   **pick New**
   prompt   `Create New Dimension Style dialogue box`
   respond   *a*) alter New Style Name: A3DIM
           b) Start with: ISO-25 or similar
           c) Use for: All dimensions – Figure 18.1
           d) Pick Continue
   prompt   `New Dimension Style: A3DIM dialogue box`
   with   six tab options for selection
   respond   **continue with the tab selections which follow.**

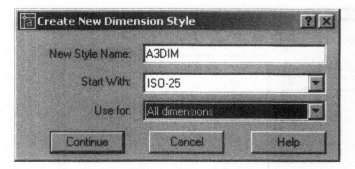

Figure 18.1 The Create New Dimension Style dialogue box.

2 *Lines and Arrows*
   respond   pick Lines and Arrows tab
   prompt   Lines and Arrows tab dialogue box
   alter   a) Dimension Lines
              1. Color: By Block
              2. Lineweight: By Block
              3. Baseline Spacing: 10
              4. Suppress: both not active
           b) Extension Lines
              1. Color: By Block
              2. Lineweight: By Block
              3. Extend beyond dim lines: 2.5
              4. Offset from origin: 2.5
              5. Suppress: Both not active
           c) Arrowheads
              1. 1st; 2nd; Leader: Closed Filled
              2. Arrow size: 4
              3. Center Marks for Circle:
                  a) Type: Mark
                  b) Size: 2
   and   dialogue box similar to Figure 18.2.

3 *Text*
   respond   pick Text tab
   prompt   Text tab dialogue box
   alter   a) Text Appearance
              1. Text style: STANDARD – this will be altered later
              2. Text color: By Block
              3. Fill Color: None
              4. Text height: 4
              5. Draw frame around text: not active
           b) Text Placement
              1. Vertical: Above
              2. Horizontal: Centered
              3. Offset from dim line: 1.5
           c) Text Alignment
              1. ISO Standard active
   and   dialogue box similar to Figure 18.3.

4 *Fit*
   respond   pick Fit tab
   prompt   Fit tab dialogue box

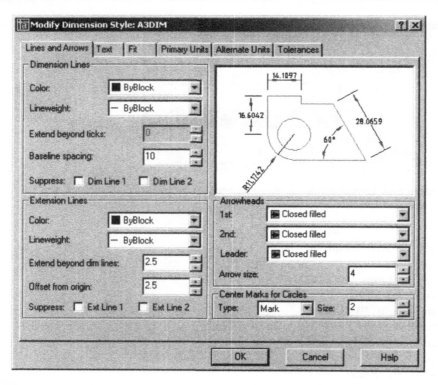

Figure 18.2 The lines and Arrows tab dialogue box.

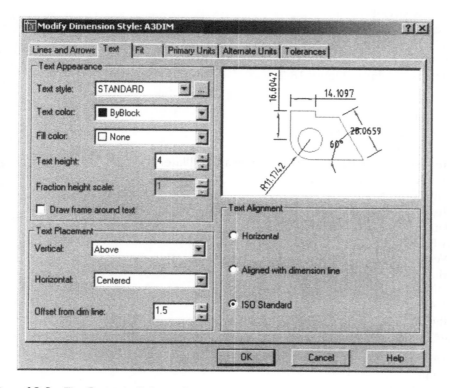

Figure 18.3 The Text tab dialogue box.

alter	a)	Fit Options
		1. Either the text or the arrows, whichever fits best: active – i.e. black dot
		2. Suppress arrows: not active
	b)	Text Placement
		1. Beside the dimension line active
	c)	Scale for Dimension Features
		1. Use overall scale of: 1
		2. Scale dimensions to layout (paperspace): not active
	d)	Fine Tuning
		1. Place text manually when dimensioning: not active
		2. Always draw dim line between ext lines: not active
and		dialogue box similar to Figure 18.4.

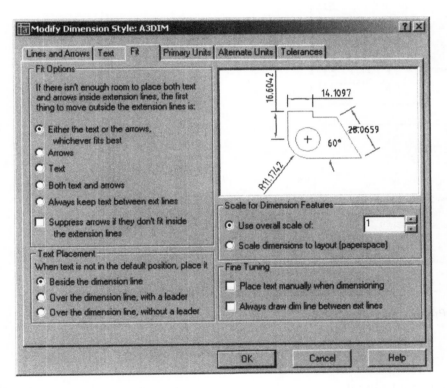

Figure 18.4 The Fit tab dialogue box.

5 *Primary Units*
   respond   pick Primary Units tab
   prompt   `Primary Units tab dialogue box`
   alter   a) Linear Dimensions
                   1. Unit Format: Decimal
                   2. Precision: 0.00
                   3. Decimal separator: '.' (Period)
                   4. Round off: 0
                   5. Measurement Scale
                       a) Scale factor: 1
                       b) Apply to layout dimension only: not active
                   6. Zero Suppression
                       a) Leading: not active
                       b) Trailing: active – i.e. tick in box

b) Angular Dimensions
   1. Units Format: Decimal Degrees
   2. Precision: 0.0
   3. Zero Suppression
      a) Leading: not active
      b) Trailing active

and dialogue box similar Figure 18.5.

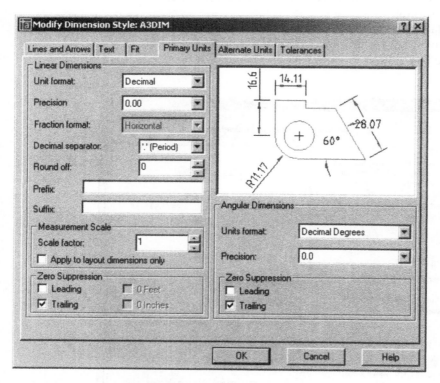

Figure 18.5 The Primary Units dialogue box.

6 *Alternate Units*
   respond pick Alternate Units tab
   prompt   `Alternate Units tab dialogue box`
   respond Display alternate units: not active – i.e. no tick (no alterations to this dialogue box).

7 *Tolerances*
   respond pick Tolerances tab
   prompt   `Tolerances tab dialogue box`
   respond a) Tolerance Format
      1. Method: None and no alterations to this dialogue box.

8 *Continue*
   respond **pick OK from New Dimension Style: A3DIM dialogue box**
   prompt   `Dimension Style Manager dialogue box`
   with   a) A3DIM added to styles list
              b) Preview of the A3DIM style
              c) Description of the A3DIM style

respond a) pick A3DIM – becomes highlighted
b) pick Set Current – note description
c) scroll at List and pick: Styles in use
d) dialogue box similar to Figure 18.6
e) pick Close
f) note that if you have been altering other dimension styles during this process, the AutoCAD Alert message about overrides may be displayed. If it is, respond as you require.

9 *Note*: We have still to 'set' a text style for our A3DIM dimension style. This will be covered in a later chapter and for the present, our A3DIM dimension style is 'complete'. It will be used for all future dimensioning work.

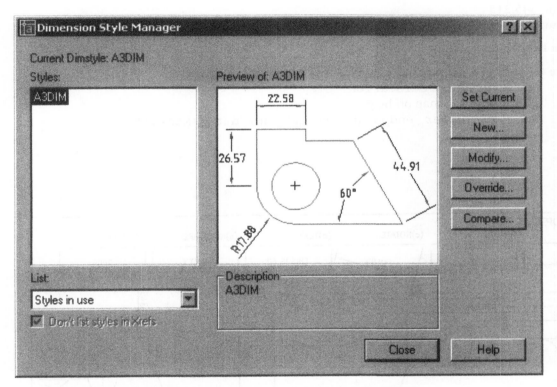

Figure 18.6 The Dimension Style Manager dialogue box for A3DIM.

Using the A3DIM dimension style

1 Make layer DIMS current and freeze layer OUT.

2 Refer to Figure 18.7 and add the displayed dimensions:
   a) linear with baseline
   b) linear with continue
   c) diameter
   d) radius
   e) angular
   f) leader with text as displayed.

3 With layer TEXT current, add suitable text, the height being at your discretion.

4 When all the dimensions have been added, save the drawing as C:\BEGIN\WORKDRG.

Ordinate dimensioning

This type of dimensioning is very popular with many companies, and although mentioned in the last chapter, it was not discussed in detail. We will now investigate how it is used.

1   Continue using WORKDRG and refer to Figure 18.7.

2   Layer OUT current and toolbars to suit.

3   Draw the following objects using absolute co-ordinate entry:

   LINE
   First point: 315,130
   Next point: 405,165
   Next point: 345,215
   Next point: close

   CIRCLE
   Centre: 370,175
   Radius: 10

4   Make the DIMS layer current.

5   Select the **ORDINATE dimension icon** from the Dimension toolbar and:
   prompt   Specify feature location
   respond   **pick point A** – snap on helps
   prompt   Specify leader endpoint or [Xdatum/Ydatum/Mtext/Text/Angle]

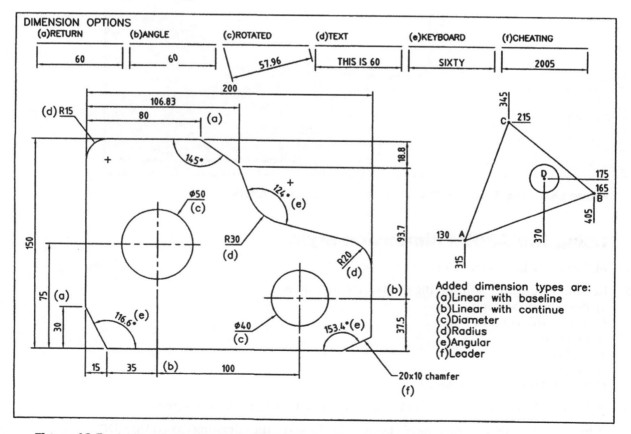

Figure 18.7 Adding dimensions to WORKDRG using dimension style A3DIM.

enter	**X <R>**	– the Xdatum option
prompt		`Specify leader endpoint or [Xdatum/Ydatum...`
enter	**@0,–10 <R>**	
prompt		`Dimension text=315`
and		command line returned.

6 Menu bar with **Dimension-Ordinate** and:
prompt   `Specify feature location`
respond   **pick point A**
prompt   `Specify leader endpoint or [Xdatum/Ydatum...`
enter   **Y <R>**
prompt   `Specify leader endpoint or [Xdatum/Ydatum...`
enter   **@–10,0 <R>**
prompt   `Dimension text=130`
and   command line returned.

7 Repeat the Ordinate dimension command and:
   a) select point B as the feature location
   b) with snap on, drag vertically downwards and click to suit.

8 With the Ordinate dimension active:
   a) select point B as the feature location
   b) drag horizontally to the right and click to suit to position the ordinate dimension.

9 Now add ordinate dimensions to the point C of the triangular shape and point D, the circle centre. The displayed dimensions should be the same as your co-ordinate entries from step 3.

10 The result should be similar to Figure 18.7.

11 Save the drawing and dimensions as C:\BEGIN\WORKDRG-1 as we will recall it later in this chapter.

Saving the A3DIM dimension style to the standard sheet

The dimension style which we have created will be used for all future work when dimensioning is required. This means that we want to have this style incorporated into our A3SHEET standard drawing sheet. We could always open the standard sheet and re-define the A3DIM dimension style, but this seems a waste of time. We will therefore use the existing layout, so:

1 Make layer OUT current.

2 Erase all objects from the screen EXCEPT the black border.

3 Menu bar with **File-Save As** and:
prompt   `Save Drawing As dialogue box`
respond   *a)* ensure type: AutoCAD 2004 Drawing (*.dwg)
          b) scroll and pick: C:\BEGIN named folder
          c) enter file name: A3SHEET
          d) pick Save
prompt   `AutoCAD message`
with   `C:\BEGIN\A3SHEET.dwg already exists`
           `Do you want to replace it?`
respond   **pick Yes**.

4 We now have the A3SHEET saved as a drawing file with the A3DIM dimension style defined. This means that we do not have to set layers or dimension styles every time we start a new drawing.

5 The A3SHEET standard drawing sheet should be used for all new work.

Quick dimensioning

AutoCAD 2005 has a quick dimensioning option and we will use our WORKDRG to discuss how it is used. This topic should really have been investigated in the last chapter, but I decided to leave it until we had discussed dimension styles. To demonstrate the topic:

1 Open the C:\BEGIN\USEREX-A drawing used to demonstrate the dimension concept and erase all the dimensions.

2 Make layer DIMS current and refer to Figure 18.8.

3 Menu bar with **Dimension-Quick Dimension** and:
   prompt   Select geometry to dimension
   respond **window the complete shape then right-click**
   prompt   Specify dimension line position or [Continuous/Staggered/Baseline/Ordinate...
   enter   **B <R>** – the baseline option

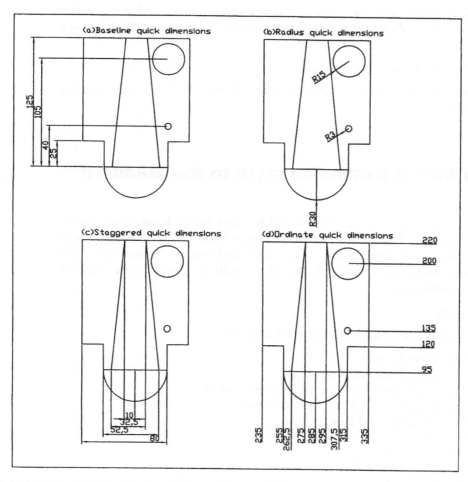

Figure 18.8 Using quick dimensioning with USEREX.

	prompt	`Specify dimension line position or...`
	respond	**pick any point to the left of the shape**.

4. All vertical dimensions will be displayed from the horizontal baseline of the shape as Figure 18.8(a). The user can now delete any dimensions which are unwanted.

5. Erase all the dimensions.

6. Select the **QUICK DIMENSION icon** from the Dimension toolbar and:
   - *prompt*   `Select geometry to dimension`
   - *respond*   **window the shape then right-click**
   - *prompt*   `Specify dimension line position or [Continuous/Staggered/Baseline...`
   - *enter*   **R <R>** – the radius option
   - *prompt*   `Specify dimension line location`
   - *respond*   **pick any point to suit**.

7. All radius dimensions for the component will be displayed as Figure 18.8(b).

8. Erase the radius dimensions then use the Quick Dimension command to add:
   - *a)* staggered dimensions to the component as Figure 18.8(c)
   - *b)* ordinate dimensions as Figure 18.8(d).

9. If Quick Dimension is being used, then it may be necessary to use the command several times to obtain the required dimension effect. QDIM is another method of adding dimensions to a drawing and it is the user's decision as to whether to use quick dimensions or 'ordinary' dimensions.

10. The command line entry to activate the Quick Dimension command is QDIM.

11. Do not save this part of the dimension exercise.

Dimension options

When using the dimension commands, the user may be aware of various options when the prompts are displayed. To investigate these options:

1. Re-open C:\BEGIN\WORKDRG-1.

2. Make layer OUT current and draw five horizontal lines of length 60 at the top of the screen.

3. Make layer DIMS current and refer again to Figure 18.7.

4. *The RETURN option*
   Select the **LINEAR DIMENSION icon** and:
   - *prompt*   `Specify first extension line origin or <select object>`
   - *respond*   **press the RETURN/ENTER key**
   - *prompt*   `Select object to dimension`
   - *respond*   **pick the first line**
   - *prompt*   `Specify dimension line location`
   - *respond*   **pick below the line** – Figure 18.7(a).

5. *The ANGLE option*
   Select the LINEAR DIMENSION icon, press RETURN, pick the second line and:
   - *prompt*   `Specify dimension line location`
   - *and*   `[Mtext/Text/Angle...`
   - *enter*   **A <R>** – the angle option
   - *prompt*   `Specify angle of dimension text`
   - *enter*   **15 <R>**

prompt   `Specify dimension line location`
respond   **pick a point to suit** – Figure 18.7(b).

6 *The ROTATED option*
LINEAR dimension icon, right-click, pick third line and:
prompt   `Specify dimension line location`
and   `[Mtext/Text/Angle...`
enter   **R <R>** – the rotated option
prompt   `Specify angle of dimension line<0>`
enter   **15 <R>**
prompt   `Specify dimension line location`
respond   **pick to suit** – Figure 18.7(c).

7 *The TEXT option*
LINEAR dimension icon, right-click, pick the fourth line and:
prompt   `Specify dimension line location`
and   `[Mtext/Text/Angle...`
enter   **T <R>** – the text option
prompt   `Enter dimension text<60>`
enter   **THIS IS 60 <R>**
prompt   `Specify dimension line location`
respond   **pick to suit** – Figure 18.7(d).

8 *Dimensioning with keyboard entry*
At the command line enter **DIM <R>** and:
prompt   `Dim`
enter   **HOR <R>** – horizontal dimension
prompt   `Specify first extension line origin`
respond   **right-click and pick the next line**
prompt   `Specify dimension line location`
respond   **pick to suit**
prompt   `Enter dimension text<60>`
enter   **SIXTY <R>** – Figure 18.7(e)
prompt   Dim returned at command line. Press ESC to end command.

9 *Dimension cheating* – not recommended
Enter **DIM <R>** at the command line and:
prompt   `Dim`
enter   **HOR <R>** – horizontal dimension
prompt   `Specify first extension line origin`
respond   **right-click and pick the last line**
prompt   `Specify dimension line location`
respond   **pick to suit**
prompt   `Enter dimension text<60>`
enter   **2005 <R>** – Figure 18.7(f)
prompt   Dim returned. Press ESC to end command.

10 This exercise is now complete and does not need to be saved.

Assignments

As dimensioning is an important concept, I have included three activities which will give you practice with:

1 using the standard sheet with layers

2 using the draw commands with co-ordinates

3   adding suitable text

4   adding dimensions with the A3DIM dimension style.

In each activity the procedure is the same:

1   Open the A3SHEET standard drawing sheet

2   Using layers correctly, complete the drawings

3   Save the completed work as C:\BEGIN\ACT11, etc.

Activity 11

Two simple components to be drawn and dimensioned. The sizes are more awkward than usual.

Activity 12

Two components created mainly from circles and arcs. Use offset as much as possible. The signal arm is interesting.

Activity 13

A component which is much easier to complete than it would appear. Offset and fillet will assist.

Summary

1   Dimension styles are created by the user to 'individual' standards.

2   Different dimension styles can be created and 'made current' at any time.

3   The standard A3SHEET sheet has a 'customised' dimension style named A3DIM which will be used for all future drawing work including dimensioning.

4   Dimension styles will be discussed again in a later chapter.

Chapter **19**

Modifying objects

The draw and modify commands are probably the most commonly used of all the AutoCAD commands and we have already used several of each. The modify commands discussed previously were Erase, Offset, Trim, Extend, Fillet and Chamfer. In this chapter we will use C:\BEGIN\WORKDRG to investigate several other modify commands as well as some additional selection set options.

Getting ready

1. Open C:\BEGIN\WORKDRG to display the red component with green centre lines. There should be no dimensions displayed.
2. Layer OUT current, with the Draw, Modify and Object Snap toolbars displayed.
3. Freeze layer CL – you will find out why shortly.
4. Menu bar with **View-Zoom-All** and the drawing should 'fill the screen'.

Copy

Allows objects to be copied to other parts of the screen and the command can be used for single or multiple copies. The user specifies a start (base) point and a displacement (or second point).

1. Refer to Figure 19.1.
2. Select the **COPY OBJECT icon** from the Modify toolbar and:
   - *prompt* Select objects
   - *enter* **C <R>** – the crossing selection set option
   - *prompt* Specify first corner and: **pick a point P1**
   - *prompt* Specify opposite corner and: **pick a point P2**
   - *but* **DO NOT RIGHT-CLICK YET**
   - *prompt* 11 found – note objects not highlighted
   - *then* Select objects – i.e. any more objects to be included in the selection
   - *enter* **A <R>** – the add selection set option
   - *prompt* Select objects
   - *respond* **pick objects D1, D2 and D3 then right-click**
   - *and* note command line as each object is selected
   - *prompt* Specify base point or displacement
   - *enter* **50,50 <R>** – note copy image as mouse is moved!
   - *prompt* Specify second point of displacement or <use first point as displacement>
   - *enter* **300,300 <R>**
   - *prompt* Specify second point of displacement
   - *respond* **right-click**.
3. The original component will be copied to another part of the screen and MAY not all be visible (if at all?).

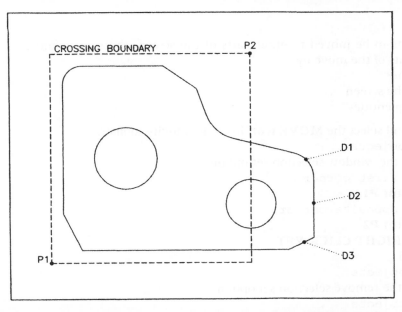

Figure 19.1 WORKDRG with selection information for the COPY command.

4 Do not panic! Select from the menu bar **View-Zoom-All** to 'see' the complete copied effect – Figure 19.2. You may have to reposition your toolbars.

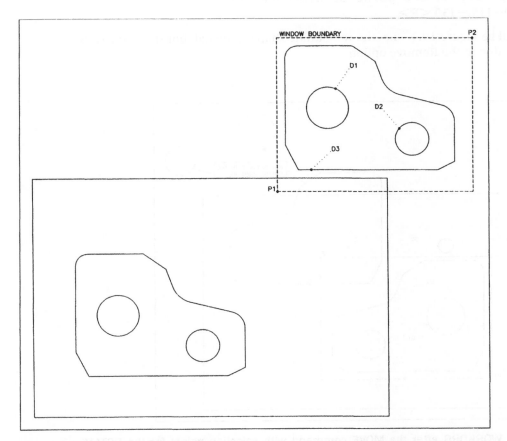

Figure 19.2 WORKDRG after the COPY command with selection points for the MOVE command.

Move

Allows selected objects to be moved to other parts of the screen, the user defining the start and end points of the move by:
a) entering co-ordinates
b) picking points on the screen
c) referencing existing entities.

1. Refer to Figure 19.2 and select the **MOVE icon** from the Modify toolbar:
    prompt   Select objects
    enter   **W <R>** – the window selection set option
    prompt   Specify first corner
    respond **pick a point P1**
    prompt   Specify opposite corner
    respond **pick a point P2**
    but   **DO NOT RIGHT-CLICK YET**
    prompt   14 found
    and   Select objects
    enter   **R <R>** – the remove selection set option
    prompt   Remove objects
    respond **pick circles D1 and D2 then right-click**
    and   note command line as each object is selected
    prompt   Specify base point or displacement
    respond **Endpoint icon and pick line D3 at end indicated** (note image as cursor is moved)
    prompt   Specify second point of displacement or...
    enter   **@–115,–130 <R>**.

2. The result will be Figure 19.3 – i.e. the red outline shape is moved, but the two circles do not move, due to the Remove option.

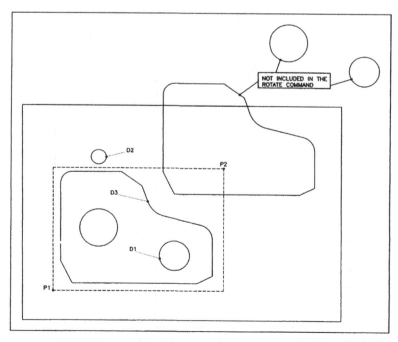

Figure 19.3 WORKDRG after the MOVE command with selection points for the ROTATE command.

3   *Task*: Using the Layer Properties Manager dialogue box, THAW layer CL and the centre lines are still in their original positions. They have not been copied or moved. This can be a problem with objects which are on frozen or off layers.

4   Now:
    a) freeze layer CL
    b) erase the copied–moved objects to leave the original WORKDRG
    c) View-Zoom-All to 'restore' the original screen.

Rotate

Selected objects can be 'turned' about a designated point in either a clockwise (−ve angle) or a counter-clockwise (+ve angle) direction. The base point for the rotation can be selected as a point on the screen, entered as a co-ordinate or referenced to existing objects.

1   Refer to Figure 19.3 and draw a circle of radius 10 with the centre point at 100,220.

2   Menu bar with **Modify-Rotate** and:
    prompt   Current positive angle in UCS: ANGDIR=counterclockwise ANGBASE=0.0
    then   Select objects
    respond **window from P1 to P2 – but NO right-click yet**
    prompt   14 found
    then   Select objects
    enter   **R <R>** – the remove option
    prompt   Remove objects
    respond **pick circle D1**
    prompt   1 found, 1 removed, 13 total
    then   Remove objects
    enter   **A <R>** – the add option
    prompt   Select objects
    respond **pick circle D2**
    prompt   1 found, 14 total
    then   Select objects
    respond **right-click** to end selection sequence
    prompt   Specify base point
    respond **Endpoint icon and pick line D3** – at 'lower end'
    prompt   Specify rotation angle or [Reference]
    enter   **−90 <R>**.

3   The selected objects will be rotated as displayed in Figure 19.4.

4   Note that two selection set options were used in this single rotate sequence:
    a) R – removed a circle from the selection set
    b) A – added a circle to the selection set.

Scale

The scale command allows selected objects or complete 'shapes' to be increased/decreased in size, the user selecting/entering the base point and the actual scale factor.

1   Refer to Figure 19.4 and erase the two circles 'outside the shape'.

2   Make layer DIMS current and with the Dimension command (icon or menu bar):
    a) linear dimension line AB
    b) diameter dimension the circle.

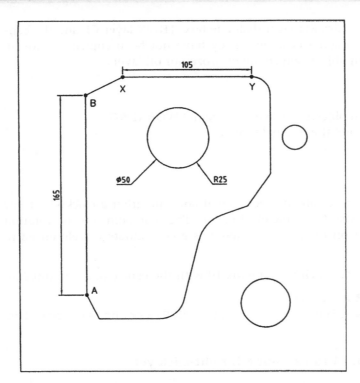

Figure 19.4 WORKDRG after the ROTATE command with dimensions for the SCALE command.

3   At the command line enter **DIM <R>** and:
    prompt   Dim
    enter   **HOR <R>**
    prompt   Specify first extension line origin
    respond   **Endpoint icon and pick line at X**
    prompt   Specify second extension line origin
    respond   **Endpoint icon and pick line at Y**
    prompt   Specify dimension line location
    respond   **pick above line**
    prompt   Enter dimension text<105>
    enter   **105 <R>**
    prompt   Dim
    enter   **RAD <R>**
    prompt   Select arc or circle
    respond   **pick the circle**
    prompt   Enter dimension text<25>
    enter   **R25 <R>**
    prompt   Specify dimension text location
    respond   **pick to suit**
    prompt   Dim and **ESC** to end sequence.

4   Make layer OUT current.

5   Select the **SCALE icon** from the Modify toolbar and:
    prompt   Select objects
    respond   **window the complete shape with dimensions then right-click**
    prompt   Specify base point

respond	**Endpoint of line AB at 'B' end**
prompt	Specify scale factor or [Reference]
enter	**0.5 <R>**.

6   The complete shape will be scaled as Figure 19.5.

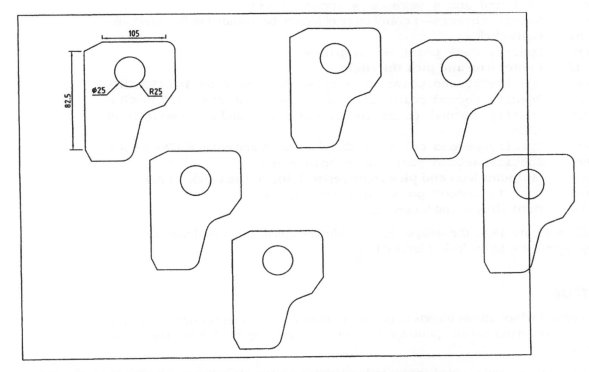

Figure 19.5 WORKDRG after the SCALE and MULTIPLE COPY commands.

7   Note the dimensions:
   a) the vertical dimension of 165 is now 82.5
   b) the diameter value of 50 is now 25
   c) the horizontal dimension of 105 is still 105
   d) the radius of 25 is still 25.

8   *Questions*
   a) Why have two dimensions been scaled by 0.5 and two have not?
   Answer: The scaled dimensions are those which used the icon or menu bar selection and those not scaled had their dimension text values entered from the keyboard.
   b) Which of the resultant dimensions are correct? Is it the 82.5 and diameter 25, or the 105 and R25?
   Answer: I will let you reason this one for yourself, but I can assure you that it causes a great deal of debate.

Multiple copy

The COPY command allows the user to create multiple copies. This is an option of the COPY command and does what it says – it produces as many copies of selected objects as you want.

1   Refer to Figure 19.5.

2   Using the Layer Properties Manager dialogue box, lock the DIMS layer.

3   From menu bar select **Modify-Copy** and:
    prompt   Select objects
    respond   **window the shape and dimensions then right-click**
    prompt   17 found and 4 were on a locked layer
    then   Select objects – i.e. any more objects to be included in the selection
    respond   **right-click**
    prompt   Specify base point or displacement
    respond   **Center icon and pick the circle**
    prompt   Specify second point of displacement and enter: **145,150 <R>**
    prompt   Specify second point of displacement and enter: **@170,10 <R>**
    prompt   Specify second point of displacement and enter: **@170<−90 <R>**
    prompt   Specify second point of displacement and enter: **@275<0 <R>**
    prompt   Specify second point of displacement
    respond   **Midpoint icon and pick right vertical line of the black 'border'**
    prompt   Specify second point of displacement
    respond   **right-click** to end sequence.

4   Result as Figure 19.5, the 'shape' being multiple-copied, but the dimensions not being copied, due to the locked layer effect.

Mirror

A command which allows objects to be mirror imaged about a line designated by the user. The command has an option for deleting the original set of objects – the source objects.

1   Erase the five multiple-copied shapes to leave the original scaled shape with dimensions. Unlock layer DIMS.

2   Move the shape and dimensions:
    Base point: **Centre of circle**
    Second point: **@120,0**.

3   Refer to Figure 19.6 and with layer OUT current, draw the following four lines:

    AB   MN   PQ   XY
    first: 140,270   first: 260,200   first: 240,150   first: 155,150
    next: @0,−100   next: @0,70   next: @50<45   next: @60<0

4   With layer TEXT current, add the following text items each with 0 rotation and using:
    a) centred on 210,257; height: 6; item: AutoCAD
    b) centred on 195,180; height: 7; item: R2005.

5   Make layer OUT current and select the **MIRROR icon** from the Modify toolbar and:
    prompt   Select objects
    respond   **window the shape and dimensions then right-click**
    prompt   Specify first point of mirror line
    respond   **Endpoint icon and pick point A**
    prompt   Specify second point of mirror line
    respond   **Endpoint icon and pick point B**
    prompt   Delete source objects [Yes/No]<N>
    enter   **N <R>**.

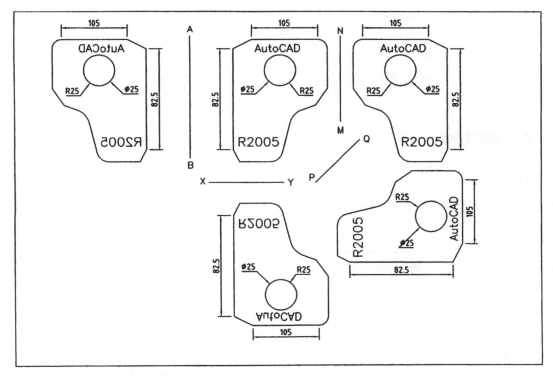

Figure 19.6 WORKDRG after the MIRROR command.

6. The selected objects are mirrored about the line AB. The added text items may or may not be mirrored, this being dependant on the system variable MIRRTEXT which is either 0 or 1 and:
   a) if MIRRTEXT: 1, then text items are mirrored
   b) if MIRRTEXT: 0, text items are not mirrored.

7. At the command line enter **MIRRTEXT <R>**
   prompt   Enter new value for MIRRTEXT<?>
   enter   0 <R>.

8. Menu bar with **Modify-Mirror** and:
   prompt   Select objects
   enter   **P <R>** – previous selection set option
   prompt   19 found
   then   Select objects
   respond **right-click** to end selection
   prompt   Specify first point of mirror line and: **pick endpoint of point M**
   prompt   Specify second point of mirror line and: **pick endpoint of point N**
   prompt   Delete source objects and enter: **N <R>**.

9. The shape is mirrored. Are the text items also mirrored?

10. At the command line enter **MIRROR <R>** and:
    a) select objects: enter P <R><R> – yes two returns – why?
    b) first point – pick point P
    c) second point – pick point Q
    d) delete – N.

11 Set the MIRRTEXT variable to 1 and mirror the original shape about line XY without deleting the source objects.

12 The final result should be Figure 19.6.

13 *Task*: Before leaving the exercise, thaw layer CL. The circle centre lines are still in their original positions.

Reference option

The rotate and scale modify commands have a reference option, and we will use the rotate command to demonstrate how this option is used.

1 Close any existing drawing (no save) and open the original WORKDRG. Refer to Figure 19.7.

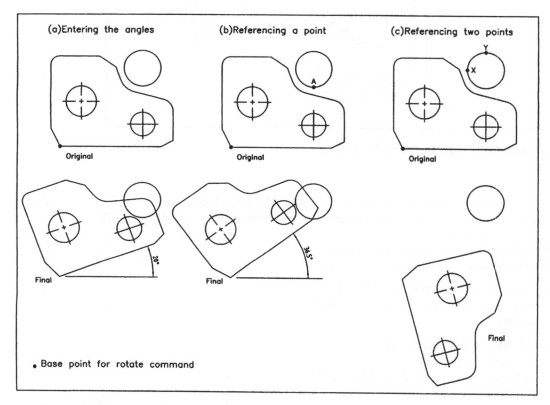

Figure 19.7 Using the ROTATE (reference option) command with WORKDRG.

2 Draw a 30-radius circle with centre at 200,180.

3 Activate the **rotate** command, window the shape (not the circle) right-click and:
    prompt   Specify base point
    respond   **Endpoint icon and pick as indicated**
    prompt   Specify rotation angle or [Reference]
    enter   **R <R>** – the reference option
    prompt   Specify the reference angle and enter **70 <R>**
    prompt   Specify the new angle and enter **90 <R>**
    and   the selected shape will be rotated through an angle of 20 degrees as Figure 19.7(a)
                This 20 degree angle is equivalent to (new−old).

4   Undo the rotate effect to restore the original WORKDRG.

5   With the rotate command:
    a) objects: window the WORKDRG shape then right-click
    b) base point: endpoint icon and pick as indicated
    c) rotation angle: quadrant icon and pick pt A on circle
    d) the selected shape will be rotated about the base point as Figure 19.7(b). The actual angle of rotation is from the horizontal to a line from the base point through the selected quadrant of the circle. I dimensioned this angle and it was 36.5 degrees.

6   Undo the rotate effect.

7   With the rotate command again:
    a) objects: window the shape and right-click
    b) base point: endpoint icon and pick as indicated
    c) rotation angle: enter **R <R>**, the reference option, and:
       prompt   Specify the reference angle
       respond **quadrant icon and pick pt X as indicated**
       prompt   Specify second point
       respond **quadrant icon and pick pt Y as indicated**
       prompt   Specify new angle
       enter   −30 <R>
    d) the selected shape will be rotated about the base point as Figure 19.7(c). The actual rotation angle is equivalent to −75 degrees and is obtained from the (new−old). I will let you work out how this (new−old) gives −75.

8   This exercise does not need to be saved.

Assignments

Five assignments are given for this chapter as the Modify commands are used extensively in CAD. Some of these assignments are new while others have been used successfully in previous books. Each activity should be completed on the A3SHEET standard drawing sheet and layers should be used correctly. The components do not need to be dimensioned, unless you are feeling adventurous. Remember to save your completed drawings as C:\BEGIN\ACT?? etc. and always use your discretion for sizes which are not given.

Activity 14: A vent cover plate

The original component has to be drawn and then copied into the rectangular plate. Commands could be rotate and multiple copy, but that is your decision.

Activity 15: Designs

You have to create the basic shapes then multiple copy them into the geometric outline. Relatively simple, but involves extensive use of the TRIM command. The geometric shapes are your own sizes.

Activity 16: A template

This is easier to complete than you may think. Draw the quarter template using the sizes, then MIRROR. The text item is to be placed at your discretion. Multiple copy and scale by half, third and quarter. Additional: add the given dimensions to a quarter template.

Activity 17: An Aztec pattern

With grid and snap set to 5, create the basic shape (or design your own). The shape is to 'fit into' a 75-side square.

Activity 18: Two well-known objects

a) With the simple steps given, the SUNGLASSES are easier than you would expect. Really a drawing exercise, the only modify command being mirror to get the complete effect. Some sizes of your own in step 3.

b) The screwdriver can be completed as a half shape then mirrored. Use offset as much as possible. Add the dimensions to the completed shape and check dimension A. It should be 21.54.

Summary

1. The commands COPY, MOVE, ROTATE, MIRROR and SCALE can all be activated:
   a) from the menu bar with **Modify-Copy**, etc.
   b) by selecting the icon from the Modify toolbar
   c) by entering the command at the command line, e.g. SCALE <R>.

2. The selection set is useful when selecting objects, especially the add/remove options.

3. All of the commands require a base point, and this can be:
   a) entered as co-ordinates
   b) referenced to an existing object
   c) picked on the screen.

4. Objects on layers which are OFF or FROZEN are not affected by the modify commands. This could cause problems!!

5. MIRRTEXT is a system variable with a 0 or 1 value and:
   a) value 1: text is mirrored – this is the default
   b) value 0: text is not mirrored.

6. Keyboard entered dimensions are not scaled.

Chapter **20**

Grips

Grips are an aid to the draughting process, offering the user a limited number of modify commands. In an earlier chapter we 'turned grips off' with the command line entry of GRIPS: 0. This was to allow the user to become reasonably proficient with using the draw and modify commands. Now that this has been achieved, we will investigate how grips are used.

Toggling grips on/off

Grips can be toggled on/off using:
a) the Selection tab from the Options dialogue box
b) command line entry.

1 Open your A3SHEET standard sheet with layer OUT current.

2 From the menu bar select **Tools-Options** and:
   prompt   Options dialogue box
   respond **pick Selection tab**
   prompt   Selection tab dialogue box
   with   four distinct parts, two of which refer to grips:
              a) Grip Size
              b) Grips
   respond refer to Figure 20.1 and:
              a) set grip size to suit
              b) set the three grip colours to:
                   1. unselected grip: blue
                   2. selected grip: red
                   3. hover grip: green
              c) Enable grips: active – i.e. tick in box
              d) Enable grips within blocks: not active
              e) Enable grip tips: active
              f) Object selection limit: 100
              g) pick Apply then OK.

3 The drawing screen is returned, with the grip box 'attached' to cursor cross-hairs.

4 At the command line enter GRIPS <R> and:
   prompt   Enter new value for GRIPS<1>
   respond **<RETURN>** – i.e. leave the 1 value.

5 The command entry to toggle grips on/off is GRIPS: 1 are ON; GRIPS: 0 are OFF.

6 *Notes*
   a) the grip box attached to the cross-hairs should not be confused with the pick box used with modify commands. Although similar in appearance they are entirely different
   b) when any command is activated (e.g. LINE), the grip box will disappear from the cross-hairs, and re-appear when the command is terminated.

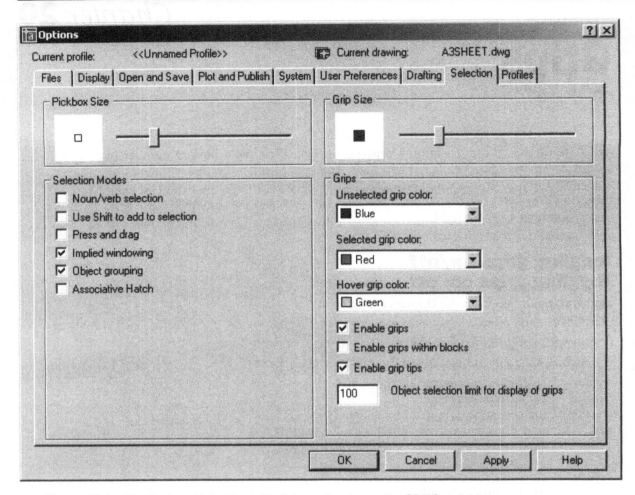

Figure 20.1 The Options (Selection tab) dialogue box to set the GRIPS variables.

What do grips do and how do they work?

1 Grips are small solid-filled squares displayed at strategic points on objects selected by the user.

2 Grips provide the user with five modify commands: stretch, move, rotate, scale and mirror.

3 Grips work in the 'opposite sense' from normal command selection:
   a) the usual sequence is to activate the command then select the objects, e.g. MOVE, then pick object to be moved
   b) with grips, the user selects the objects and then activates one of the five grip options.

4 To demonstrate how grips work:
   a) Draw a line, circle, arc and text item anywhere on the screen.
   b) Ensure grips are on, i.e. GRIPS: 1 at command line.
   c) Refer to Figure 20.2 and move the cursor to each object and 'pick them' with the grip box and:
      1. solid blue grip boxes appear at each object 'snap point' and the selected object is highlighted as Figure 20.2(a)
      2. move the grip box on the cursor cross-hairs over any solid blue box and leave for a second. The solid blue box will change to a green solid box. This is the **hover grip**

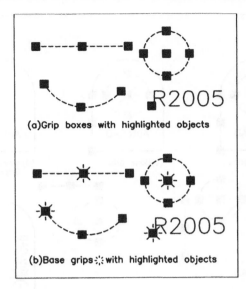

Figure 20.2 Grip 'states' or types.

3. move the grip box to any one of the solid blue boxes and 'pick it'. The solid blue box changes to a solid red box and the selected object is still highlighted. The red box is the **BASE GRIP** – Figure 20.2(b)
4. press the ESC key – solid blue grips with highlighted object
5. ESC again to cancel the grips operation and restore the selected objects to their original appearance.

Grips exercise 1

This demonstration is relatively simple but rather long. It is advisable to work through the exercise without missing out any of the steps.

1 Erase all objects from the screen, or re-open A3SHEET.

2 Refer to Figure 20.3 and draw the original shape using the sizes given as Figure 20.3(a). Make the lower left corner at the point (100,100) and ensure grips are on.

3 Move the cursor to the circle and pick it, then move to the right vertical line and pick it. Solid blue grip boxes appear and the two objects are highlighted as Figure 20.3(b).

4 Move the cursor grip box over the grip box at the circle centre and the green hover grip will be displayed, then left-click, i.e. pick the circle centre grip. The selected box will be displayed in red as it is now the base grip as Figure 20.3(c).

5 Observe the command line:
    prompt   ** STRETCH **
                 Specify stretch point or [Base point/Copy/Undo/eXit]
    respond **with a <RETURN>**
    prompt   ** MOVE **
                 Specify move point or [Base point/Copy/Undo/eXit]
    enter   **@25,25 <R>**.

6 The following should have happened:
    a) the circle and line are moved
    b) the command prompt line is returned
    c) the grips are still active
    d) there is no base grip – Figure 20.3(d).

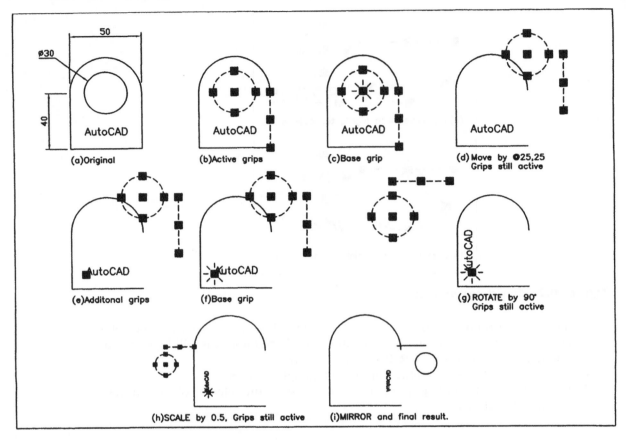

Figure 20.3 Grips exercise 1.

7 Move the cursor and pick the text item to add it to the grip selection – Figure 20.3(e).

8 Make the left grip box of the text item the base grip, by moving the cursor pick box onto it, left-clicking as Figure 20.3(f) and:
   prompt   ** STRETCH **
              Specify stretch point or [Base point/Copy...
   respond **with a <RETURN>**
   prompt   ** MOVE **
              Specify move point or [Base point/Copy...
   respond **<RETURN>**
   prompt   ** ROTATE **
              Specify rotation angle or [Base point/Copy...
   enter   **90 <R>**.

9 The circle, line and text item will be rotated and the grips are still active – Figure 20.3(g).

10 Make the same text item grip box the base grip (easy!) and:
    prompt   ** STRETCH **
               Specify stretch point or...
    enter   **SC <R>** – the scale grip option
    prompt   ** SCALE **
               Specify scale factor or [Base point...
    enter   **0.5 <R>**.

11 The three objects are scaled and the grips are still active as Figure 20.3(h).

12 Make the right box on the line the base grip and:
    prompt   ** STRETCH **
    enter   **MI <R>** – the mirror option
    prompt   ** MIRROR **
              Specify second point or [Base point...
    enter   **B <R>** – the base point option
    prompt   Specify base point
    respond   **Midpoint icon and pick the original horizontal line**
    prompt   ** MIRROR **
              Specify second point or...
    respond   **Midpoint icon and pick the arc**.

13 The three objects are mirrored about the selected 'line' and the grips are still active as Figure 20.3(i).

14 Press ESC – removes the grips and ends the sequence.

15 The exercise is now complete. Do not exit yet.

Selection with grips

Selecting individual objects for use with grips can be tedious. It is possible to select a window/crossing option when grips are on.

1 Does your screen still display the line, circle and text item after the grips exercise has been completed?

2 Refer to Figure 20.4(a), move the cursor and pick a first point 'roughly' where indicated. Move the cursor down and to the right, pick a second point and all complete objects within the window will display solid blue grip boxes with highlighted objects.

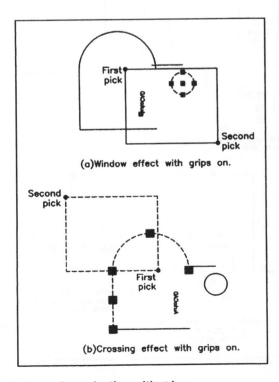

Figure 20.4 Window/crossing selection with grips.

3   ESC to cancel the grip effect.

4   Move the cursor to about the same point as step 2, and pick a first point as Figure 20.4(b), then move the cursor upwards and to the left and pick a second point. All objects within or which cross the boundary will display solid blue grip boxes with highlighted objects.

5   ESC to cancel grip selection.

6   The effect can be summarised as:
    a) window effect to the right of first pick
    b) crossing effect to the left of the first pick.

Grips exercise 2

1   Open C:\BEGIN\USEREX to display the component used in the offset and text exercises and refer to Figure 20.5.

2   Ensure only the red objects are displayed – Figure 20.5(a).

3   Grips on?

4   Pick a point 1 and drag out a window and pick a point 2. Five lines will display solid blue grip boxes and be highlighted as Figure 20.5(b).

5   Make the rightmost grip the base grip and:
    prompt   ** STRETCH **
    respond   **right-click** to display the grip pop-up menu options

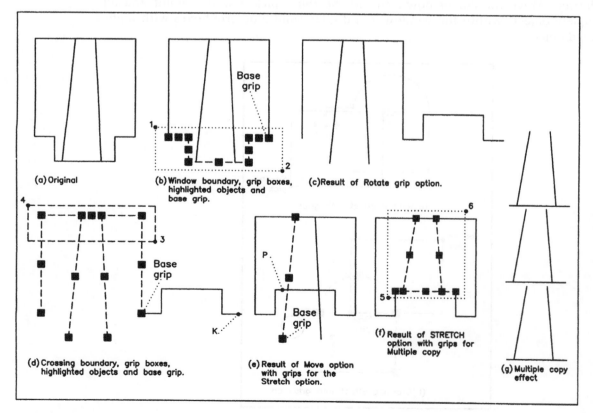

Figure 20.5   Grips exercise 2.

then	**pick Rotate**
prompt	Specify rotation angle or...
enter	**180 <R>**.

6. Now ESC to cancel the grips – Figure 20.5(c).

7. Pick a point 3 and drag out a crossing-window and pick a point 4 to display five highlighted lines with solid blue grip boxes as Figure 20.5(d).

8. Make the lowest grip box on the right vertical line hot and:
   prompt   ** STRETCH **
   respond   **right-click** to display the grip options menu
   then   **pick Move** – Figure 20.6
   prompt   Specify move point or...
   respond   **Endpoint icon and pick line K.**

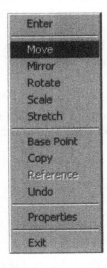

Figure 20.6 The GRIPS shortcut menu.

9. The lines are moved 'onto' the rotated lines – Figure 20.5(e).

10. ESC to cancel the grips.

11. Pick one of the inclined lines and:
    a) make the lowest grip box the base grip
    b) activate the STRETCH grips option
    c) stretch the grip box perpendicular to line P
    d) repeat for the other inclined line as Figure 20.5(f).

12. Finally:
    a) drag a window from point 5 to point 6 – three highlighted lines
    b) make the leftmost grip hot
    c) activate the MOVE grips option
    d) enter C <R> to activate the copy option
    e) copy the selected objects to three parts of the screen – Figure 20.5(g)
    f) right-click and enter to end grip options
    g) ESC to end command.

13. *Note*: This exercise could have been completed using the modify commands, e.g. rotate and trim. It demonstrates that there is no one way to complete a drawing.

14. Save this exercise if required, but not as USEREX.

Assignments

Activity 19 involves the re-positioning of a robotic arm and using the arm to re-position a component. This is an updated activity which has proved relatively successful in my previous 'Beginning books'.

1. Refer to Activity 19(a) and:
   a) create the robotic arm in the original position using the sizes given
   b) use your discretion for sizes not given
   c) position the two pallets (A and B) and the component C.

2. Refer to Activity 19(b) and:
   Fig(a): the original positions
   Fig(b): the upper and lower arms MIRRORED about a vertical line through the large lower arm circle with the hot grip as the large circle centre
   Fig(c): the upper arm MOVED, STRETCHED and ROTATED onto the component
   Fig(d): the upper arm ROTATED to a horizontal position and STRETCHED with component attached
   Fig(e): the upper arm STRETCHED and ROTATED to place the component on pallet B
   Fig(f): the robot home position with the lower arm vertical and the upper arm horizontal.

3. *Notes*
   a) I have added the upper arm aligned dimensions to assist with the stretch operations
   b) In the activity figure, I have trimmed the robot for appearance purposes only.

4. Save when complete.

Summary

1. Grips allow the user access to the STRETCH, MOVE, ROTATE, SCALE and MIRROR modify commands without icon or menu bar selection.

2. Grips work in the 'opposite sense' from the normal AutoCAD commands, i.e. select object first then the command.

3. Grips **do not have to be used**. They give the user another draughting aid.

4. Grips are toggled on/off using the Grips dialogue box or by keyboard entry. The dialogue box allows the grip box colours and size to be altered.

5. If grips are not being used, I would always recommend that they be toggled off, i.e. GRIPS: 0 at the command line.

6. When a grip box is the base grip, the options can be activated by:
   a) return at the keyboard
   b) entering SC, MO, MI, etc.
   c) right-click the mouse to display the grip option menu.

Chapter 21
Drawing assistance

Up until now, all objects have been created by picking points on the screen, entering co-ordinate values or by referencing existing objects, e.g. midpoint, endpoint, etc. There are other methods which enable objects to be positioned on the screen, and in this chapter we will investigate three new concepts, these being:
a) point filters
b) construction lines
c) ray lines.

Point filters

These allow objects to be positioned by referencing the X and Y co-ordinate values of existing objects.

Example 1

1 Open the A3SHEET standard sheet and refer to Figure 21.1.

2 Draw a 50-sided square, lower left corner at 20,220. Ensure that grips are not active – i.e. off.

3 Multiple copy this square to three other positions.

4 A circle of diameter 30 has to be created at the 'centre' of each square and this will be achieved by four different methods:
   a) *Co-ordinates*
      Activate the circle command with centre: 45,245; radius: 15.
   b) *Object snap midpoint*
      1. draw in a diagonal of the square
      2. select the CIRCLE icon from the Draw toolbar and:
         a) centre point: **Snap to Midpoint** of diagonal
         b) radius: enter 15.
   c) *Object snap from*
      1. select the CIRCLE icon from the Draw toolbar and:
         a) centre: pick the **Snap from icon**
         b) base point: **Endpoint icon** and pick left end of line AB
         c) offset: enter @25,25
         d) radius: enter 15.
   d) *Point filters*
      Activate the circle command and:
      prompt   Specify center point
      enter   **.X <R>**
      prompt   of

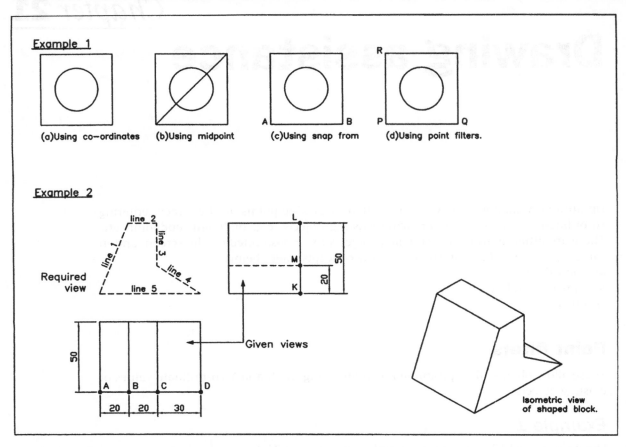

Figure 21.1 Point filter examples.

respond	**Midpoint icon and pick line PQ**
prompt	`(need YZ)`
respond	**Midpoint icon and pick line PR**
prompt	`Specify radius` and enter: **15 <R>**.

Example 2

Figure 21.1 displays the top, end and isometric views of a shaped block. The front view has to be created from the two given views, and we will use the point-filter technique to achieve this.

1. Draw the top and end views using the sizes given. Use the lower part of the screen. Draw with the snap on.

2. Select the Line icon from the Draw toolbar and:
prompt	`Specify first point`
enter	**.X <R>**
prompt	`of`
respond	**Intersection icon and pick point A**
prompt	`(need YZ)`
respond	**Intersection icon and pick point K**
and	cursor 'snaps' to a point on the screen
prompt	`Specify next point`
enter	**.X <R>**
prompt	`of`

respond	**Intersection icon and pick point B**	
prompt	`(need YZ)`	
respond	**Intersection icon and pick point L** – line 1 is drawn	
prompt	`Specify next point`	
enter	**.X <R>** and **pick Intersection of point C**	
prompt	`(need YZ)` and **pick Intersection of point L** – line 2 is drawn	
prompt	`Specify next point`	
enter	**.X <R>** and **pick Intersection of point C**	
prompt	`(need YZ)` and **pick Intersection of point M** – line 3 is drawn	
prompt	`Specify next point`	
enter	**.X <R>** and **pick Intersection of point D**	
prompt	`(need YZ)` and **pick Intersection of point K** – line 4 is drawn	
prompt	`Specify next point`	
enter	**C <R>** to draw line 5.	

3 The front view of the shaped block is now complete.

4 This exercise does not need to be saved.

5 *Notes*
   a) The point-filter method of creating objects can be rather 'cumbersome' to use.
   b) Point filters are another draughting aid.
   c) I would suggest that Object Snap Tracking is 'easier' than point filters for creating the front view in this exercise. Why not try this for yourself by setting the object snap to the Endpoint mode and acquiring the endpoints of the various lines with Object Snap Tracking on.

Construction lines

Construction lines are lines that extend to infinity in both directions from a selected point on the screen. They can be referenced to assist in the creation of other objects.

1 Open the A3SHEET standard sheet, layer OUT current and display toolbars Draw, Modify and Object Snap. Refer to Figure 21.2(a).

2 With layer OUT current, draw:
   a) a 100-sided square, lower left corner at 50,50
   b) a circle, centred on 250,220 with radius 50.

3 Make a new layer (**Format-Layer**) named CONLINE, colour to suit and with a DASHED linetype, and make it current. Set the LTSCALE value to your requirements.

4 Menu bar with **Draw-Construction Line** and:
   prompt   `Specify a point or [Hor/Ver/Ang/Bisect/Offset]`
   enter   **50,50 <R>**
   and   line 'attached' to cursor through the entered point 'rotates' as the mouse is moved
   prompt   `Specify through point`
   enter   **80,200 <R>**
   prompt   `Specify through point`
   respond   **Center icon and pick the circle**
   prompt   `Specify through point` and right-click.

5 At the command line enter **XLINE <R>** and:
   prompt   `Specify a point or...`
   enter   **H <R>** – the horizontal option
   prompt   `Specify through point`

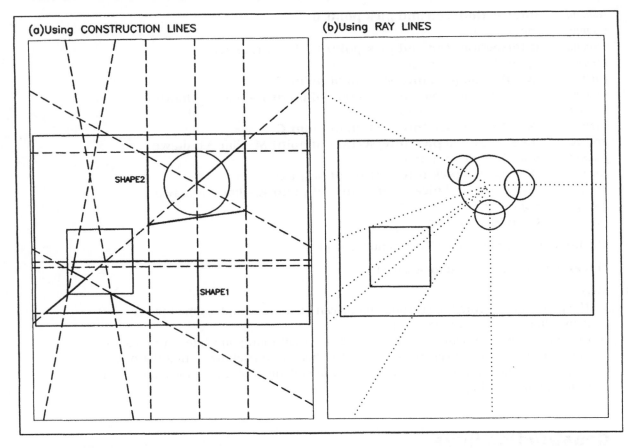

Figure 21.2 Construction and ray lines.

    enter   **100,20 <R>**
    prompt   Specify through point
    respond   **Midpoint icon and pick a vertical line of square**
    prompt   Specify through point
    respond   **Quadrant icon and pick top of circle**
    prompt   Specify through point and right-click.

6   Select the **CONSTRUCTION LINE icon** from the Draw toolbar and:
    prompt   Specify a point or...
    enter   **V <R>** – the vertical option
    prompt   Specify through point
    respond   **Center icon and pick the circle**.
    prompt   Specify through point and right-click.

7   Activate the construction line command and:
    prompt   Specify a point or...
    enter   **O <R>** – the offset option
    prompt   Specify offset distance or [Through] and enter: **75 <R>**
    prompt   Select a line object
    respond   **pick the vertical line through the circle centre**
    prompt   Select side to offset
    respond   **offset to the right**
    prompt   Select a line object
    respond   **offset the same line to the left**
    prompt   Select a line object and right-click.

8 Construction line command and at prompt:
   enter   **A <R>** – the angle option
   prompt   Enter angle of xline (0.0) or [Reference]
   enter   **−30 <R>**
   prompt   Specify through point
   respond **Center icon and pick the circle**
   prompt   Specify through point
   enter   **135,40 <R> then right-click**.

9 Construction line command for last time and at prompt:
   enter   **B <R>** – the bisect option
   prompt   Specify angle vertex point
   respond **Midpoint icon and pick top line of square**
   prompt   Specify angle start point
   respond **pick lower left vertex of square**
   prompt   Specify angle end point
   respond **Midpoint icon and pick right vertical line of square**
   prompt   Specify angle end point and right-click.

10 Construction lines can be copied, moved, trimmed, etc. and object snap referenced to create other objects. SHAPE1 and SHAPE2 in Figure 21.2(a) have been 'drawn' by referencing the existing construction lines. Try some shape creation for yourself.

11 *Notes*
   a) Think of how construction lines could have been used to create the front view of the point filters example completed earlier in this chapter.
   b) I would recommend that construction lines are created on their own layer, and that this layer is frozen to avoid 'screen clutter' when not in use. I would also recommend that they are given a colour and linetype not normally used.

12 *Task* – try the following:
   a) At the command line **enter LIMITS <R>** and:
      prompt   Specify lower left corner and enter: **0,0 <R>**
      prompt   Specify upper right corner and enter: **10000,10000 <R>**.
   b) From menu bar select **View-Zoom-All** and:
      1. drawing appears very small at bottom of screen
      2. the construction lines 'radiate outwards' to the screen edges.
   c) Using **LIMITS <R>** enter the following values:
      lower left corner: −10000,−10000
      upper right corner: 0,0.
   d) **View-Zoom-All** to 'see' the construction lines.
   e) Return limits to 0,0 and 420,297 then **View-Zoom-All** to restore the original drawing screen.

13 The construction line exercise is now complete. The drawing can be saved if required, but we will not use it again.

Rays

Rays are similar to construction lines, but they only extend to infinity in one direction from the selected start point.

1 Make a new layer called RAYLINE, colour to suit and with a dotted linetype. Make this layer current and refer to Figure 21.2(b).

2 Freeze layer CONLINE and erase any objects to leave the original square and circle.

3   Menu bar with **Draw-Ray** and:
    prompt   Specify start point
    respond   **Center icon and pick the circle**
    prompt   Specify through point and enter: **@100<150 <R>**
    prompt   Specify through point and enter: **@100<0 <R>**
    prompt   Specify through point and enter: **@100<−90 <R>**
    prompt   Specify through point and right-click.

4   With layer OUT current, draw three 25-radius circles at the intersection of the ray lines and circle.

5   Make layer RAYLINE current and at the command line enter RAY <R> and:
    prompt   Specify start point
    respond   **pick the circle centre point**
    prompt   Specify through point
    respond   **Intersection icon and pick the four vertices of the square then right-click**.

6   Use the LIMITS command and enter:
    a) lower left: −10000,−10000
    b) upper right: 0,0.

7   Menu bar with **View-Zoom-All** and note the position of 'our drawing'.

8   Return limits to 0,0 and 420,297, then **Zoom-Previous** to restore the original drawing screen.

9   *Note*: As with construction lines, it is recommended that ray lines be drawn on 'their own layer' with a colour and linetype to suit. This layer can then be frozen when not in use.

Summary

1   Point filters allow the user a method of creating objects by referencing existing object co-ordinates.

2   Point filters are activated by keyboard entry.

3   Construction and ray lines are aids to draughting. They allow lines to be created from a selected point and:
    a) construction lines extend to infinity in both directions from the selected start point
    b) ray lines extend to infinity in only one direction from the selected start point.

4   Both construction and ray lines can be activated from the menu bar or by keyboard entry. XLINE is the construction line entry and RAY the ray line entry.

5   The default Draw toolbar displays only the Construction Line icon, although this can be altered to include the ray line icon.

6   Both construction and ray lines are very useful aids for the CAD operator, but it is personal preference if they are to be used.

Chapter 22
Viewing a drawing

Viewing a drawing is important as the user may want to enlarge a certain part of the screen for more detailed work or return the screen to a previous display. AutoCAD allows several methods of altering the screen display including:
a) the scroll bars
b) the Pan and Zoom commands
c) the Aerial view option
d) Realtime pan and zoom.

In this chapter we will investigate several view options.

Getting ready

1 Open WORKDRG to display the red outline, two red circles, four green centre lines and a black border.

2 With layer TEXT current, menu bar with **Draw-Text-Single Line Text** and:
 a) start point: **centred** on 90,115
 b) height: 0.1 – yes 0.1
 c) rotation: 0
 d) text item: AutoCAD.

3 With layer OUT current, draw a circle centred on 89.98,115.035 and radius 0.01 – the awkward co-ordinate entries are deliberate.

4 These two objects cannot yet be 'seen' on the screen.

5 Ensure the Zoom toolbar is displayed and positioned to suit.

Pan

Allows the graphics screen to be 'moved', the movement being controlled by the user:
a) with co-ordinate entry
b) by selecting points on the screen
c) using the mouse-tracker wheel (if available).

1 From the Standard toolbar select the **PAN REALTIME icon** and:
 prompt Press Esc or Enter to exit, or right-click to display shortcut menu
 and   cursor changes to a hand

respond a) hold down the left button of mouse
b) move mouse and complete drawing moves
c) note that scroll bars also move
d) move image roughly back to original position
e) right-click and:
   1. pop-up shortcut menu displayed
   2. Pan is active – tick
   3. pick Exit.

2 Menu bar with **View-Pan-Point** and:
prompt   Specify base point or displacement and enter: **0,0 <R>**
prompt   Specify second point and enter: **500,500 <R>**.

3 No drawing on screen – do not panic!

4 Use PAN REALTIME to pan down and to the left, and 'restore' the drawing roughly to its original position, then right-click and exit.

5 Menu bar with **View-Zoom-All**.

Zoom

This is one of the most important and widely used of all the AutoCAD commands. It allows parts of a drawing to be magnified/enlarged on the screen. The command has several options, and it is these options which will now be investigated. To assist us in investigating the zoom options:
a) use the COPY command, window the shape and copy the component from 50,50 to 210,260
b) refer to Figure 22.1.

Zoom All

Displays a complete drawing including any part of the drawing which is 'off' the current screen.

1 From menu bar select **View-Zoom-All**.

2 The two components and the black border are displayed – Figure 22.1(a).

Zoom Window

Perhaps the most useful of the zoom options. It allows areas of a drawing to be 'enlarged' for clarity or more accurate work.

1 Select the **ZOOM WINDOW icon** from the Zoom toolbar and:
prompt   Specify first corner
respond   **window the original left circle** – Figure 22.1(b).

2 At the command line enter **ZOOM <R>** and:
prompt   Specify corner of window, enter a scale factor...
enter   **W <R>** – the window option
prompt   Specify first corner and enter: **89.6,114.9 <R>**
prompt   Specify opposite corner and enter: **90.4,115.2 <R>**.

3 The text item and circle will now be displayed – Figure 22.1(c).

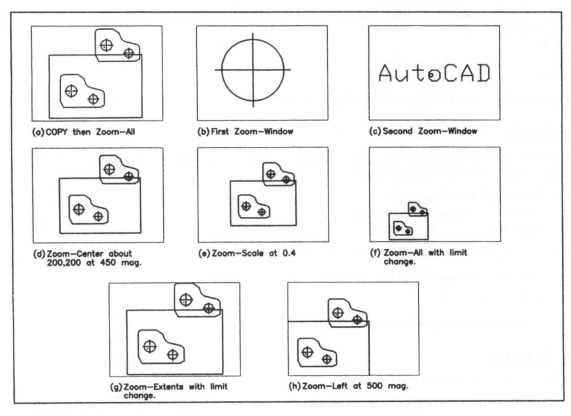

Figure 22.1 Various zoom options.

Zoom Previous

Restores the drawing screen to the display before the last view command.

1 Menu bar with **View-Zoom-Previous** – restores Figure 22.1(b).
2 At command line enter **ZOOM <R>** then **P <R>** – restores Figure 22.1(a).

Zoom Center

Allows a drawing to be centred about a user-defined centre point.

1 Select the **ZOOM CENTER icon** and:
   prompt   Specify center point and enter: **200,200 <R>**
   prompt   Enter magnification or height<?> and enter **450 <R>**.
2 The complete drawing is centred on the drawing screen about the entered point – Figure 22.1(d).
3 Notes
   a) The 'size' of the displayed drawing depends on the magnification/height value entered by the user and is relative to the displayed **default<?> value** and:
      1. a value less than the default – magnifies drawing on screen
      2. a value greater than the default – reduces the size of the drawing on the screen.
   b) Toggle the grid ON and note the grid effect with this centre option, then toggle the grid off.

4   Menu bar with **View-Zoom-Center** and enter the following centre points, all with 1000 magnification:
    a) 200,200
    b) 0,0
    c) 500,500.

5   Now Zoom Previous four times to restore Figure 22.1(a).

Zoom Scale

Centres a drawing on the screen at a scale factor entered by the user. It is similar (and easier?) than the Zoom Center option.

1   Select the **ZOOM SCALE icon** and:
    prompt   Enter a scale factor (nX or nXP)
    enter   **0.4 <R>**.

2   The drawing is displayed centred and scaled – Figure 22.1(e).

3   Menu bar with **View-Zoom-Scale** and enter a scale factor of 0.25.

4   Zoom to a scale factor of 1.5.

5   Zoom Previous three times to restore Figure 22.1(a).

Zoom Extents

Zooms the drawing to extent of the current limits.

1   At command line enter **LIMITS <R>** and:
    prompt   Specify lower left corner and enter: **0,0 <R>**
    prompt   Specify upper right corner and enter: **1000,1000 <R>**.

2   Menu bar with **View-Zoom-All** – Figure 22.1(f).

3   Select the ZOOM EXTENTS icon – Figure 22.1(g).

4   Menu bar with **Format-Drawing Limits** and set the limits back to 0,0 and 420,297.

5   Zoom All to restore Figure 22.1(a).

Zoom Dynamic

This option will not be discussed. The other zoom options and the realtime pan and zoom should be sufficient for all user needs.

Zoom Realtime

1   Menu bar with **View-Zoom-Realtime** and:
    prompt   Press ESC or ENTER to exit, or right-click to display
              shortcut menu
    and   cursor changes to a magnifying glass with a + and –
    respond   a) hold down left button of mouse and move upwards to give a magnification effect
              b) move downwards to give a decrease in size effect
              c) left–right movement – no effect
              d) right-click to display pop-up menu with Zoom active
              e) pick Exit.

2   **View-Zoom-All** to display Figure 22.1(a).

3   The Zoom Realtime icon is in the Standard Toolbar.

Zoom Left

This option is not available as an icon and does not appear as a command line option. It is an old zoom option that can still be activated.

1   At the command line enter **ZOOM <R>** and:
    prompt   Specify corner of window or...
    enter   **L <R>** – the left option
    prompt   Lower left corner point and enter: **0,0 <R>**
    prompt   Enter magnification or height and enter: **500 <R>**.

2   Complete drawing is moved to left of drawing screen and displayed at the entered magnification – Figure 22.1(h).

3   Zoom Previous to restore Figure 22.1(a).

Aerial view

The aerial view is a navigation tool (AutoCAD expression) and allows the user to pan and zoom a drawing interactively. This view option will be investigated when we construct a large-scale drawing.

Transparent zoom

A transparent command is one which can be activated while using another command and zoom has this facility.

1   Restore the original Figure 21.2(a) – Zoom All?

2   Select the **LINE icon** and:
    prompt   Specify first point
    enter   **0,0 <R>**
    prompt   Specify next point
    enter   **'ZOOM <R>** – the transparent zoom command
    prompt   All/Center/Dynamic... – the zoom options
    enter   **W <R>** – the window zoom option
    prompt   Specify corner of window and enter: **89.6,114.9 <R>**
    prompt   Specify opposite corner and enter: **90.4,115.2 <R>**
    prompt   Specify next point
    respond **centre icon and pick the circle created at the start of the chapter**
               (which should now be displayed)
    prompt   Specify next point
    enter   **'Z <R> then A <R>** – the zoom all option
    prompt   Specify next point
    respond **right-click-enter** to end line sequence.

3   The transparent command was activated by entering 'ZOOM <R> or 'Z <R> at the command line.

4   Only certain commands have transparency, generally those which alter a drawing display, e.g. GRID, SNAP and ZOOM.

5   Do not save this drawing modification.

6 *Note*: In this exercise we entered Z <R> to activate the zoom command and this is an example of using an AutoCAD Alias or Abbreviation. Many of the AutoCAD commands have an alias which allows the user to activate the command from the keyboard by entering one or two letters.
Typical aliases are:
L line   C circle   E erase   M move   CP Copy

Summary

1 Pan and Zoom are VIEW commands usually activated by icons.

2 Pan does 'not move a drawing' – it 'moves' the complete drawing screen.

3 The zoom command has several options, the most common being:
   All   displays the complete drawing
   Window   allows parts of a drawing to be displayed in greater detail
   Center   centres a drawing about a user-defined point at a user-specified magnification
   Previous   displays the drawing screen prior to the last pan/zoom command
   Extents   zooms the drawing to the existing limits.

4 Both the pan and zoom commands have a REALTIME option.

Chapter **23**

Hatching

AutoCAD 2005 has four 'types' of associated boundary hatching:
a) predefined, i.e. AutoCAD's stored hatch patterns
b) user-defined
c) custom – not considered in this book
d) gradient.

When applying hatching (section detail) the user has two methods of defining the hatch pattern boundary:
a) by selecting objects which make the boundary
b) by picking a point within the boundary.

There are two hatch commands available:
a) HATCH: command line entry only
b) BHATCH: activated by icon, menu bar or command line entry. This command allows access to a dialogue box.

In this chapter we will consider all of the above options, with the exception of custom hatch patterns.

Getting ready

1. Open the A3SHEET standard sheet and refer to Figure 23.1.

2. Draw a 50-unit square with four lines (do NOT use the Rectangle command) on layer OUT and multiple copy it to ten other parts of the screen. Add the other lines within the required squares (any size). Note that I have included additional squares to indicate appropriate object selection and to demonstrate the 'before and after' effect.

3. Make layer SECT current.

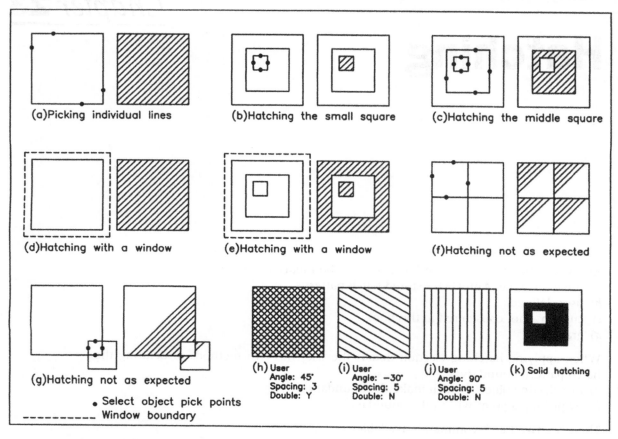

Figure 23.1 User-defined hatching using the Select Objects method.

User-defined hatch patterns – Select Objects method

User-defined hatching consists of straight line patterns, the user specifying:
a) the hatch angle relative to the horizontal
b) the distance between the hatch lines – spacing
c) whether single or double (cross) hatching.

1 At the command line enter **HATCH <R>** and:
   prompt   Enter a pattern name or [?/Solid/User defined]<ANGLE>
   enter   U <R> – user-defined option
   prompt   Specify angle for crosshatch lines<0.0>
   enter   45 <R>
   prompt   Specify spacing between lines<1.0000>
   enter   3 <R>
   prompt   Double hatch area [Yes/No]<N>
   enter   N <R>
   prompt   Select objects
   respond   **pick the four lines of first square and right-click**
   and   hatching added to the square – Figure 23.1(a).

2 **Immediately** after the hatching has been added, right-click and:
   prompt   Shortcut pop-up menu
   respond   **pick Repeat HATCH**
   prompt   Select objects
   respond   **pick four lines indicated in second square then right-click**
   and   hatching added to the small square – Figure 23.1(b).

3   Select the **HATCH icon** from the Draw toolbar and:
    prompt   Boundary Hatch and Fill dialogue box as Figure 23.2
    with   Three tabs: Hatch (current), Advanced, Gradient
    and   a) Type: User-defined
             b) Swatch: Bylayer
             c) Angle: 45
             d) Spacing: 3 – i.e. our previous entries from step 1
             e) Draw Order: Send behind boundary
             f) Composition: Associative
             g) Double: not active
    respond **pick Select Objects**
    and   dialogue box disappears
    prompt   Select objects at command line
    respond **pick eight lines in third square then right-click and pick Enter**
    prompt   Boundary Hatch and Fill dialogue box returned
    respond **pick Preview**
    and   drawing screen displayed with hatching added
    respond either a) ENTER or right-click if hatching correct
             or   b) ESC to return to dialogue box and reselect points
    and   hatching added as Figure 23.1(c).

4   Menu bar with **Draw-Hatch** and:
    prompt   Boundary Hatch and Fill dialogue box – settings as before?
    respond **pick Select Objects**
    prompt   Select objects
    enter   **W <R>** – the window selection option

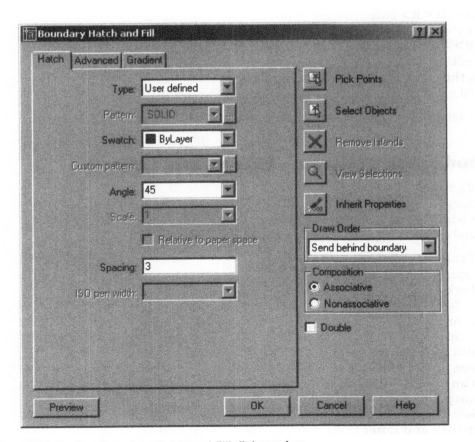

Figure 23.2 The Boundary Hatch and Fill dialogue box.

	then	window the fourth square, right-click and Enter
	prompt	Boundary Hatch and Fill dialogue box
	respond	**Preview and right-click**
	and	square hatched – Figure 23.1(d).

5. At the command line enter **HATCH <R>** then:
   a) pattern name: U
   b) angle: 45
   c) spacing: 3
   d) double: N
   e) Select objects: window the fifth square then right-click
   f) hatching as Figure 23.1(e).

6. Using HATCH from the keyboard, accept the four defaults of U, 45, 3 and N, and pick the four lines of the sixth square to give hatching as Figure 23.1(f). Not as expected?

7. Repeat the hatch command and select the four lines indicated in the seventh square – Figure 23.1(g).

8. With HATCH <R> at the command line, add hatching to the next three squares using the following entries:

	I	II	III
Pattern	U	U	U
Angle	45	−30	90
Spacing	3	5	5
Double	Y	N	N
Figure 23.1	(h)	(i)	(j)

9. Finally enter **HATCH <R>** and:

prompt	Enter a pattern name...
enter	**S <R>** – the solid option
prompt	Select objects
respond	**pick the eight lines as (c) then right-click**
and	a solid hatch pattern is added – Figure 23.1(k).

10. This drawing is complete but does not need to be saved.

User-defined hatch patterns – Pick Points method

1. Refer to Figure 23.3 and:
   a) erase all hatching
   b) erase squares which are not required
   c) add squares and circles as required – size not important.

2. With layer SECT current, pick the **HATCH icon** and:

prompt	Boundary Hatch and Fill dialogue box
respond	a) check/alter:
	1. Type: User-defined
	2. Angle: 45
	3. Spacing: 3
	b) **pick Pick Points**
prompt	Select internal point
respond	**pick any point within first square**
prompt	Various prompts at command line
then	Select internal point – i.e. any more internal points for hatching
respond	**right-click and Enter**

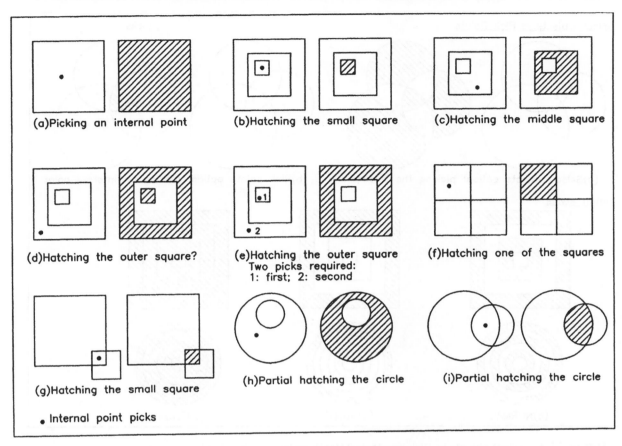

Figure 23.3 User-defined hatching using the Pick Points method.

    prompt   Boundary Hatch and Fill dialogue box
    respond **Preview-right-click**
    and   hatching added to square – Figure 23.3(a).

3 Using the **HATCH icon** with the Pick Points option from the Boundary Hatch and Fill dialogue box, add hatching using the points indicated in Figure 23.3. The only 'problem' is hatching the outer square in Figure 23.3(e) – two pick points are required.

Select Objects vs Pick Points

With two options available for hatching, new users to AutoCAD may be confused as to whether they should use Select Objects or Pick Points. In general the Pick Points option is simpler to use, and will allow complex shapes to be hatched with a single pick within the area to be hatched. To demonstrate the effect:

1 Erase all objects from the screen and refer to Figure 23.4.

2 Draw two sets of intersecting circles – any size.

3 We want to hatch the intersecting area of the three circles.

4 Select the HATCH icon and from the Boundary Hatch dialogue box:
    a) set: User-defined, Angle of 45, Spacing of 3
    b) pick the Select Objects option

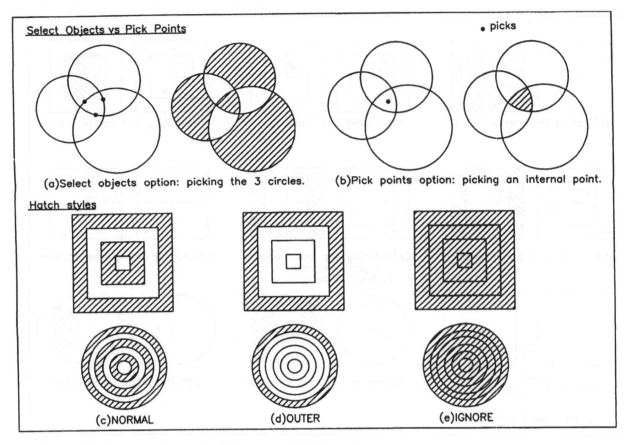

Figure 23.4 Select objects vs Pick points and Hatch styles.

    c) pick the three circles
    d) preview-right-click-OK and hatching as Figure 23.4(a).
5 Select the HATCH icon again and:
    a) pick Pick Points option
    b) pick any point within the area to be hatched
    c) preview-right-click-OK and hatching as Figure 23.4(b).

Hatch style

AutoCAD has a hatch style (called Island detection) option which allows the user to control three 'variants' of the hatch command. To demonstrate the hatch style, refer to Figure 23.4 and draw:
    a) a 70-sided square with four smaller squares 'inside it'
    b) six concentric circles, smallest radius being 5
    c) copy the squares and circles to two other areas of the screen.

1 Select the **HATCH icon** and:
```
prompt    Boundary Hatch and Fill dialogue box
ensure    User-defined, Angle: 45, Spacing: 3
then      pick Advanced tab
prompt    Boundary Hatch and Fill dialogue box with Advanced tab
          active
```

note	a)	Island detection style 'picture'
	b)	Style: Normal active
	c)	Island detection method: Flood active – Figure 23.5
respond		**pick Select Objects**
prompt		Select objects at the command line
respond		**window the first square and circle then right-click**
prompt		shortcut menu as Figure 23.5
respond	a)	pick Preview
	b)	right-click
and		Hatching added as Figure 23.4(c).

2 Repeat the HATCH icon and with the Advanced tab:
   a) alter Island detection style to Outer
   b) pick Select Objects and window the second square and circle
   c) preview then OK – hatching as Figure 23.4(d).

3 Using the Advanced tab of the Boundary Hatch and Fill dialogue box, alter the island detection style to Ignore, and add hatching to the third square and circle – Figure 23.4(e).

4 *Notes*
   a) The user can leave the style at 'Normal' and then control the hatch area by picking the relevant points within the area to be hatched.
   b) Islands are objects within a hatch boundary.
   c) Flood is the recommended island detection method and will recognise objects within a hatch boundary.

5 Save if required, but we will not use this drawing again.

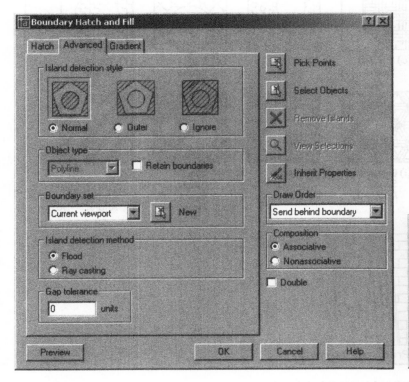

Figure 23.5 The Boundary Hatch and Fill (Advanced tab) dialogue box and Shortcut menu.

Predefined hatch patterns

AutoCAD 2005 has several stored hatch patterns which can be accessed using the command line HATCH entry or from the Boundary Hatch dialogue box which is slightly easier. With predefined hatch patterns, the user specifies:
a) the scale of the patterns
b) the angle of the patterns.

1. Open your standard sheet, refer to Figure 23.6(A) and:
   a) draw a 50-unit square
   b) multiple copy the square eleven times
   c) add other lines and circles as displayed.

2. Select the **HATCH icon** and:
   prompt   Boundary Hatch and Fill dialogue box
   respond *a*) Advanced tab and set:
                1. Island detection style: Normal
                2. Island detection method: Flood
              b) Hatch tab and set:
                1. Type: Predefined
                2. Pattern: pick the [. . .] button
   prompt   Hatch Pattern Palette dialogue box
   with   four tab selections:
              ANSI, ISO, Other Predefined, Custom

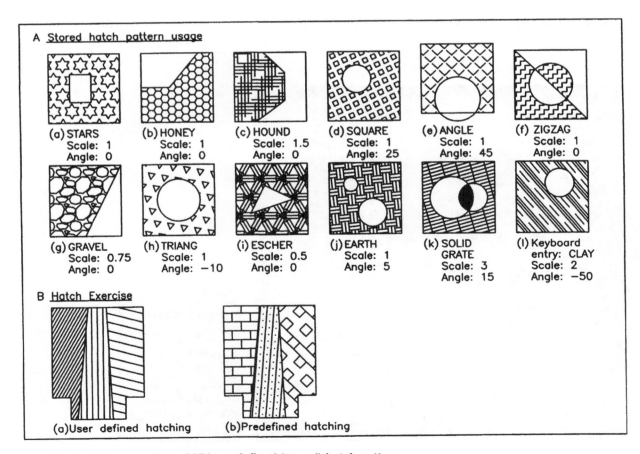

Figure 23.6 Using AutoCAD's predefined (stored) hatch patterns.

respond	*a)*	ensure Other Predefined tab active
	b)	scroll until LINE to ZIGZAG displayed – Figure 23.7
	c)	pick STARS then OK
prompt		Boundary Hatch and Fill dialogue box
with	*a)*	Pattern: STARS
	b)	Swatch: display of stars hatch pattern
respond	*a)*	Angle: 0
	b)	Scale: 1 – dialogue box as Figure 23.8
	c)	select Pick Points
	d)	select any internal point in first square
	e)	right-click and pick Preview from pop-up menu
	f)	right-click to accept the hatch pattern
and		Hatching added as Figure 23.6A(a).

3 Repeat the HATCH icon selection and using the following hatch pattern names, scales and angles, add hatching to the other squares using the Pick Points option:

Figure 23.6A	Pattern	Scale	Angle	Figure 23.6A	Pattern	Scale	Angle
b	HONEY	1	0	h	TRIANG	1	−10
c	HOUND	1.5	0	i	ESCHER	0.5	0
d	SQUARE	1	25	j	EARTH	1	5
e	ANGLE	1	45	k	SOLID		
f	ZIGZAG	1	0	k	GRATE	3	15 – i.e. two hatch patterns for figure 23.6A(k)
g	GRAVEL	0.75	0				

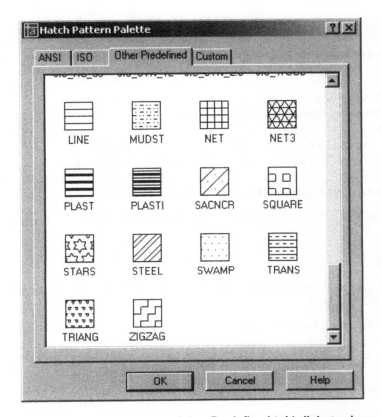

Figure 23.7 The Hatch Pattern Palette (Other Predefined tab) dialogue box.

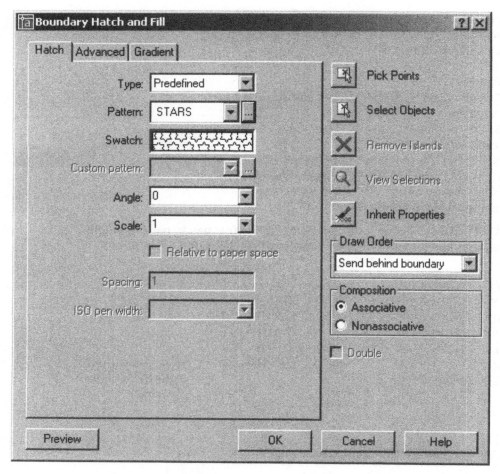

Figure 23.8 The Boundary Hatch and Fill dialogue box for the STARS predefined pattern.

4 Finally enter **HATCH <R>** at the command line and:
   prompt   Enter a pattern name or [?/Solid/...
   enter   **CLAY <R>**
   prompt   Specify a scale for the pattern and enter: **2 <R>**
   prompt   Specify an angle for the pattern and enter: **−50 <R>**
   prompt   Select objects
   respond **pick lines and circle then right-click** – Figure 23.6A(l).

5 Exercise is complete. Save it as C:\BEGIN\HATCH-A for the associative hatch exercise.

Hatch exercise

1 Open the USEREX and copy the component to another part of the screen. Refer to Figure 23.6(B). Layer SECT current.

2 Add the following hatching using the Pick Points option:

   User-defined
   a) angle: 60, spacing: 2
   b) angle: 90, spacing: 4
   c) angle: −15, spacing: 6

   Predefined
   a) BRICK, scale: 1, angle: 0
   b) SACNCR, scale: 2, angle: 40
   c) BOX, scale: 1, angle: 45

3 Save this exercise if required, but not as USEREX.

Associative hatching

AutoCAD 2005 has associative hatching, i.e. if the hatch boundary is altered, the hatching within the boundary will be 'regenerated' to fill the new boundary limits. Associative hatching is applicable to both user-defined and predefined hatch patterns, irrespective of whether the Select Objects or Pick Points option was used. We will demonstrate the effect by example so:

1. *a)* Re-open the C:\BEGIN\HATCH-A drawing from the predefined exercise and refer to Figure 23.9.
   b) Erase squares which are not included in Figure 23.9.
   c) Note that I have included two sets of each square to demonstrate the 'before and after' effect.

2. Make layer SECT current.

3. Select the MOVE icon and:
   a) window the small square in the first square
   b) move it to another position in the square
   c) hatching 'changes' – Figure 23.9(a), i.e. it is associative.

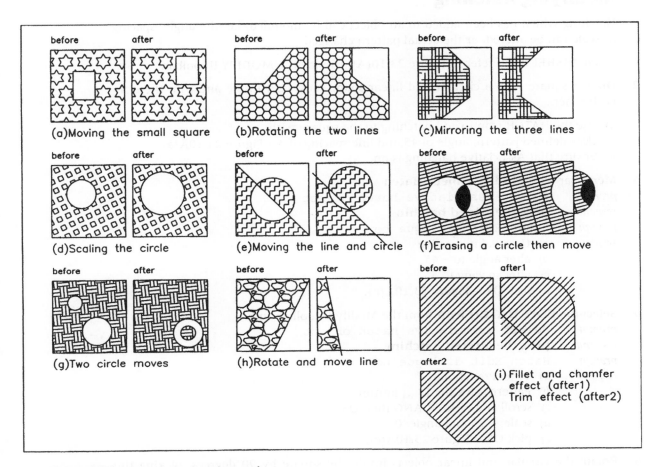

Figure 23.9 Associative hatch exercise.

4   Figure 23.9 displays some associative hatching effects, these being:

Figure 23.9	Effect
b	rotating the two lines by 90 degrees about midpoint of top line
c	mirroring the three lines about midpoints of top and bottom lines
d	scaling the circle
e	moving the line and circle
f	erasing a circle, then moving the other circle
g	moving the two circles using centre snap
h	rotating and moving the line

5   Try the above operations and some other modify commands of your own, you may find that associated hatching does not always work as expected.

6   *Hatch trim*
AutoCAD 2005 allows hatching to be trimmed. To demonstrate the concept, refer to Figure 23.9(i) and:
 a) draw a square, size to suit
 b) hatch the square
 c) fillet one corner and chamfer another corner – no associative effect (after1)
 d) trim the hatch to the fillet curve and chamfer line (after2).

7   This exercise is complete and can be saved.

Modifying hatching

Hatching which has been added to an object can be modified, i.e. the angle, spacing or scale can be altered, or the actual pattern changed.

1   Open A3SHEET and refer to Figure 23.10(A). Display the MODIFY II toolbar.

2   Draw a square with a circular and linear shape (any size) inside and trim these 'to each other'.

3   *a*) Use the HATCH icon to add hatching to the square with:
       User-defined pattern, angle of 45 and line spacing of 3 – Figure 23.10A(a).
   b) Ensure that Associative hatching is on.

4   Menu bar with **Modify-Object-Hatch** and:
   prompt   Select associative hatch object
   respond **pick the added hatching**
   prompt   Hatch Edit dialogue box
   respond *a*) User-defined active
              b) alter angle to −45
              c) alter spacing to 5
              d) pick OK – Figure 23.10A(b).

5   Select the **EDIT HATCH icon** from the Modify II toolbar and:
   prompt   Select associative hatch object
   respond **pick the altered hatching**
   prompt   Hatch Edit dialogue box
   respond *a*) Type: Predefined
              b) Pattern: pick the [...] button
              c) scroll and pick TRIANG then OK
              d) scale: 0.75 and angle: 0
              e) pick OK – Figure 23.10A(c).

6   Rotate the circular and linear objects inside the square by 90 degrees, picking the base point 'roughly' at the square 'centre' – Figure 23.10A(d).

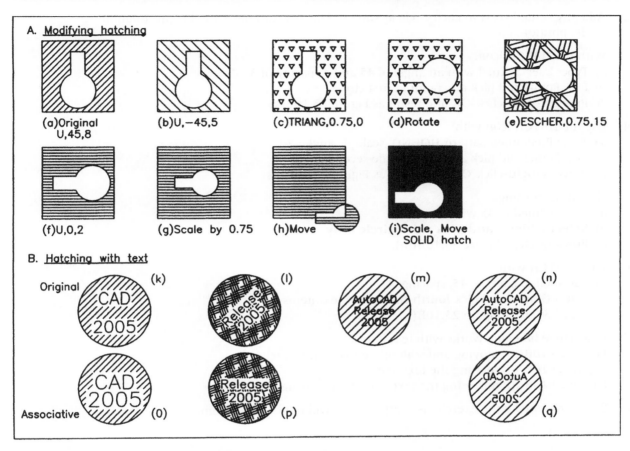

Figure 23.10 Modifying hatching and hatching with text.

7 Using the Edit Hatch icon:
   a) Type: Predefined ESCHER
   b) scale: 0.6 and angle: 15 – Figure 23.10A(e).

8 Edit the hatching to:
   User-defined, angle: 0, spacing: 2 – Figure 23.10A(f).

9 Scale the circular and linear objects inside the square by 1.5, again picking the square 'centre' as the base point – Figure 23.10A(g). I suggest a window selection for the objects.

10 Move the circular and linear objects using the circle centre as the base point and the lower right vertex of the square as the second point – Figure 23.10A(h).

11 Finally, select the circular and linear objects and:
   a) scale them by 1.15 using the circle centre as base point
   b) move them anywhere into the square
   c) change hatch pattern to SOLID
   d) Figure 23.10A(i).

Text and hatching

Text which is placed in an area to be hatched can be displayed with a 'clear border' around it.

1 Refer to Figure 23.10(B) and draw four 50-diameter circles.

2   Add any suitable text to each circle as shown. This text can have any height, rotation angle, position, etc.

3   With the HATCH icon:
    a) Type: User-defined with an angle of 45 and a spacing of 3
    b) Pick Points and pick any point in first circle
    c) Preview-right-click-OK and result as Figure 23.10B(k).

4   Use the HATCH icon with:
    a) Type: Predefined pattern HOUND, scale: 1, angle: 20
    b) Pick Points and pick any point inside second circle
    c) Preview-right-click-OK and result as Figure 23.10B(l).

5   HATCH icon using:
    a) User-defined at 45 with 4 spacing
    b) **Select Objects and pick third circle only**
    c) Preview, etc. – Figure 23.10B(m).

6   Final HATCH with:
    a) User-defined, angle: 45, spacing: 4
    b) Select Objects and pick fourth circle AND text items
    c) Preview, etc. – Figure 23.10B(n).

7   Associative hatching works with text items:
    Figure 23.10B(o): moving and scaling the two items of text
    Figure 23.10B(p): rotating the text item
    Figure 23.10B(q): mirroring the text item, then erasing part of it.

8   Save if required – the exercise is complete and will not be used again.

Gradient fill

A gradient fill is a solid hatch fill that gives the blended colour effect of a surface with light on it. You can use gradient fills to suggest a solid form in two-dimensional drawings. Gradient fill is applied in the same way as hatching and is associative and can be modified as previously described. To demonstrate using gradient fill:

1   Open your A3SHEET standard sheet and draw a 50-sided square then copy it to eight other places.

2   Menu bar with **Draw-Hatch** and:
    prompt   Boundary Hatch and Fill dialogue box
    respond **pick the Gradient tab**
    prompt   Gradient tab display
    with   the following option effects:
        a) one colour with shade to tint control
        b) two colour effect
        c) centred pattern
        d) angle of pattern
        e) nine fill options, these being linear, spherical or radial
        f) Figure 23.11 displays the gradient tab dialogue box
    respond *a)* One color active
        b) Full tint – i.e. slider bar to right
        c) Centered active
        d) Angle: scroll and pick 45
        e) pick Pick Points
        f) add the gradient fill to the first square.

3   Figure 23.12 displays the gradient fill effects and two applications.
4   Try some of these for yourself.
5   The hatching exercises are now all complete.

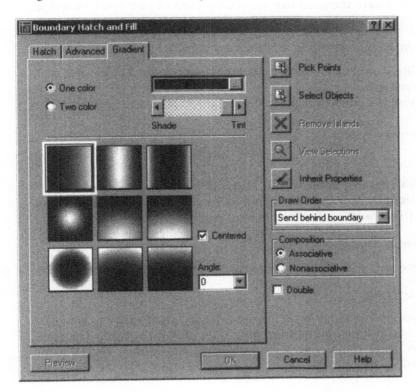

Figure 23.11   The Boundary Hatch and Fill (Gradient tab) dialogue box.

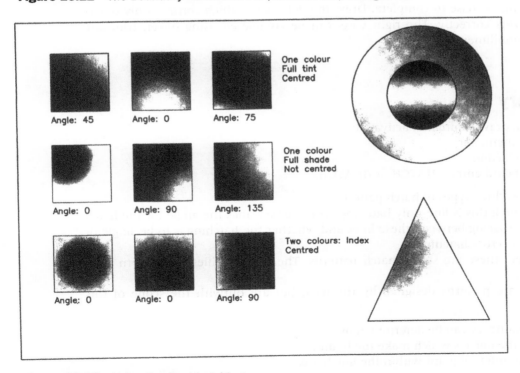

Figure 23.12   Using Gradient hatching.

Assignments

Some AutoCAD users may not use hatching in their draughting work, but they should still be familiar with the process. The Pick Points option makes hatching fairly easy and for this reason I have included five interesting exercises for you to attempt. These should test all your existing CAD draughting skills.

Activity 20: Three different types of component to be drawn and hatched

a) A small engineering component – the hatching is user-defined, use your own angle and spacing values.
b) A pie chart – the text is to have a height of 8 and the hatch names and variables are given.
c) A model airport runway system.

Activity 21: Cover plate

A relatively simple drawing to complete. The MIRROR command is useful and the hatching should not trouble. Add text and then dimensions. A gradient fill effect has been added with the Draw order set to 'Send behind boundary'.

Activity 22: Protected bearing housing

Four views to complete, two with hatching. A fairly easy drawing to complete.

Activity 23: Steam expansion box

This activity has proved very popular in my previous books and is easier to complete than it would appear. Create the outline from lines then fillet the corners. The complete component uses many commands, e.g. offset, fillet, mirror, etc. The hatching should not be mirrored – why?

Activity 24: Gasket cover

An interesting exercise to complete. Draw the 'left view' which consists only of circles, use layers correctly. The right view can be completed using offset, trim and mirror. Do not dimension.

Summary

1 Hatching is a draw command, activated:
   a) from the menu bar
   b) by icon selection
   c) with keyboard entry – HATCH or BHATCH.

2 AutoCAD has three types of hatch pattern:
   a) User-defined: this is line-only hatching. The user specifies the angle for the hatch lines, the spacing between these lines and whether the hatching is to be single or double, i.e. cross-hatching.
   b) Predefined: these are stored hatch patterns. The user specifies the pattern scale and angle.
   c) Custom: are patterns designed by the user, but are outwith the scope of this book.

3 The hatch boundary can be determined by:
   a) selecting the objects which make the boundary
   b) picking an internal point within the hatch area.

4   Hatching is a single object but can be exploded.

5   AutoCAD 2005 has associative hatching which allows added hatching to be altered when the hatch boundary changes.

6   Added hatching can be edited.

7   AutoCAD 2005 has a gradient-fill hatch effect.

8   Hatching can be trimmed with Release 2005.

Chapter 24

Point, polygon and solid

These are three useful draw commands which will be demonstrated by example, so:
a) open A3PAPER standard sheet with layer OUT current
b) activate the Draw, Modify and Object Snap toolbars
c) refer to Figure 24.1.

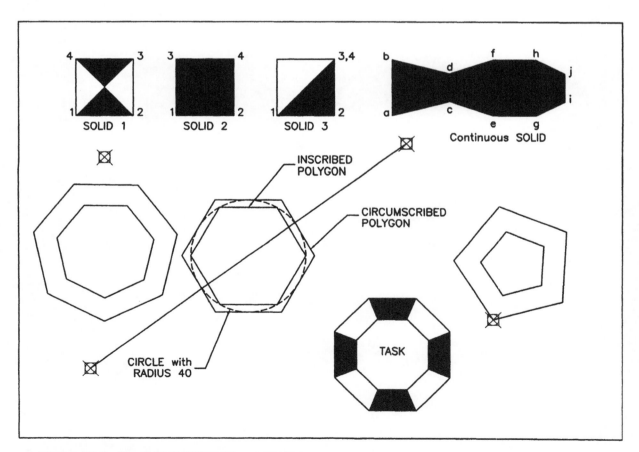

Figure 24.1 The POINT, POLYGON and SOLID draw commands.

Point

A point is an object whose size and appearance is controlled by the user.

1. From the Draw toolbar select the **POINT icon** and:
    prompt   Current point modes: PDMODE=0 PDSIZE=0.00
                  Specify a point
    enter   **50,50 <R>**
    prompt   Specify a point and enter: **60,200 <R>**
    prompt   Specify a point and ESC to end the command.

2. Two point objects will be displayed in red on the screen. You may have to toggle the grid off to 'see' these points properly.

3. From the menu bar select **Format-Point Style** and:
    prompt   Point Style dialogue box as Figure 24.2
    respond a) pick point style indicated
                  b) set Point Size to 5%
                  c) Set Size Relative to Screen active
                  d) pick OK.

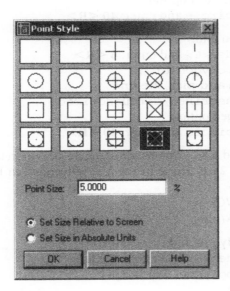

Figure 24.2 The Point Style dialogue box.

4. The screen will be displayed with the two entered points regenerated to this new point style.

5. Menu bar with **Draw-Point-Multiple Point** and:
    prompt   Specify a point and enter: **330,85 <R>**
    prompt   Specify a point and enter: **270,210 <R>**
    prompt   Specify a point and ESC.

6. The screen will now display two additional points in the selected style.

7. Select the **LINE icon** and:
    prompt   Specify first point
    respond **Snap to Node icon and pick lower left point**
    prompt   Specify next point
    respond **Snap to Node icon and pick upper right point**
    prompt   Specify next point and right-click then Enter.

8   *Tasks*
   a) set a new point style and size
   b) restore the previous point style and size.

Polygon

A polygon is a multi-sided figure, each side having the same length and can be drawn by the user specifying:
   a) a centre point and an inscribed or circumscribed radius
   b) the endpoints of an edge of the polygon.

1 Select the **POLYGON icon** from the Draw toolbar and:
   prompt   Enter number of sides<4> and enter: **6 <R>**
   prompt   Specify center of polygon or [Edge]
   respond   **Snap to Midpoint icon and pick the line**
   prompt   Enter an option [Inscribed in circle/Circumscribed about circle]<I>
   enter   **I <R>** – the inscribed (default) option
   prompt   Specify radius of circle and enter: **40 <R>**.

2 Repeat the **POLYGON icon** selection and:
   prompt   Enter number of sides<6> and enter: **6 <R>**
   prompt   Specify center of polygon or...
   respond   **Snap to Midpoint icon and pick the line**
   prompt   Enter an option [Inscribed/Circumscribed...
   enter   **C <R>** – the circumscribed option
   prompt   Specify radius of circle and enter: **40 <R>**.

3 The screen will display an inscribed and circumscribed circle drawn relative to a 40-radius circle as shown in Figure 24.1. These hexagonal polygons can be considered as equivalent to:
   a) inscribed: ACROSS CORNERS (A/C)
   b) circumscribed: ACROSS FLATS (A/F).

4 From the menu bar select **Draw-Polygon** and:
   prompt   Enter number of sides<6> and enter: **5 <R>**
   prompt   Specify center of polygon or [Edge]
   enter   **E <R>** – the edge option
   prompt   Specify first endpoint of edge
   respond   **Snap to Node icon and pick lower right point**
   prompt   Specify second endpoint of edge
   enter   **@50<15 <R>**.

5 At the command line enter POLYGON <R> and:
   a) Number of sides: 7
   b) Edge/Center: E
   c) First point: 60,100
   d) Second point: @30<25.

6 Using the OFFSET icon, set an offset distance of 15 and:
   a) offset the 5-sided polygon 'inwards'
   b) offset the 7-sided polygon 'outwards'.

7 *Note*: A polygon is a POLYLINE type object and has the 'properties' of polylines. Polylines are discussed in the next chapter.

Solid (or more correctly 2D Solid)

A command which 'fills-in' lined shapes, the appearance of the final shape being determined by the pick point order.

1. With the snap on, draw three squares of side 40, towards the top part of the screen – Figure 24.1.

2. Menu bar with **Draw-Surfaces-2D Solid** and:
   - *prompt*   Specify first point and **pick point 1 of SOLID 1**
   - *prompt*   Specify second point and **pick point 2**
   - *prompt*   Specify third point and **pick point 3**
   - *prompt*   Specify fourth point and **pick point 4**
   - *prompt*   Specify third point and right-click to end command.

3. At the command line enter SOLID <R> and:
   - *prompt*   Specify first point and **pick point 1 of SOLID 2**
   - *prompt*   Specify second point **and pick point 2**
   - *prompt*   and pick points 3 and 4 in order displayed then right-click.

4. Activate the SOLID command and with SOLID 3 pick points 1–4 in the order given, i.e. 3 and 4 are the same points.

5. The three squares demonstrate how 3- and 4-sided shapes can be solid filled.

6. Using the SOLID command and Snap on:
First point	pick a point a
Second point	pick a point b
Third point	pick point c
Fourth point	pick point d
Third point	pick point e
Fourth point	pick point f
Third point	pick point g
Fourth point	pick point h
Third point	pick point i
Fourth point	pick point j
Third point	right-click.

7. These last entries demonstrate how continuous filled 2D shapes can be created.

8. *Task 1*
   - a) Draw an 8-sided polygon, centre at 260,60 and circumscribed in a 40-radius circle.
   - b) Offset the polygon inwards by a distance of 15.
   - c) Use the SOLID command to produce the effect in Figure 24.1.
   - d) A running object snap to Endpoint will help.

9. *Task 2*
   What are the minimum and maximum number of sides allowed with the POLYGON command?

10. *Task 3*
    - a) At the command line enter **FILL <R>** and:
      - *prompt*   Enter mode [ON/OFF]<ON>
      - *enter*   **OFF <R>**.
    - b) At command line enter REGEN <R>.
    - c) The solid fill effect is not displayed.
    - d) Turn FILL back on then REGEN the screen.

11. This exercise is complete and can be saved if required.

Assignments

Two activities have been included with this chapter.

Activity 25: A backgammon board

This is a nice simple drawing to complete. The filled triangles can be created with the 2D SOLID command or by hatching with the predefined SOLID pattern. The multiple copy and mirror commands are useful.

Activity 26: Four exercises

These four exercises have to be created using your own sizes. No help given.

Summary

1 A point is an object whose appearance depends on the selection made from the Point Style dialogue box.

2 Only ONE point style can be displayed on the screen at any time.

3 The Snap to Node icon is used with points.

4 A polygon is a multi-sided figure having equal sides.

5 A polygon is a polyline type object.

6 Line shapes can be 'solid filled', the appearance of the solid fill being dependant on the order of the pick points.

7 Only 3- and 4-sided shapes can be solid filled.

Chapter **25**

Polylines

A polyline is a single object which can consist of line and arc segments and can be drawn with varying widths. It has its own editing facility and can be activated by icon selection, from the menu bar or by keyboard entry.

A polyline is a very useful and powerful object yet it is probably one of the most under-used draw commands. The demonstration which follows is quite long and several keyboard options are required.

Creating polyines

1 Open the A3SHEET standard drawing sheet with layer OUT current and refer to Figure 25.1. Display the Draw, Modify, Object Snap and Modify II toolbars.

2 Select the **POLYLINE icon** from the Draw toolbar and:
   prompt   Specify start point
   enter   15,220 <R>
   prompt   Current line-width is 0.00

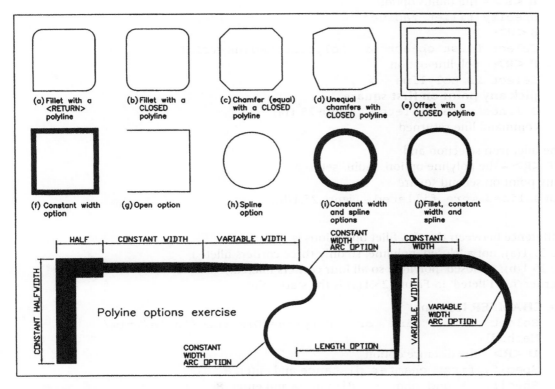

Figure 25.1 Polyline demonstration exercise.

then	`Specify next point or [Arc/Close/Halfwidth/Length/Undo/Width]`	
enter	**@50,0 <R>**	
prompt	`Specify next point or [Arc/Close/Halfwidth/Length/Undo/Width]`	
enter	**@0,50 <R>**	
prompt	`Specify next point or...` and enter **@−50,0 <R>**	
prompt	`Specify next point or...` and enter **@0,−50 <R>**	
prompt	`Specify next point or...`	
respond	**right-click-Enter** to end command.	

3. From menu bar select **Draw-Polyline** and:

prompt	`Specify start point` and enter **90,220 <R>**	absolute entry
prompt	`Specify next point` and enter **@50<0 <R>**	relative polar entry
prompt	`Specify next point` and enter **140,270 <R>**	absolute entry
prompt	`Specify next point` and enter **@−50,0 <R>**	relative absolute entry
prompt	`Specify next point` and enter **C <R>**	close option.

4. Select the **COPY icon** and:

prompt	`Select objects`
respond	**pick any point on SECOND square**
and	all four lines are highlighted with one pick
then	**right-click**
prompt	`Specify base point`
and	multiple copy the square to eight other parts of the screen.

5. Select the **FILLET icon** and:

prompt	`Select first object or [Polyline/Radius/Trim]`
enter	**R <R>** – the radius option
prompt	`Specify fillet radius`
enter	**8 <R>**
prompt	`Select first object or [Polyline/Radius/Trim]`
enter	**P <R>** – polyline option
prompt	`Select 2D polyline`
respond	**pick any point on first square**
prompt	`3 lines were filleted` – Figure 25.1(a)
and	command line returned.

6. Repeat the fillet icon selection and:
    a) enter P <R> – the polyline option, radius set to 8
    b) pick any point on second square
    c) prompt `4 lines were filleted` – Figure 25.1(b).

7. *Notes*
    a) The difference between the two fillet operations is:
        Figure 25.1(a): not a 'closed' polyline, so only three corners filleted
        Figure 25.1(b): a 'closed' polyline, so all four corners filleted.
    b) The corner 'not filleted' in Figure 25.1(a) is the start point.

8. Select **the CHAMFER icon** and:

prompt	`Select first line or [Polyline/Distance/Angle/Trim/Method]`
enter	**D <R>** – the distance option
prompt	`Specify first chamfer distance and enter:` **8**
prompt	`Specify second chamfer distance and enter:` **8**
prompt	`Select first line or [Polyline/Distance/Angle/Trim/Method]`

enter	**P <R>** – the polyline option	
prompt	`Select 2D polyline`	
respond	pick any point on third square	
prompt	`4 lines were chamfered` – Figure 25.1(c)	
and	command line returned.	

9 *Task 1*
   a) set chamfer distances to 12 and 5
   b) chamfer the fourth square remembering to enter P <R> to activate the polyline option
   c) result is Figure 25.1(d)
   d) note the orientation of the 12 and 5 chamfer distances.

10 *Task 2*
   a) set an offset distance to 5 and offset the fifth square 'outwards'
   b) set an offset distance to 8 and offset the fifth square 'inwards'
   c) the complete square is offset with a single pick – Figure 25.1(e).

11 Select the **EDIT POLYLINE icon** from the Modify II toolbar and:

prompt	`Select polyline or [Multiple]`
respond	**pick the sixth square**
prompt	`Enter an option [Open/Join/Width/Edit vertex/Fit/Spline/Decurve...`
enter	**W <R>** – the width option
prompt	`Specify new width for all segments`
enter	**4 <R>**
prompt	`Enter an option [Open/Join/Width/Edit vertex/Fit/Spline/Decurve...`
respond	right-click-Enter to end command – Figure 25.1(f).

12 Menu bar with **Modify-Object-Polyline** and:

prompt	`Select polyline or [Multiple]`
respond	**pick the seventh square**
prompt	`Enter an option [Open/Join/Width/Edit vertex/Fit/Spline/Decurve...`
enter	**O <R>** – the open option
prompt	`Enter an option [Close/Join/Width/Edit vertex/Fit/Spline/Decurve...`
enter	right-click-Enter
and	square displayed with 'last segment' removed: Figure 25.1(g).

13 At the command line enter **PEDIT <R>** and:

prompt	`Select polyline` and **pick the eighth square**
prompt	options and enter **S <R>** – the spline option
prompt	options and enter **X <R>** – exit command
and	square displayed as a splined curve, in this case a circle – Figure 25.1(h).

14 Activate the polyline edit command, pick the ninth square then:
   a) enter W <R> then 5 <R>
   b) enter S <R>
   c) enter X <R> – Figure 25.1(I).

15 *Task*
   a) set a fillet radius to 9
   b) fillet the tenth square – remember P
   c) use the polyline edit command with options: width of 3, then spline, then exit – Figure 25.1(j).

16 *Notes*
   a) If a polyline is drawn with the close option, then when the edit polyline command is used, the option is: Open
   b) If the polyline was not closed, then the option is: Close

Polyline options

The polyline command has several options displayed at the prompt line when the start point is selected and these options are activated by entering the capital letter corresponding to the option. The options are:

Arc	draws an arc segment
Close	closes a polyline shape to the start point
Halfwidth	user enters start and end halfwidths
Length	length of line segment entered
Undo	undoes the last option entered
Width	user enters the start- and end-widths.

To demonstrate these options, at the command line enter PLINE <R> and enter the options at the command prompt:

prompt	*enter*	*comment*
Start point	40,10 <R>	or pick any point
Next point/options	H <R>	halfwidth option
Starting half-width<0.00>	8 <R>	
Ending half-width<8.00>	8 <R>	
Next point/options	@0,80 <R>	line segment endpoint
Next point/options	@30,0 <R>	line segment endpoint
Next point/options	W <R>	width option
Starting width<16.00>	8 <R>	note default <16>
Ending width<8.00>	8 <R>	
Next point/options	@80,0 <R>	line segment endpoint
Next point/options	W <R>	width option
Starting width<8.00>	8 <R>	
Ending width<8.00>	3 <R>	
Next point/options	@75,0 <R>	line segment endpoint
Next point/options	A <R>	arc option
Endpoint arc/options	@0,−30 <R>	arc segment endpoint
Endpoint arc/options	@0,−50 <R>	arc segment endpoint
Endpoint arc/options	L <R>	back to line option
Next point/options	L <R>	line length option
Length of line	80 <R>	
Next point/options	W <R>	width option
Starting width<3.00>	3 <R>	
Ending width<3.00>	10 <R>	
Next point/options	@0,85 <R>	line segment endpoint
Next point/options	@50,0 <R>	line segment endpoint
Next point/options	A <R>	arc option
Endpoint arc/options	W <R>	width option
Starting width<10.00>	10 <R>	
Ending width<10.00>	0 <R>	
Endpoint arc/options	@0,−80 <R>	arc segment endpoint
Endpoint arc/options	right-click-Enter	to end the command

Task

Before leaving this exercise:

1 MOVE the complete polyline shape with a single pick from its start point by @25,25.

2 With FILL <R> at the command line, toggle fill off.

3 REGEN the screen.

4 Turn FILL on then REGEN the screen.

5 This exercise is now complete. Save if required.

Line and arc segments

A continuous polyline object can be created from a series of line and arc segments of varying width. In the demonstration which follows we will use several of the options and the final shape will be used in the next chapter. The exercise is given as a rather LONG list of options and entries, but persevere with it.

1 Open A3SHEET, layer OUT current, toolbars to suit.

2 Refer to Figure 25.2, select the polyline icon and:

prompt	enter	ref	comment
start point	50,50	pt 1	
next pt/options	L		length option
length of line	45	pt 2	
next pt/options	W		width option
starting width	0		
ending width	10		
next pt/options	@120,0	pt 3	
next pt/options	A		arc option
arc end/options	@50,50	pt 4	
arc end/options	L		back to line option
next pt/options	W		width option
starting width	10		
ending width	0		
next pt/options	@0,100	pt 5	
next pt/options	220,220	pt 6	
next pt/options	W		width option
starting width	0		
ending width	5		
next pt/options	@30<-90	pt 7	
next pt/options	A		arc option
arc end/options	@-40,0	pt 8	
arc end/options	@-50,30	pt 9	
arc end/options	L		back to line option
next pt/options	W		width option
starting width	5		
ending width	0		
next pt/options	50,180	pt 10	
next pt/options	C	pt 1	back to start

3 If your entries are correct the polyshape will be the same as that displayed in Figure 25.2. Mistakes with polylines can be rectified as each segment is being

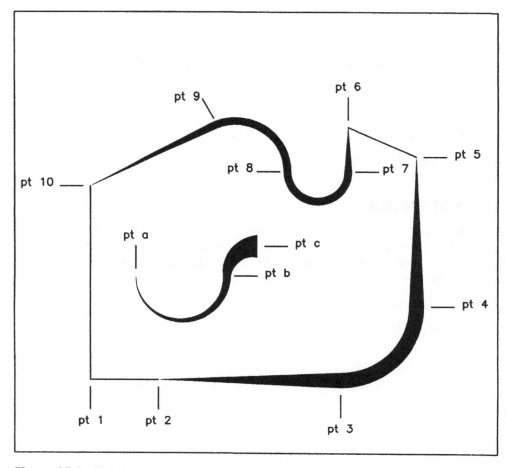

Figure 25.2 Polyline shape exercise.

constructed with the U (undo) option. Both the line and arc segments have their own command line option entries.

4 Repeat the polyline icon selection and:

prompt	enter	ref	comment
start point	80,120	pt a	
next pt/options	A		arc option
arc end/options	W		width option
starting width	0		
ending width	5		
arc end/options	@60<0	pt b	
arc end/options	W		width option
starting width	5		
ending width	15		
arc end/options	@20,20	pt c	
arc end/options	L		back to line option
next pt/options	right-click and Enter		

5 Save the drawing layout as C:\BEGIN\POLYEX for the next chapter which demonstrates how polylines can be edited.

Polyline tasks

Polyline shapes can be used with the modify commands. To demonstrate their use, open your A3SHEET standard sheet and refer to Figure 25.3.

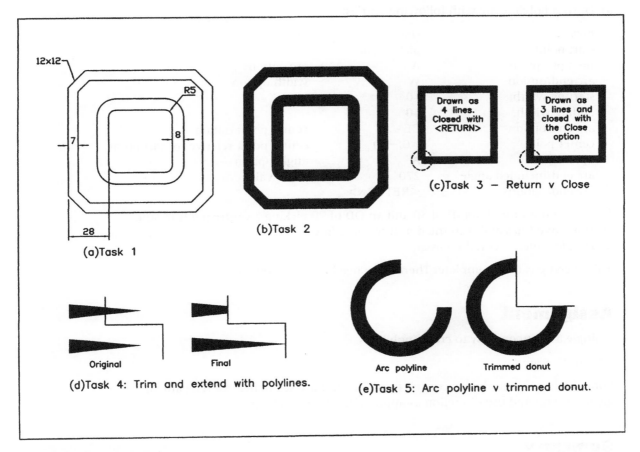

Figure 25.3 Polyline tasks.

Task 1: Figure 25.3(a)

a) Draw a 100-sided closed polyline square.
b) Use the sizes given to complete the component – it is easier than you may think.
c) Commands are OFFSET, CHAMFER and FILLET.

Task 2: Figure 25.3(b)

a) Draw a 100-sided closed polyline square with constant width 7.
b) Use the fillet, chamfer and offset sizes from task 1 to complete the component.

Task 3: Figure 25.3(c)

a) Draw a 50-sided polyline square of width 5 with four lines ended with <RETURN>.
b) Draw a 50-sided polyline square of width 5 with three lines and closed with close option.
c) Note the difference at the polyline start point.

Task 4: Figure 25.3(d)

a) Polylines can be trimmed and extended.
b) Try these operations with an 'arrowhead' type polyline with starting width 10 and ending width 0.

Task 5: Figure 25.3(e)

a) Draw a polyline arc with following entries:

prompt	enter	comment
start point	pick to suit	
next pt/options	A	arc option
arc end/options	W	width option
starting width	10	
ending width	10	
arc end/options	CE	centre point option
center point	@0,−30	centre point relative to start point
options	A	angle option
arc end/included angle	270	angle value
arc end/options	<RETURN>	

b) Draw a donut with an ID of 50 and an OD of 70 picking a centre point to suit.
c) Draw two lines and trim the donut to these lines.
d) Decide which method is easier.

This exercise is now complete. There is no need to save these tasks.

Assignment

A single activity on how to create polylines.

Activity 27: Shapes

Some basic polyline shapes created from line and arc segments. All relevant sizes are given for you and use discretion as appropriate. Snap on helps.

Summary

1 A polyline is a single object which can consist of line and arc segments of varying width.

2 A polyshape which is to be closed should be completed with the Close option.

3 Polyline shapes can be chamfered, filleted, trimmed, extended, copied, scaled, etc.

4 Polylines have their own edit command – next chapter.

Chapter 26
Modifying polylines

Polylines have their own editing facility which gives the user access to several extra options in addition to the existing modify commands.

Editing line and arc segments

1 Open the polyline exercise POLYEX and refer to Figure 26.1 which displays a scaled version of each option.

2 Select the EDIT POLYLINE icon from the Modify II toolbar and:
prompt   Select polyline
respond   **pick any point on outer polyline shape**
prompt   Enter an option [Open/Join/Width/Edit vertex/Fit/Spline/Decurve/Ltype gen/Undo]

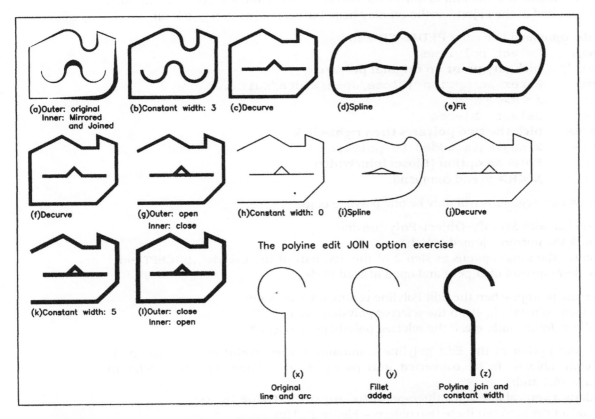

Figure 26.1   Editing polylines using POLYEX.

prompt	enter	ref	comments
Open/Join ...	W		constant width
new width	3	Figure 26.1(b)	width value
options	D	Figure 26.1(c)	decurve
options	S	Figure 26.1(d)	spline
options	F	Figure 26.1(e)	fit option
options	D	Figure 26.1(f)	decurve again
options	O	Figure 26.1(g)	open from start point
options	W		constant width
new width	0	Figure 26.1(h)	width value
options	S	Figure 26.1(i)	spline
options	D	Figure 26.1(j)	decurve
options	W		constant width
new width	5	Figure 26.1(k)	width value
options	c	Figure 26.1(l)	close shape
options	ESC		end command

The join option

This is a very useful option as it allows several individual polylines to be 'joined' into a single polyline object. Refer to Figure 26.1 again and:

1. Use the MIRROR command to mirror the polyarc shape about a vertical line through the right end of the object – ortho on may help, but remember to toggle it off.

2. At the command line enter **PEDIT <R>** and:
   - prompt   Select polyline
   - respond   **pick any point on original polyarc**
   - prompt   Enter an option [Close/Join/Width/Edit ...
   - enter   **J <R>**
   - prompt   Select objects
   - respond   **pick the two polyarcs then right-click**
   - prompt   2 segments added to polyline
   - then   **Enter an option [Close/Join/Width** ...
   - enter   **X <R>** to end command.

3. The two arc segments will now be one polyline object.

4. Menu bar with **Modify-Object-Polyline** and:
   a) pick the mirrored joined polyshape
   b) enter the same options as step 2 of the first part of the exercise, EXCEPT enter C(lose) instead of O(pen) and open instead of close.

5. Note the prompt when the Edit Polyline command is activated:
   a) Open/Join/Width, etc. if the selected polyshape is 'closed'
   b) Close/Join/Width, etc. if the selected polyshape is 'opened'.

6. The join option of the Edit polyline command is very useful as it allows 'non-polyline objects' to be converted into polylines. To demonstrate this, refer to Figure 26.1 and:
   a) draw a vertical line and a three-point arc, any size as Figure 26.1(x)
   b) add a fillet between these two objects – Figure 26.1(y)
   c) activate the Edit polyline command and:
      - prompt   Select polyine
      - respond   **pick the line**

prompt	Object selected is not a polyline
	Do you want to turn it into one <Y>
enter	**Y <R>**
prompt	Enter an option
enter	**J <R>** – the join option
prompt	Select objects
respond	**window the line, fillet and arc then right-click**
prompt	Enter an option
enter	**W <R>** then **3 <R>** – Figure 26.1(z)
prompt	Enter an option
respond	**X <R>** to end the command.

7   This exercise is now complete and can be saved.

Edit vertex option

The options available with the Edit Polyline command usually 'redraws' the selected polyshape after each entry, e.g. if S is entered at the options prompt, the polyshape will be redrawn as a splined curve. Using the Edit vertex option is slightly different from this. When E is entered as an option, the user has another set of options. Refer to Figure 26.2 and:

1   Erase all objects from the screen or open A3SHEET.

2   Draw an 80-sided closed square polyshape and multiple copy it to three other areas of the screen.

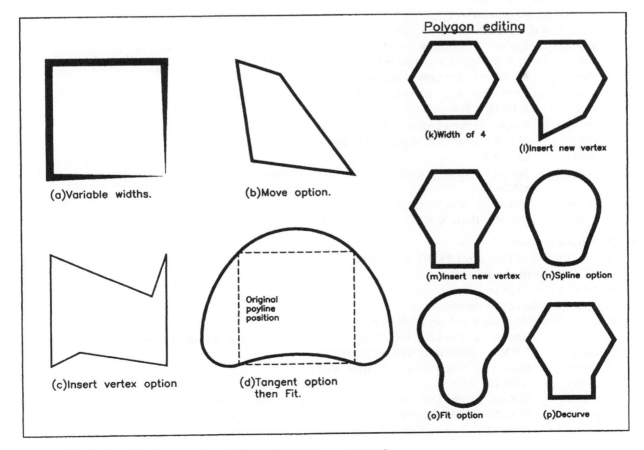

Figure 26.2   Edit vertex option of the Edit Polyline command.

3   *Variable widths*
    Menu bar with **Modify-Object-Polyline** and:
prompt	Select polyline or [Multiple]
respond	**pick the first square**
prompt	Enter an option [Open/Join/Width...
enter	**W <R>** and then **5 <R>**
prompt	Enter an option [Open/Join/Width...
enter	**E <R>** – the edit vertex option
prompt	Enter a vertex editing option [Next/Previous/Break/Insert/...
and	an X is placed at the start vertex, e.g. lower left
enter	**W <R>**
prompt	Specify starting width for next segment and enter **5 <R>**
prompt	Specify ending width for next segment<5> and enter **0 <R>**
prompt	Enter a vertex editing option [Next/Previous/Break...
enter	**N <R>** and X moves to next vertex (lower right for me)
prompt	Enter a vertex editing option [Next/Previous/Break...
enter	**W <R>**
prompt	Specify starting width for next segment and enter **0 <R>**
prompt	Specify ending width for next segment and enter **5 <R>**
prompt	Enter a vertex editing option [Next/Previous/Break...
enter	**X <R>** – to exit the edit vertex option
then	**X <R>** – to exit the Edit polyline command – Figure 26.2(a).

4   *Moving a vertex*
    Select the Edit Polyline icon and:
    a) pick the second square
    b) enter a constant width of 2
    c) enter **E <R>** – the edit vertex option
prompt	Enter a vertex editing option [Next/Previous/Break...
enter	**N <R>** until X at lower left vertex – probably is?
then	**M <R>** – the move option
prompt	Specify new location for marked vertex
enter	**@10,10 <R>**
prompt	Enter a vertex editing option [Next/Previous/Break...
enter	**N <R>** until X at diagonally opposite vertex
then	**M <R>**
prompt	Specify new location for marked vertex and enter **@−50,−10 <R>**
prompt	Enter a vertex editing option [Next/Previous/Break...
enter	**X <R>** then **X <R>** – Figure 26.2(b).

5   *Inserting a new vertex*
    At the command line enter PEDIT <R> and:
    a) pick the third square
    b) enter a constant width of 1
    c) pick the edit vertex option
    d) enter **N <R>** until X at lower left vertex then:
prompt	Enter a vertex editing option[Next/Previous/Break...
enter	**I <R>** – the insert new vertex option
prompt	Specify location of new vertex
enter	**@20,10 <R>**
prompt	Enter a vertex editing option [Next/Previous/Break...
prompt	**N <R>** until X at opposite corner then enter **I <R>**
prompt	Specify location of new vertex

enter	@−10,−30 <R>
then	X <R> and X <R> – Figure 26.2(c).

6. *Tangent-Fit Options*
   Activate the Edit polyline command, select the fourth square, set a constant width of 2, select the edit vertex option with the X at lower left vertex and:

prompt	Enter a vertex editing option [Next/Previous/Break...
enter	**T <R>** – the tangent option
prompt	Specify direction of vertex tangent
enter	**20 <R>**
and	**note arrowed line direction**
prompt	Enter a vertex editing option [Next/Previous/Break...
enter	**N <R>** until X at lower right vertex
then	**T <R>**
prompt	Specify direction of vertex tangent
enter	**−20 <R>** – note arrowed line direction
prompt	Enter a vertex editing option [Next/Previous/Break...
enter	**X <R>** – to end the edit vertex options
prompt	Enter an option [Open/Join...
enter	**F <R>** – the fit option
then	**X <R>** – to end command and give Figure 26.2(d)
and	the final fit shape 'passes through' the vertices of the original polyline square.

Editing a polygon

A polygon is a polyline and can therefore be edited with the Edit Polyline command. To demonstrate the effect:

1. Draw a 6-sided polygon inscribed in a 40-radius circle towards the right of the screen.

2. Activate the Edit Polyline command, pick the polygon then enter the following option sequence:

enter	ref	comment
W then 4	Figure 26.2(k)	constant width
E		edit vertex option
N		until X at lower left vertex
I then @0,−20	Figure 26.2(l)	new vertex inserted
I then @40,0	Figure 26.2(m)	new vertex inserted
X		end edit vertex option
S	Figure 26.2(n)	spline curve
F	Figure 26.2(o)	fit option
D	Figure 26.2(p)	decurve option
X		end command

Assignments

Two polyline (and perhaps polyline edit) activities to complete. The first is several smallish type components, the second an activity to demonstrate the use of polylines.

Activity 28: Hooks and eyes

Three shapes have to be drawn as polylines. You can create them as lines and arcs, then use the JOIN polyline edit option.

Activity 29: Printed circuit board

A practical use for polylines in a drawing and is harder to complete than you may think, especially with the dimensions being given as ordinates. Use your discretion when positioning the various 'lines' and add the dimensions.

Summary

Polylines can be edited and modified with their own modify commands.

Chapter 27

Divide, measure and break

These are three useful commands, divide and measure being Draw commands while break is a Modify command. We will demonstrate their use by example, so:

1 Open your standard sheet with layer OUT current.

2 Activate the Draw, Modify, Dimension and Object Snap toolbars.

3 Refer to Figure 27.1 and set the point style and size indicated with **Format-Point Style** from the menu bar.

4 Draw the following objects:
   a) LINE   from 35,25 to @65<15
   b) CIRCLE   centre 190,250 with radius: 25
   c) POLYLINE 1. line segment from 260,270 to: @50,0
                     2. arc segment to @0,−50
                     3. line segment to @−30,0.

5 Copy the four objects below the originals.

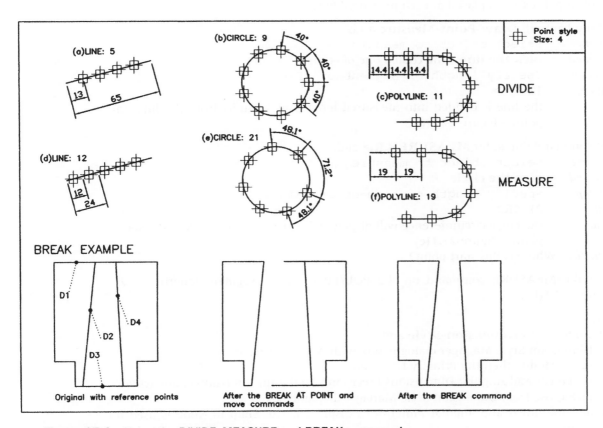

Figure 27.1 Using the DIVIDE, MEASURE and BREAK commands.

Divide

A selected object is 'divided' into an equal number of segments, the user specifying this number. The current point style is 'placed' at the division points.

1. Menu bar with **Draw-Point-Divide** and:
   prompt   Select object to divide
   respond   **pick the line**
   prompt   Enter the number of segments or [Block]
   enter   **5 <R>**
   and   the line will be divided into five equal parts, and a point is placed at the end of each segment length as Figure 27.1(a).

2. At the command line enter **DIVIDE <R>** and:
   prompt   Select object to divide
   respond   **pick the circle**
   prompt   Enter the number of segments or [Block]
   enter   **9 <R>**
   and   the circle will be divided into nine equal arc lengths and nine points placed on the circle circumference as Figure 27.1(b).

3. Divide the POLYLINE into 11 segments – Figure 27.1(c).

Measure

A selected object is 'divided' into a number of user-specified equal lengths and the current point style is placed at each measured length.

1. Menu bar with **Draw-Point-Measure** and:
   prompt   Select object to measure
   respond   **pick the line** – the copied one of course!
   prompt   Specify length of segment or [Block]
   enter   **12 <R>**
   and   the line is divided into measured lengths of 12 units from the line start point – Figure 27.1(d).

2. At command line enter **MEASURE <R>** and:
   prompt   Select object to measure
   respond   **pick the circle**
   prompt   Specify length of segment or [Block]
   enter   **21 <R>**
   and   the circle circumference will display points every 21 units from the start point – Figure 27.1(e)
   Question: Where is this start point?

3. Use the MEASURE command on the POLYLINE with a segment length of 19 – Figure 27.1(f).

4. *Task*
   Menu bar with **Dimension-Style** and:
   a) Dimension Style Manager dialogue box with A3DIM current
   b) select Modify then the Primary Units tab
   c) set Linear and Angular Dimensions Precision to 0.0 if this precision is not active
   d) pick Close from Dimension Style Manager dialogue box
   e) add the linear and angular dimensions displayed. The Snap to Node is used for points.

5. This completes the exercise – no need to save.

Break

This Modify command allows a selected object to be broken:
a) at a specified point
b) between two specified points with an erase effect.

1 Open USEREX and refer to Figure 27.1. Ensure that the top and bottom horizontal lines of the object have been created as single objects. Copy the complete shape to another part of the screen.

2 We want to modify the shape:
a) by splitting it into two parts
b) by erasing part of it.

3 Select the **BREAK AT POINT icon** from the Modify toolbar and:
   prompt   Select object
   respond **pick line D1**
   prompt   Specify first break point
   respond **Snap to Endpoint of 'top' of line D2**
   and   command line returned.

4 Select the **BREAK AT POINT icon** and:
   prompt   Select object
   respond **pick line D3**
   prompt   Specify first break point
   respond **Snap to Endpoint of 'lower' end of line D2**
   and   command line returned.

5 Now move line D2 and the other lines to the left of D2 (six lines in total) to a suitable point.

6 Select the **BREAK icon** from the Modify toolbar and:
   prompt   Select object
   respond **pick line D1 of the copied shape**
   prompt   Specify second point or [First point]
   enter   **F <R>** – the first point option
   prompt   Specify first break point
   respond **Snap to Endpoint of 'top' of line D2**
   prompt   Specify second break point
   respond **Snap to Endpoint of 'top' of line D4**.

7 The segment of line D1 between the two snapped endpoints will be erased.

8 The exercise is now complete, but do not save.

9 Note that when using BREAK, **an entry of @ at the second break point prompt** ensures that the first and second break points are the same.

Summary

1 Objects can be 'divided' into:
   a) an equal number of parts – DIVIDE
   b) equal segment lengths – MEASURE.
   Both commands have a Block option which will be investigated in a later chapter.

2 The BREAK command allows objects to be 'broken' at a single point or between two specified points.

3 The three commands can be used on line, circle, arc and polyline objects.

Chapter 28

Lengthen, align and stretch

Three useful modify commands which can be used to increase drawing efficiency. The three commands will be demonstrated with worked examples.

Lengthen

This command will alter the length of objects (including arcs) but cannot be used with CLOSED objects.

1 Open your A3SHEET standard sheet and draw a horizontal line of length 80 then multiple copy it below the original four times as Figure 28.1.

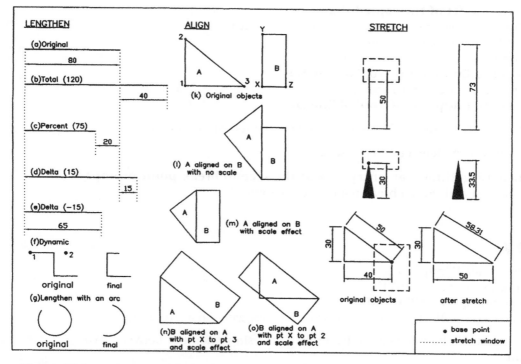

Figure 28.1 Using the lengthen, align and stretch commands.

2   Menu bar with **Modify-Lengthen** and:
    prompt   Select an object or [Delta/Percent/Total/Dynamic]
    enter   **T <R>** – the total option
    prompt   Specify total length or [Angle]<1.00>
    enter   **120 <R>**
    prompt   Select an object to change or [Undo]
    respond **pick the second line at the right end**
    then   right-click and Enter – Figure 28.1(b).

3   Menu bar with **Modify-Lengthen** and:
    prompt   Select an object or [Delta/Percent...
    enter   **P <R>** – the percent option
    prompt   Enter percentage length<100.00>
    enter   **75 <R>**
    prompt   Select an object to change or [Undo]
    respond **pick the next line**
    then   right-click and Enter – Figure 28.1(c).

4   At the command line enter **LENGTHEN <R>** and:
    enter   **DE <R>** – the delta option
    enter   **15 <R>** – the delta length
    respond pick the next line – Figure 28.1(d).

5   Activate the LENGTHEN command and:
    prompt   Select an object or...
    respond **pick the fifth line**
    prompt   Current length: 80.00
    then   Select an object or...
    enter   **DE <R>**
    prompt   Enter delta length or [Angle]
    enter   **−15 <R>**
    prompt   Select an object to change or...
    respond pick the fifth line – Figure 28.1(e).

6   a) Draw a Z shape as Figure 28.1(f)
    b) Activate the lengthen command and:
       1. enter DY <R> – the dynamic option
       2. pick point 1 on the line as indicated
       3. move the cursor and pick a point 2 then right-click-Enter
       4. the original shape will be altered as shown.

7   a) Draw a three-point arc as Figure 28.1(g)
    b) Activate the lengthen command and:
       prompt   Select an object or...
       respond **pick the arc**
       prompt   Current length: 70.87, included angle: 291.0 (these values depend on the arc you have drawn)
       prompt   Select an object or...
       enter   **T <R>** – the total option
       prompt   Specify total length or [Angle]
       enter   **A <R>** – the angle option
       prompt   Specify total angle
       enter   **180 <R>**
       prompt   Select an object to change or...
       respond **pick the arc**.

8   *Notes*
    a) An object is 'lengthened' at the end of the line selected.
    b) If an object is picked before an option is entered, the length of the object is displayed. With arcs, the arc length and included angle are displayed.
    c) When the angle option is used with arcs, the entered total angle is relative to the arc start point.

9   *Task*: Dimension the five lines to check if the various options have been performed correctly.

Align

A very powerful command which combines the move and rotate commands into one operation. The command is mainly used with 3D objects, but can also be used in 2D.

1   Draw a right-angled triangle and a **rectangle** (own sizes) and copy the two objects [Figure 28.1(k)] to four other parts of the screen – the snap on will help with this operation.

2   We want to align:
    a) side 23 of the triangle onto side XY of the rectangle
    b) side XY of the rectangle onto side 23 of the triangle.

3   Menu bar with **Modify-3D Operation-Align** and:
    prompt   Select objects
    respond   **pick the three lines of triangle A then right-click**
    prompt   Specify first source point
    respond   **snap to endpoint and pick point 3**
    prompt   Specify first destination point
    respond   **snap to endpoint and pick point X**
    and   a line is drawn between points 3 and X
    prompt   Specify second source point
    respond   **snap to endpoint and pick point 2**
    prompt   Specify second destination point
    respond   **snap to endpoint and pick point Y**
    and   a line is drawn between points 2 and Y
    prompt   Specify third source point or...
    respond   **right-click** as no more selections are needed
    prompt   Scale objects based on alignment points? [Yes/No]<N>
    enter   **N <R>**.

4   The triangle is moved and rotated onto the rectangle with sides 23 and XY in alignment – Figure 28.1(l).

5   Repeat the Align selection and:
    a) pick the three lines of the next triangle then right-click
    b) pick the same source and destination points as step 3
    c) enter Y <R> at the Scale prompt
    d) the triangle will be aligned onto the rectangle, and side 23 scaled to side XY – Figure 28.1(m).

6   At the command line enter **ALIGN <R>** and:
    prompt   Select objects
    respond   **pick the next rectangle then right-click**
    prompt   Specify first source point
    respond   **Snap to Midpoint icon and pick line XY**

prompt		Specify first destination point
respond		**Snap to Midpoint icon and pick line 23**
prompt		Specify second source point and **pick point X**
prompt		Specify second destination point and **pick point 3**
prompt		Specify third source point and right-click
prompt		Scale objects based on alignment points
enter		**Y <R>**.

7. The selected rectangle will be aligned onto the triangle as Figure 28.1(n) and scaled to suit.

8. Repeat the align command and:
    a) pick the last rectangle then right-click
    b) midpoint of side XY as first source point
    c) midpoint of side 23 as first destination point
    d) point X as second source point
    e) point 2 as second destination point
    f) right-click at third source point prompt
    g) enter Y to scale option – Figure 28.1(o).

9. Note that the orientation of the aligned object is dependent on the order of selection of the source and destination points.

Stretch

This command does what it says – it 'stretches' objects. If hatching and dimensions have been added to the object to be stretched, they will both be affected by the command – remember that hatch and dimensions are associative.

1. Select a clear area of the drawing screen and draw:
    a) a vertical dimensioned line
    b) a dimensioned variable width polyline
    c) a dimensioned triangle.

2. Select the **STRETCH icon** from the Modify toolbar and:
prompt	Select object to stretch by crossing-window or crossing-polygon
then	Select objects
enter	**C <R>** – the crossing option
prompt	Specify first corner
respond	**window the top of vertical line and dimension**
prompt	2 found
then	Select objects
respond	**right-click**
prompt	Specify base point or displacement
respond	**pick top end of line (endpoint)**
prompt	Specify second point of displacement
enter	**@0,23 <R>**.

3. The line and dimension will be stretched by the entered value.

4. Menu bar with **Modify-Stretch** and:
prompt	Select objects
enter	**C <R>** – crossing option
prompt	First corner
respond	**window the top of polyline and dimension**

then **right-click**
prompt Specify base point and pick top end of polyline
prompt Specify second point and enter: **@0,3.5 <R>**.

5 The polyline and dimension are stretched by the entered value.

6 At the command line enter STRETCH <R> and:
 a) enter C <R> for the crossing option
 b) window the vertex of triangle indicated then right-click
 c) pick indicated vertex as the base point
 d) enter @10,0 as the displacement.

7 The triangle is stretched as are the appropriate dimensions.

Stretch example

1 Open your WORKDRG and refer to Figure 28.2.

2 *a)* erase the centre lines
 b) add hatching – own selection
 c) linear dimension the two lines and angle indicated.

3 Activate the STRETCH command and:
 a) enter C <R> – crossing option
 b) first corner: pick a point P1

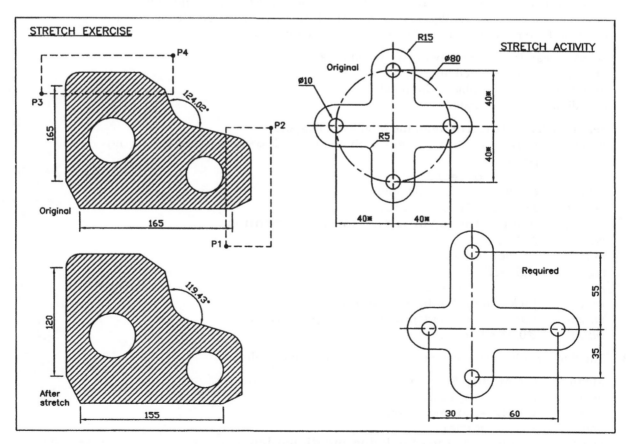

Figure 28.2 Using the STRETCH command with WORKDRG.

c) opposite corner: pick a point P2 then right-click
   d) base point: pick any suitable point
   e) second point: enter @−10,0 <R>.

4. Repeat the STRETCH command and:
   a) activate the crossing option
   b) pick a point P3 for the first corner
   c) pick a point P4 for the opposite corner
   d) pick a suitable base point
   e) enter @0,15 as the second point.

5. The component and dimensions will be stretched relative to the entered values.

Stretch activity

Refer to Figure 28.2 and:

1. Draw the original component as shown adding the four 40(*) dimensions.

2. Using the STRETCH command only, produce the modified component.

3. There is no need to save this activity.

Summary

1. Lengthen will increase/decrease the length of lines, arcs and polylines. There are several options available.

2. Dimensions are not lengthened with the command.

3. Align is a powerful command which combines move and rotate into one operation. The order of selecting points is important.

4. Stretch can be used with lines, polylines and arcs. It does not affect circles.

5. Dimensions and hatching are stretched due to association.

6. The lengthen and align commands are activated from the Modify menu bar, while the stretch command can be activated from the Modify menu bar and toolbar. All three commands can be activated from the command line.

Chapter 29
Obtaining information from a drawing

Drawings contain information which may be useful to the user, e.g. co-ordinate data, distances between points, area of shapes, etc. We will investigate how this information can be obtained, so:
a) open C:\BEGIN\USEREX and refer to Figure 29.1
b) draw a circle, centre at 190,140 and radius: 30
c) activate the Inquiry toolbar.

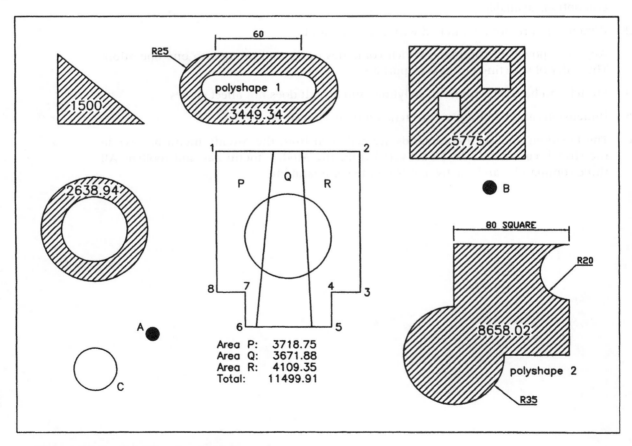

Figure 29.1 Obtaining drawing information.

Point identification

This command displays the co-ordinates of a selected point.

1. Select the **LOCATE POINT icon** from the Inquiry toolbar and:
   prompt   Specify point
   respond   **Snap to Midpoint icon and pick line 23**.

2. The command line area displays:
   X = 240.00   Y = 150.00   Z = 0.00.

3. Menu bar with **Tools-Inquiry-ID Point** and:
   prompt   Specify point
   respond   **Snap to Center icon and pick the circle**.

4. Command line display is X = 190.00   Y = 140.00   Z = 0.00.

5. The command can be activated with ID <R> at the command line.

Distance

Returns information about a line between two selected points including the distance and the angle to the horizontal.

1. Select the **DISTANCE icon** from the Inquiry toolbar and:
   prompt   Specify first point
   respond   **pick point 8** (snap to endpoint or intersection)
   prompt   Specify second point
   respond   **pick point 2**.

2. The command prompt area will display:
   Distance = 141.42, Angle in XY plane = 45.0, Angle from XY plane = 0.0
   Delta X = 100.00, Delta Y = 100.00, Delta Z = 0.00.

3. Menu bar with **Tools-Inquiry-Distance** and:
   prompt   Specify first point
   respond   **snap to centre of circle**
   prompt   Specify second point
   respond   **pick midpoint of line 23**.

4. The display at the command prompt is:
   Distance = 50.99, Angle in XY plane = 11.3, Angle from XY plane = 0.0
   Delta X = 50.00, Delta Y = 10.00, Delta Z = 0.00.

5. Using the DISTANCE command, select point 2 as the first point and point 8 as the second point. Is the displayed information any different from step 1?

6. Entering DIST <R> at the command line will activate the command.

List

A command which gives useful information about a selected object.

1. Select the **LIST icon** from the Inquiry toolbar and:
   prompt   Select objects
   respond   **pick the circle then right-click**
   prompt   AutoCAD Text Window with information about the circle
   respond   **F2 to flip back to drawing screen**.

2   Menu bar with **Tools-Inquiry-List** and:
    prompt   Select objects
    respond   **pick line 12 then right-click**
    prompt   AutoCAD text window
    either   a) F2 to flip back to drawing screen
    or   b) cancel icon from text window title bar.

3   Figure 29.2 is a screen dump of the AutoCAD Text window display for the two selected objects.

4   LIST <R> is the command line entry.

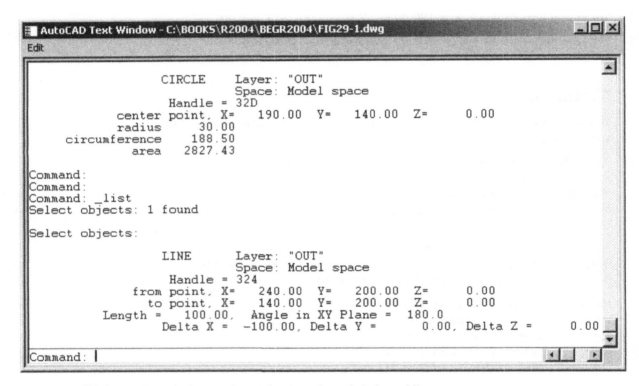

Figure 29.2 The AutoCAD text window for the selected circle and line.

Area

This command will return the area and perimeter for selected shapes or polyline shapes. It has the facility to allow composite shapes to be selected.

1   Select the **AREA icon** from the Inquiry toolbar and:
    prompt   Specify first corner point or [Object/Add/Subtract]
    enter   **O <R>** – the object option
    prompt   Select objects
    respond   **pick the circle**.

2   The command line will display:
    Area = 2827.43, Circumference = 188.50.

3   *Question*: Are these values the same as from the LIST command?

4   Menu bar with **Tools-Inquiry-Area** and:
    prompt   Specify first corner point or [Object/Add/Subtract]
    respond **Snap to endpoint and pick point 1**
    prompt   Specify next corner point or press ENTER for total
    respond **Snap to endpoint and pick point 2**
    prompt   Specify next corner point or...
    respond **Snap to endpoint and pick points 3,4,5,6,7,8**
    prompt   Specify next corner point or press ENTER for total
    respond **press <RETURN>**.

5   Command line displays:
    Area = 11500.00, Perimeter = 450.00.

6   *Question*: Are these figures correct for the shape selected?

7   At the command line enter **AREA <R>** and:
    prompt   Specify first corner point or [Object/Add/Subtract]
    enter   **A <R>** – the add option
    prompt   Specify first corner point or [Object/Subtract]
    respond **Endpoint icon and pick point 1**
    prompt   Specify next corner point or press ENTER for total (ADD mode)
    respond **Endpoint icon and pick point 2**
    prompt   Specify next corner point or press ENTER for total (ADD mode)
    respond **Endpoint icon and pick points 3,4,5,6,7,8**
    prompt   Specify next corner point or press ENTER for total (ADD mode)
    respond **press <RETURN>**
    prompt   Area=11500.00, Perimeter: 450.00
               Total area=11500.00
    then   Specify first corner point or [Object/Subtract]
    enter   **S <R>** – the subtract option
    prompt   Specify first corner point or [Object/Add]
    enter   **O <R>** – the object option
    prompt   (SUBTRACT mode) Select objects
    respond **pick the circle**
    prompt   Area: 2827.43, Circumference: 188.50
               Total area=8672.57
    then   (SUBTRACT mode) Select objects
    respond **ESC to end command**.

8   *Question*: Is the area value of 8672.57 correct for the outline area less the circle area?

Time

This command displays information in the AutoCAD Text Window about the current drawing, e.g.:
a) when it was originally created
b) when it was last updated
c) the length of time worked on it.

1   The command can be activated:
    a) from the menu bar with **Tools-Inquiry-Time**
    b) by entering **TIME <R>** at the command line.

2   The command has options of Display, On, Off and Reset.

3   A useful command or a nightmare?

Status

This command gives additional information about the current drawing as well as disk space information. Select the sequence **Tools-Inquiry-Status** to 'see' the status display in the AutoCAD text window.

Calculator

1   AutoCAD 2005 has a built-in calculator which can be used:
    a) to evaluate mathematical expressions
    b) to assist on the calculation of co-ordinate point data.

2   The mathematical operations obey the usual order of preference with brackets, powers, etc.

3   At the command line enter **CAL <R>** and:
    prompt   >>Expression:
    enter   **12.6*(8.2 + 5.1) <R>**
    prompt   167.58 – is it correct?

4   Enter **CAL <R>** and:
    prompt   >>Expression:
    enter   **(5*(7 − 4)) ^ 3.5 <R>**
    prompt   13071.3.

5   *Question*: What is answer to ((7 − 4) + (2*(8 + 1))) – a key question?

Transparent calculator

A transparent command is one which can be used 'while in another command' and is activated from the command line by entering the ' symbol. The calculator command has this transparent ability.

1   Activate the DONUT command with diameters of 0 and 10, then:
    prompt   Specify center of donut
    enter   **'CAL <R>** – the transparent calculator command
    prompt   >>Expression:
    enter   **CEN/2 <R>**
    prompt   >>Select entity for CEN snap
    respond   **pick the circle**
    prompt   (95.0 70.0 0.0)
    and   donut at position A
    prompt   Specify center of donut
    enter   **'CAL <R>**
    prompt   >>Expression:
    enter   **(MID + INT) <R>**
    prompt   >>Select entity for MID snap and: **pick line 65**
    prompt   >>Select entity for INT snap and: **pick point 8**
    prompt   (330.0 175.0 0.0)
    and   donut at position B
    prompt   Specify center of donut and right-click.

2   Activate the circle command and:
    prompt   Specify center point for circle and enter **55,45 <R>**
    prompt   Specify radius of circle
    enter   **'CAL <R>**
    prompt   >>Expression:
    enter   **rad/2 <R>**
    prompt   >>Select circle, arc or polyline segment for RAD function
    respond   **pick the original circle**
    and   circle at position C.

3   Check the donut centre points with the ID command. They should be (a) 95,70 and (b) 330,175. These values were given at the command prompt line as the donuts were being positioned.

4   *Question*: How were the donut centre co-ordinate values calculated during step 1?

Task

1   Refer to Figure 29.1 and create the following (anywhere on the screen, but use SNAP ON to help):
    a) right-angled triangle with vertical side of 50 and horizontal side of 60
    b) square of side 80 and inside this square two other squares of side 15 and 20
    c) circles: radii 23 and 37 – concentric
    d) polyshape 1: to sizes given, offset 15 'inwards'
    e) polyshape 2: to sizes given.

2   Find the areas of the shaded regions using the AREA command.

3   Obtain the areas of the three 'vertical strips' of the original USEREX, i.e. without the circle.

Summary

1   Drawings can be 'interrogated' to obtain information about:
    a) co-ordinate details
    b) distance between points
    c) area and perimeter of composite shapes
    d) the status and time of the current drawing.

2   AutoCAD has a built-in calculator which can be used transparently.

Chapter 30
Text fonts and styles

Text has been added to previous drawings without any discussion about the 'appearance' of these text items. In this chapter we will investigate:
a) text fonts and text styles
b) text control codes
c) the Express pull-down text options.

The words 'font' and 'style' are extensively used with text and they can be explained as:

Font   Defines the pattern which is used to draw characters, i.e. it is basically an alphabet's 'appearance'. AutoCAD 2005 has over 100 fonts available to the user, and Figure 30.1 displays the text item 'AutoCAD 2005' using 30 of these fonts.

Style Defines the parameters used to draw the actual text characters, i.e. the width of the characters, the obliquing angle, whether the text is upside-down, backwards, etc.

Notes

1   Text fonts are 'part of' the AutoCAD package.

2   Text styles are created by the user.

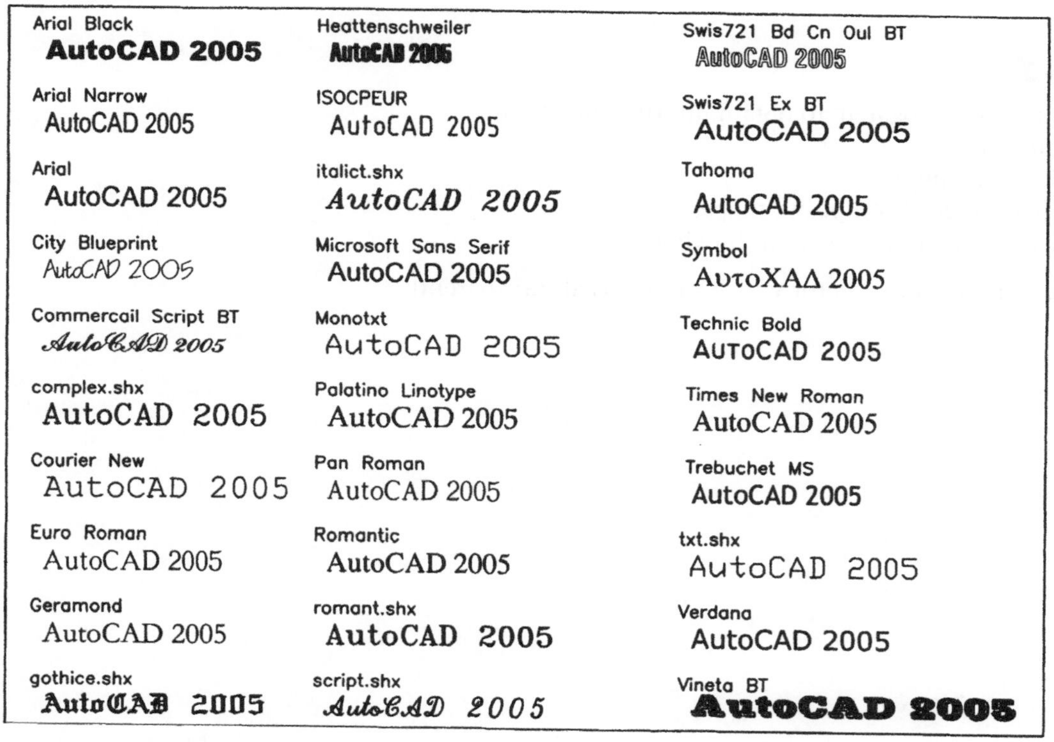

Arial Black **AutoCAD 2005**	Heattenschweiler **AutoCAD 2005**	Swis721 Bd Cn Oul BT AutoCAD 2005
Arial Narrow AutoCAD 2005	ISOCPEUR AutoCAD 2005	Swis721 Ex BT AutoCAD 2005
Arial **AutoCAD 2005**	italict.shx *AutoCAD 2005*	Tahoma AutoCAD 2005
City Blueprint AutoCAD 2005	Microsoft Sans Serif AutoCAD 2005	Symbol ΑυτοΧΑΔ 2005
Commercail Script BT *AutoCAD 2005*	Monotxt AutoCAD 2005	Technic Bold **AutoCAD 2005**
complex.shx AutoCAD 2005	Palatino Linotype AutoCAD 2005	Times New Roman AutoCAD 2005
Courier New AutoCAD 2005	Pan Roman AutoCAD 2005	Trebuchet MS AutoCAD 2005
Euro Roman AutoCAD 2005	Romantic AutoCAD 2005	txt.shx AutoCAD 2005
Geramond AutoCAD 2005	romant.shx AutoCAD 2005	Verdana AutoCAD 2005
gothice.shx AutoCAD 2005	script.shx *AutoCAD 2005*	Vineta BT **AutoCAD 2005**

Figure 30.1   Some of the AutoCAD 2005 text fonts, all of height 8.

3   Any text font can be used for many different styles.
4   A text style uses only one font.
5   If text fonts are to be used in a drawing, a text style **must be created** by the user.
6   New text fonts can be created by the user, but this is outside the scope of this book.
7   Text styles can be created
    a) by keyboard entry
    b) via a dialogue box.

Getting started

1   Open A3SHEET with layer TEXT current.

2   At the command line enter **–STYLE <R>** and
    prompt   Enter name of text style or [?]<Standard>
    enter   **? <R>** – the 'query' option
    prompt   Enter text style(s) to list<*>
    enter   **<R>**
    prompt   AutoCAD Text Window
    with   Style name: "Standard" Font files: ISOCP (or similar)
               Height: 0.00 Width Factor: 1.00 Obliquing angle: 0.0
               Generation: Normal
               Current test style: "Standard".

3   This is AutoCAD 2005's 'default' text style. It has the text style name **STANDARD** and uses the text font **ISOCP**. Realise that your system may have a different text style name and font. If it does, do not worry. It will not affect this exercise.

4   Cancel the text window.

5   With the menu bar sequence **Draw-Text-Single Line Text**, add the text item **AutoCAD 2005** at 110,275 with height 8 and rotation angle 0.

6   The command line entry **–STYLE** was to allow us to use the command line instead of a dialogue box. I thought that it would be easier to understand the concept using the command line at this stage.

Creating a text style from the keyboard

1   At the command line enter **–STYLE <R>** and:
    prompt   Enter name of text style or [?]
    enter   **ST1 <R>** – the style name
    prompt   Specify full font name or font filename (TTF or SHX)
    enter   **romans.shx <R>**
    prompt   Specify height of text and enter: **0 <R>**
    prompt   Specify width factor and enter: **1 <R>**
    prompt   Specify obliquing angle and enter: **0 <R>**
    prompt   Display text backwards? and enter: **N <R>**
    prompt   Display text upside-down? And enter: **N <R>**
    prompt   Vertical? and enter: **N <R>**
    prompt   "ST1" is now the current style.

2   The above entries of height, width factor, etc. are the parameters which must be defined for every text style created.

Creating a text style from a dialogue box

1. From the menu bar select **Format-Text Style** and
   prompt   Text Style dialogue box
   with
       a) ST1 as the Style Name
       b) romans.shx as the Font Name
       c) Height: 0.0, Width Factor: 1.00, Oblique Angle: 0.0
   respond **pick New** and:
   prompt   New Text Style dialogue box
   respond
       a) alter Style Name to **ST2**
       b) pick OK
   prompt   Text Style dialogue box
   with   ST2 as the Style Name
   respond
       a) pick the scroll arrow at right of romans.shx
       b) scroll and pick **italicc.shx**
       c) ensure that – Height: 0.0, Width Factor: 1.00, Oblique Angle: 0.0
       d) note the Preview box
       e) dialogue box as Figure 30.2
       f) pick **Apply** then **Close**.

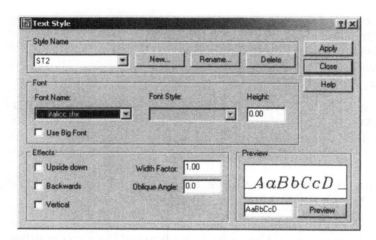

Figure 30.2 The Text Style dialogue box for the new ST2 text style.

2. With the menu bar selection **Format-Text Style**, use the Text Style dialogue box as step 1 to create the following new text styles:

					Effects		
Style name	Font name	Ht	Width factor	Obl'g angle	Backwards	Upside-down	Vert'l
ST3	gothice.shx	12	1	0	OFF	OFF	OFF
ST4	Arial Black	10	1	0	OFF	OFF	OFF
ST5	italict.shx	5	1	30	OFF	OFF	OFF
ST6	Romantic	10	1	0	OFF	ON	–
ST7	scriptc.shx	5	1	−30	OFF	OFF	OFF
ST8	monotxt.shx	6	1	0	OFF	OFF	ON
ST9	Swis721BdOulBT	12	1	0	OFF	OFF	OFF
ST10	complex.shx	5	1	0	ON	OFF	OFF
ST11	isoct.shx	5	1	0	ON	ON	–
ST12	romand.shx	5	1	0	ON	ON	ON

3. *Note*: When using the Text Style dialogue box, a TICK in a box means that the effect is on and a blank box means that the effect is off.

Using created text styles

1. Menu bar with **Draw-Text-Single Line Text** and:
    prompt Specify start point of text or [Justify/Style]
    enter   **S <R>** – the style option
    prompt Enter style name (or ?) – ST12 as default name?
    enter   **ST1 <R>**
    prompt Specify start point of text or [Justify/Style]
    enter   **20,240 <R>**
    prompt Specify height and enter: **8 <R>**
    prompt Specify rotation angle of text and enter: **0 <R>**
    prompt Enter text and enter: **AutoCAD 2005 <R>**.

2. Using the step 1 procedure and the single line text command, add the text item AutoCAD 2005 using the following information:

Style	Start pt	Ht	Rot
ST1	20,240	8	0 – already entered
ST2	225,265	8	0
ST3	15,145	NA	0 (NA: not applicable)
ST4	250,240	NA	0
ST5	25,175	NA	30
ST6	100,220	NA	0
ST7	145,195	NA	−30
ST8	215,235	NA	270 (default angle)
ST9	235,195	NA	0
ST10	365,165	NA	0
ST11	325,145	NA	0
ST12	400,140	NA	270 (default angle)

3. When completed, the screen should display 13 different text styles – the 12 created and the STANDARD default as Figure 30.3.

4. There is no need to save this drawing but:
    a) erase all the text from the screen
    b) save the 'blank' screen as **C:\BEGIN\STYLEX** – you are really saving the created text styles for future use.

Notes

Text styles and fonts can be confusing to new AutoCAD users due to the terminology, and by referring to Figure 30.3, the following may be of assistance:

1. *Effects*
    Three text style effects which can be 'set' are upside-down, vertical and backwards. These effects should be obvious to the user, and several of our created styles had these effects toggled on.

2. *Width factor*
    A parameter which 'stretches' the text characters and Figure 30.3(a) displays an item of text with six width factors. The default width factor value is 1.

3. *Obliquing angle*
    This parameter 'slopes' the text characters as is apparent in Figure 30.3(b). The default value is 0.

4. *Height*
    When the text command was used with the created text styles, only two styles prompted for a height – ST1 and ST2. The other text styles had a height value

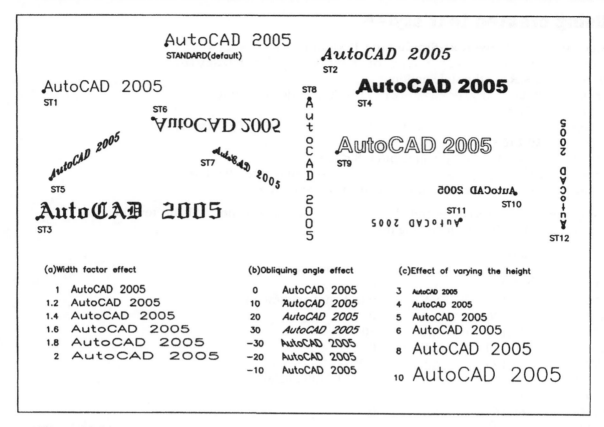

Figure 30.3 Using the created text styles.

entered when the style was created – hence no height prompt. This also means that these text styles cannot be used at varying height values. The effect of differing height values is displayed in Figure 30.3(c).

5   *Recommendation*
    I would strongly recommend that if text styles are being created, the height be left at 0. This will allow the user to enter any text height at the prompt when the text command is used.

6   The text items displayed using styles ST5 and ST7 are interesting, these items having:
    ST5   30 obliquing   30 rotation
    ST7   −30 obliquing   −30 rotation
    These styles give an 'isometric text' appearance.

Text control codes

When text is being added to a drawing, it may be necessary to underline the text item, or add a diameter/degree symbol. AutoCAD has several control codes which when used with the **Single Line Text** command will allow underscoring, overscoring and symbol insertion. The available control codes are:

%%O: toggles the OVERSCORE on/off
%%U: toggles the UNDERSCORE on/off
%%D: draws the DEGREE symbol for angle or temperature(°)
%%C: draws the DIAMETER symbol (Ø)
%%P: draws the PLUS/MINUS symbol (±)
%%%:   draws the PERCENTAGE symbol (%).

1   Open C:\BEGIN\STYLEX with the twelve created text styles.

2   Refer to Figure 30.4, select **Draw-Text-Single Line Text** and:
    prompt   Specify start point of text or [Justify/Style]
    enter   **S <R>**
    prompt   Enter style name or [?]
    enter   **ST1 <R>**
    prompt   Specify start point of text and enter: **25,250 <R>**
    prompt   Specify height and enter: **10 <R>**
    prompt   Specify rotation angle of text and enter: **−5 <R>**
    prompt   Enter text and enter **%%UAutoCAD 2005%%U <R>**
    prompt   Enter text and **<R>**.

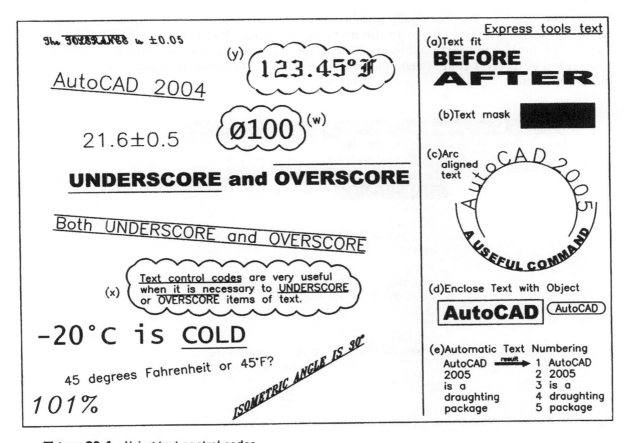

Figure 30.4 Using text control codes.

3   At the command line enter **DTEXT <R>** and:
    a) style: ST3
    b) start point: 175,255
    c) angle: 0
    d) text: 123.45%%DF.

4   Activate the single line text command and:
    a) style: ST4
    b) start point: 35,175
    c) angle: 0
    d) text: %%UUNDERSCORE%%U and %%OOVERSCORE%%O.

5   With the single line text command:
    a) style: ST9
    b) start point: 150,210
    c) angle: 0
    d) text: %%C100.

6   Refer to Figure 30.4 and add the other text items – or text items of your choice. The text style used is at your discretion.

7   *Note*: In the next chapter another method of adding text symbols will be demonstrated.

Express menu

If your 2005 package has been fully installed you will have access to the Express pull-down menu which is located in the menu bar between Modify and Window. Figure 30.5 displays the Text options available to the user and we will now investigate several of these. If your system does not have the Express menu, then proceed to the next section.

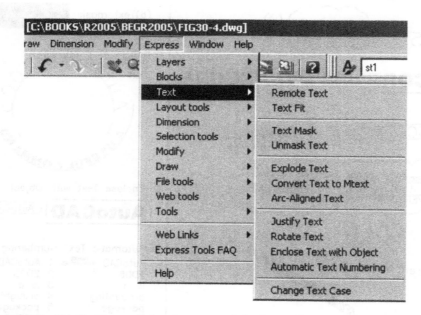

Figure 30.5   The Express pull-down menu and the Text cascade options.

1   Refer to Figure 30.4 and clear an area to the right of the sheet.

2   *Text Fit*
    Allows the user to 'stretch' an existing item of text as Figure 30.4(a).
    The user: *a*) selects the item of text
                b) specifies the end point or the start and end points.

3   *Text Mask*
    Creates a rectangular frame around a specified text item. To demonstrate the effect as Figure 30.4(b), create an item of text then select **Text Mask** and:
    prompt   Select text object to mask or [Masktype/Offset]
    enter   **M <R>** – masktype option
    prompt   Specify entity type to use for mask

enter	**S <R>** – solid option	
prompt	Select Color palette dialogue box	
respond	**enter colour number 31 then pick OK**	
prompt	Select text object to mask or [Masktype/Offset]	
enter	**O <R>** – offset option	
prompt	Enter offset factor relative to text height	
enter	**0.4 <R>**	
prompt	Select text object to mask or [Masktype/Offset]	
respond	**pick required text item** then right-click-enter.	

4 *Arc-Aligned text*
   Allows text to be aligned with an arc or trimmed circle as Figure 30.4(c) but it will not work with full circles.
   a) To demonstrate the command, **draw an arc** then activate the command and:

prompt	Select an Arc or an ArcAligned Text
respond	**pick your drawn arc**
prompt	ArcAlignedText Workshop – Create dialogue box
respond	1. style ST1 active – scroll if required
	2. Text: AutoCAD 2005
	3. C (center along the arc) active – i.e. 'pressed in'
	4. Text height: 10
	5. Width factor: 1
	6. Char spacing: 4
	7. Offset from arc: 0.5
	8. On convex side active
	9. Outward from center active – Figure 30.6
	10. pick OK.

   b) The text item will be aligned along the drawn arc.
   c) Try this command with other options from the dialogue box.

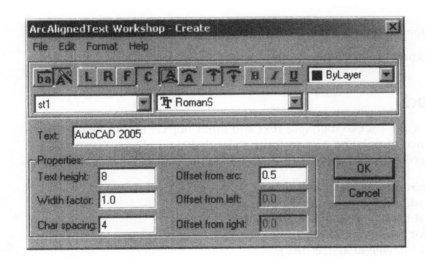

Figure 30.6 The ArcAlignedText Workshop – Create dialogue box.

5 *Justify Text*
   The user has access to the complete list of justify options when a text item is selected.

6 *Rotate Text*
   Allows a selected text item to be rotated to a new entered value.

7 *Enclose Text with Object*
   Creates a user-defined frame around a text item as Figure 30.4(d).
   a) To demonstrate this option, create an item of text, activate the command and:
prompt	Select objects
respond	**pick the text item then right-click**
prompt	Enter distance offset factor
enter	**0.35 <R>**
prompt	Enclose text with [Circles/Slots/Rectangles]
enter	**R <R>** – the rectangle option
prompt	Create rectangles of constant or variable size
enter	**C <R>** – constant size
prompt	Maintain constant rectangle [Width/Height/Both]
enter	**B <R>** – both option.
   b) The selected text item will be enclosed in a rectangle.
   c) Try the other options available.

8 *Automatic Text Numbering*
   Numbers text items according to user-specific entry – Figure 30.4(e).
   a) To demonstrate this, enter a few lines of test, activate the command from the Express Tools menu and:
prompt	Select objects
respond	**window the text items then right-click**
prompt	Sort selected objects by [X/Y/Select-order]
enter	**Y <R>**
prompt	Specify starting number and increment [Start, increment]
enter	**1,2 <R>**
prompt	Placement of number in text
enter	**P <R>** – prefix option.
   b) The windowed text items will be numbered according to the specified entries.

9 *Change Text Case*
   The user selects an item of text and then, via a dialogue box, the case type required.

   This completes the Express Tools demonstration.

Revision cloud

This feature will allow the user to highlight text and other objects in a drawing. The user creates a sequence of polyline arcs to form a cloud-shaped object. Figure 30.4 should still be displayed with various text control code items, now:

1 Menu bar with **Format-Lineweight** and:
prompt	Lineweight Settings dialogue box
respond	a) scroll at Lineweights and pick 0.40 mm
	b) units for listing: millimetres
	c) display lineweight active (tick)
	d) pick OK.

2 Draw a polyline around the 1090 diameter text.

3 Menu bar with **Draw-Revision Cloud** and:
prompt	Minimum arc length: 15 Maximum arc length: 15 Specify start point or [Arc length/Object]<Object>
enter	**O <R>** – the object option

prompt	`Select object`
respond	**pick the polyline**
prompt	`Reverse direction [Yes/No]`
enter	**N <R>** – Figure 30.4(w).

4   Repeat the Revision Cloud command and:
prompt   `Specify start point or [Arc length/Object]<Object>`
respond   a) Identify any suitable text item
　　　　　　b) pick a suitable start point
prompt   `Guide crosshairs along cloud path`
respond   **move cursor around the text item to the start point**
and   **revision cloud will be added** – Figure 30.4(x).

5   Revision cloud command again and:
prompt   `Specify start point or [Arc length/Object]<Object>`
enter   **S <R>** – the style option
prompt   `Select arc style [Normal/Calligraphy]<Normal>`
enter   **C <R>** – the calligraphy style
prompt   `Specify start point or [Arc length/Object]<Object>`
respond   a) Identify any suitable text item
　　　　　　b) pick a suitable start point
prompt   `Guide crosshairs along cloud path`
respond   **move cursor around the text item to the start point**
and   **revision cloud will be added** – Figure 30.4(y).

6   Try the revision cloud command with another item of text.

This completes the text exercises. The current drawing does not need to be saved.

Assignment

Activity 30: Club cards

Using your C:\BEGIN\STYLEX saved drawing (with the various created text styles) create several club cards of your own design. All cards have to display the welcome symbol, created from lines (and an arc) using a 5 × 5 grid. You can use the saved text styles, or create other styles as appropriate.

Summary

1   Fonts define the pattern of characters.

2   Styles define the parameters for drawing characters.

3   Text styles must be created by the user.

4   A font can be used for several text styles.

5   Every text style must use a text font.

6   The AutoCAD 2005 default text style is called STANDARD and uses the txt font.

7   Text control codes allow under/overscoring and symbols to be added to text items.

8   No mention has been made of the different types of font extension used with Auto-CAD. This is considered beyond the scope of this book.

9   The Express Tools menu allows the user several useful text command options.

Chapter 31

Multiline Text, Text Tables and Fields

Multiline is a useful draughting tool as it allows large amounts of text to be added to a pre-determined area on the screen. This added text can also be edited.

Creating the multiline text

1 Open the STYLEX drawing, saved with the created text styles. The drawing area should be blank.
2 Display toolbars to suit, including the TEXT toolbar.
3 With layer TEXT current, select the TEXT icon from the Draw toolbar and:
   prompt   Specify first corner
   enter   **10,280 <R>**
   prompt   Specify opposite corner or [Height/Justify/Line spacing/Rotation/Style/Width]
   enter   **S <R>** then **ST1 <R>** – setting the text style
   prompt   Specify opposite corner or [Height/Justify/Line spacing/Rotation/Style/Width]
   enter   **H <R>** then **5 <R>** – setting the height
   prompt   Specify opposite corner or [Height/Justify/Line spacing/Rotation/Style/Width]
   enter   **125,195 <R>**
   prompt   Multiline Text Editor dialogue box
   with   three distinct 'sections' (which can be repositioned to suit the user):
              a) the Text Formatting toolbar
              b) the Ruler with tab settings, etc.
              c) the Text Editor window
   and   cursor at start of the 'window area'
   respond a) enter the following in the Text Editor window
              b) INCLUDE the typing errors which are deliberate and have been underlined
              c) do not try to add the underline effect
              d) do not press the return key
   enter   CAD is a draughting tol with many benefits when compared to conventional draughting techniches. Some of these benefitds include incresed productivity, shorter lead tines, standardisation, acuracy amd rapid resonse to change.
   and   dialogue box as Figure 31.1
   respond **pick OK from the Text Formatting toolbar**.
4 Refer to Figure 31.2(a) for the entered text effect.

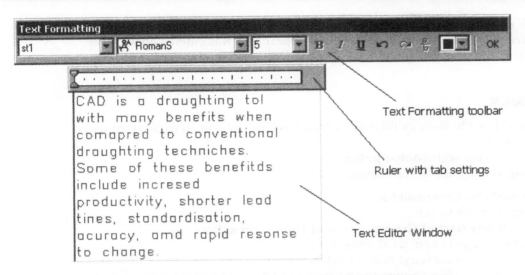

Figure 31.1 The Text Formatting toolbar and text window for multiline text.

5   *Notes*
   a) the text 'wraps around' the Text Editor window as it is entered from the keyboard
   b) the text is fitted into the width of the selected area of the screen, not the full rectangular area of the dialogue box
   c) Figure 31.2(a) displays a rectangular area equivalent to the (10,275 and 125,190) entered co-ordinates

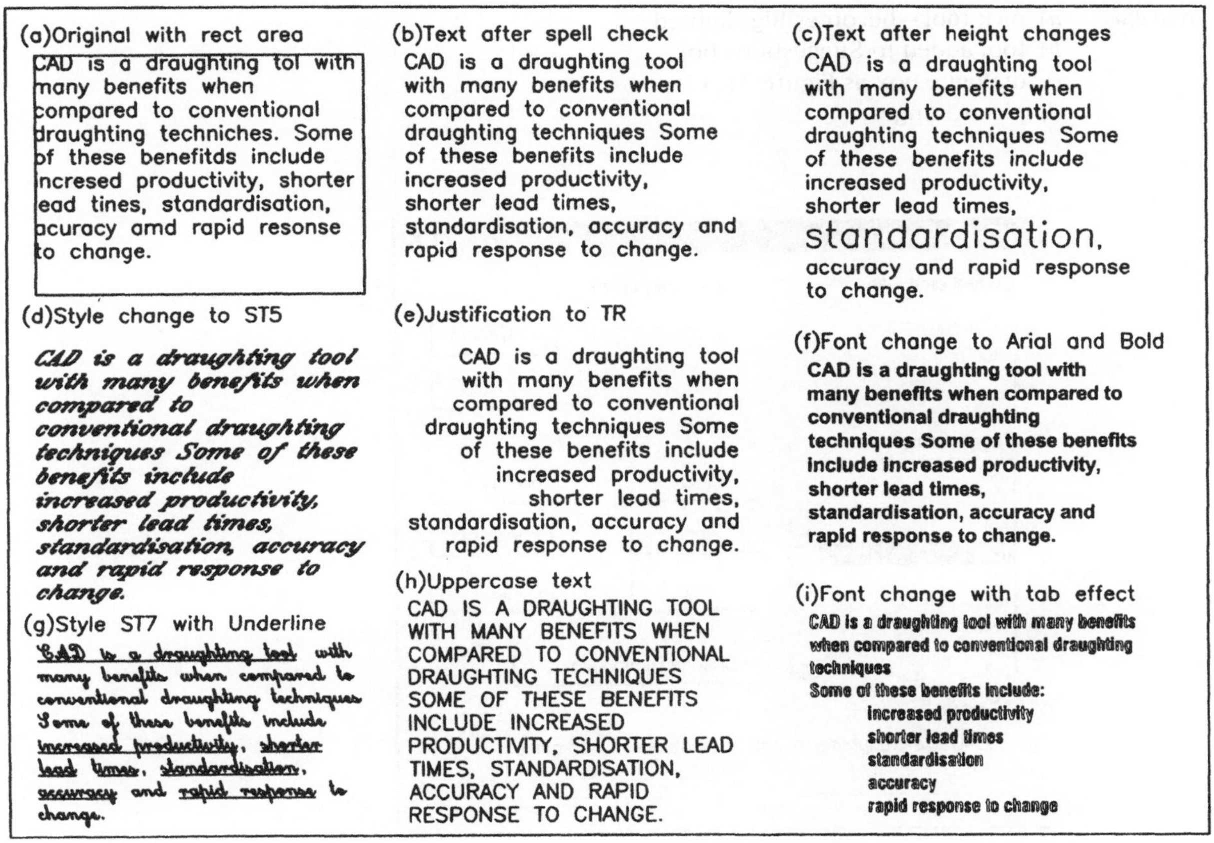

(a) Original with rect area
CAD is a draughting tol with many benefits when compared to conventional draughting techniches. Some of these benefitds include incresed productivity, shorter lead tines, standardisation, acuracy amd rapid resonse to change.

(b) Text after spell check
CAD is a draughting tool with many benefits when compared to conventional draughting techniques Some of these benefits include increased productivity, shorter lead times, standardisation, accuracy and rapid response to change.

(c) Text after height changes
CAD is a draughting tool with many benefits when compared to conventional draughting techniques Some of these benefits include increased productivity, shorter lead times, standardisation, accuracy and rapid response to change.

(d) Style change to ST5
CAD is a draughting tool with many benefits when compared to conventional draughting techniques Some of these benefits include increased productivity, shorter lead times, standardisation, accuracy and rapid response to change.

(e) Justification to TR
CAD is a draughting tool with many benefits when compared to conventional draughting techniques Some of these benefits include increased productivity, shorter lead times, standardisation, accuracy and rapid response to change.

(f) Font change to Arial and Bold
CAD is a draughting tool with many benefits when compared to conventional draughting techniques Some of these benefits include increased productivity, shorter lead times, standardisation, accuracy and rapid response to change.

(g) Style ST7 with Underline
CAD is a draughting tool with many benefits when compared to conventional draughting techniques Some of these benefits include increased productivity, shorter lead times, standardisation, accuracy and rapid response to change.

(h) Uppercase text
CAD IS A DRAUGHTING TOOL WITH MANY BENEFITS WHEN COMPARED TO CONVENTIONAL DRAUGHTING TECHNIQUES SOME OF THESE BENEFITS INCLUDE INCREASED PRODUCTIVITY, SHORTER LEAD TIMES, STANDARDISATION, ACCURACY AND RAPID RESPONSE TO CHANGE.

(i) Font change with tab effect
CAD is a draughting tool with many benefits when compared to conventional draughting techniques
Some of these benefits include:
   increased productivity
   shorter lead times
   standardisation
   accuracy
   rapid response to change

Figure 31.2 The Multiline text exercise.

d) the width is determined by the entered co-ordinates
e) the entered text in the dialogue box will not appear as it was entered above. This is normal, so do not worry about it.

Spellcheck

AutoCAD 2005 has a built-in spellchecker which has been used in an earlier chapter. It can be activated:
a) from the menu bar with **Tools-Spelling**
b) by entering SPELL <R> at the command line.

1   Activate the spell check command and:
    prompt   Select objects
    respond   **pick any part of the entered text then right-click**
    prompt   Check Spelling dialogue box
    with   *a*) current dictionary: British English (ise)
                (if not, select Change Dictionaries and select as required then Apply & Close)
           b) current word: probably *draughting*
           c) suggestions: probably draughtiness, draught
           d) context: *CAD is a draughting tol with many . . .*
    respond   **pick Ignore All**
    note   we are agreed that draughting is the correct spelling
    prompt   Check Spelling dialogue box
    with   *a*) current word: *tol*
           b) suggestions: toll, tool, to . . .
    respond   *a*) pick tool – becomes highlighted
           b) tool added to Suggestions box
           c) dialogue box as Figure 31.3
           d) pick Change

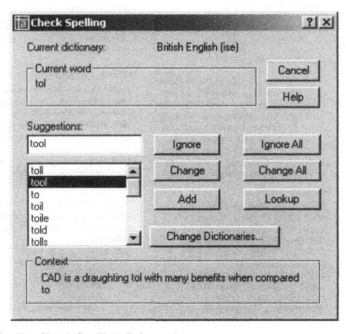

Figure 31.3   The Check Spelling dialogue box.

prompt	`Check Spelling dialogue box`
with	*a*) current word: *techniches*
	b) suggestions: tech
respond	*a*) at suggestions, alter tech to techniques
	b) pick Change
prompt	`Check Spelling dialogue box`
respond	change the following as they appear:

original	alter to
benefitds	benefits
incresed	increased
acuracy	accuracy
amd	and (manual change required)
resonse	response

then	`AutoCAD Message`
	`Spelling check complete`
respond	**pick OK**.

2 The multiline text will be displayed with the correct spelling as Figure 31.2(b).

3 One of the original spelling mistakes was 'tines' (times) and this was not highlighted with the spellcheck. This means that the word 'tines' is a 'real word' as far as the dictionary used is concerned although it is wrong to us. This can be a major problem with spellchecks.

4 Enter **DDEDIT <R>** at the command line and:
   a) pick any part of the text
   b) the Multiline Text Editor dialogue box will be displayed and the word *tines* can be manually altered to times
   c) when the alteration is complete, pick OK.

5 Multiple copy the corrected text to seven other places on the screen and refer to Figure 31.2.

Editing multiline text

1 Select the **EDIT TEXT icon** from the Text toolbar and:

prompt	`Select an annotation object or [Undo]`
respond	**pick a copied multiline text item**
prompt	`Multiline Text Editor dialogue box`
with	text displayed
respond	*a*) left-click and drag mouse over CAD
	b) alter height to 6
	c) left-click and drag over standardisation
	d) alter height to 9
	e) pick OK from the toolbar
	f) right-click-enter to end command
and	text displayed as Figure 31.2(c).

2 Menu bar with **Modify-Object-Text-Edit** and:

prompt	`Select an annotation object`
respond	**pick a copied text item**
prompt	`Multiline Text Editor dialogue box`
respond	*a*) scroll at Style, pick ST5 then pick OK from toolbar
	b) right-click-Enter to end command – Figure 31.2(d).

3 Edit Text command, select the next multiline text then:
   prompt   Multiline Text Editor dialogue box
   respond **right-click in the text window**
   prompt   shortcut menu
   respond *a*) pick Justification
              b) pick Top Right
              c) pick OK then right-click-enter
   and   text displayed as Figure 31.2(e).

4 Using the Edit Text command alter the other copied multiline text items using the following information:
   a) Highlight text, change font to Arial with Bold effect – Figure 31.2(f)
   b) Highlight text, change style to ST7 then highlight and underline indicated text – Figure 31.2(g)
   c) Highlight text, shortcut menu, UPPERCASE – Figure 31.2(h)
   d) Highlight text, change font to SWIS721BdCnOulBT and produce tab effect – Figure 31.2(i).

5 This exercise is now complete and can be saved if required.

Text modifications

To investigate some of the other text modifications, erase all items except the figure 31.2(e) – TR justification. Move this item of text to the top left of the screen and refer to Figure 31.4 for the exercise layout.

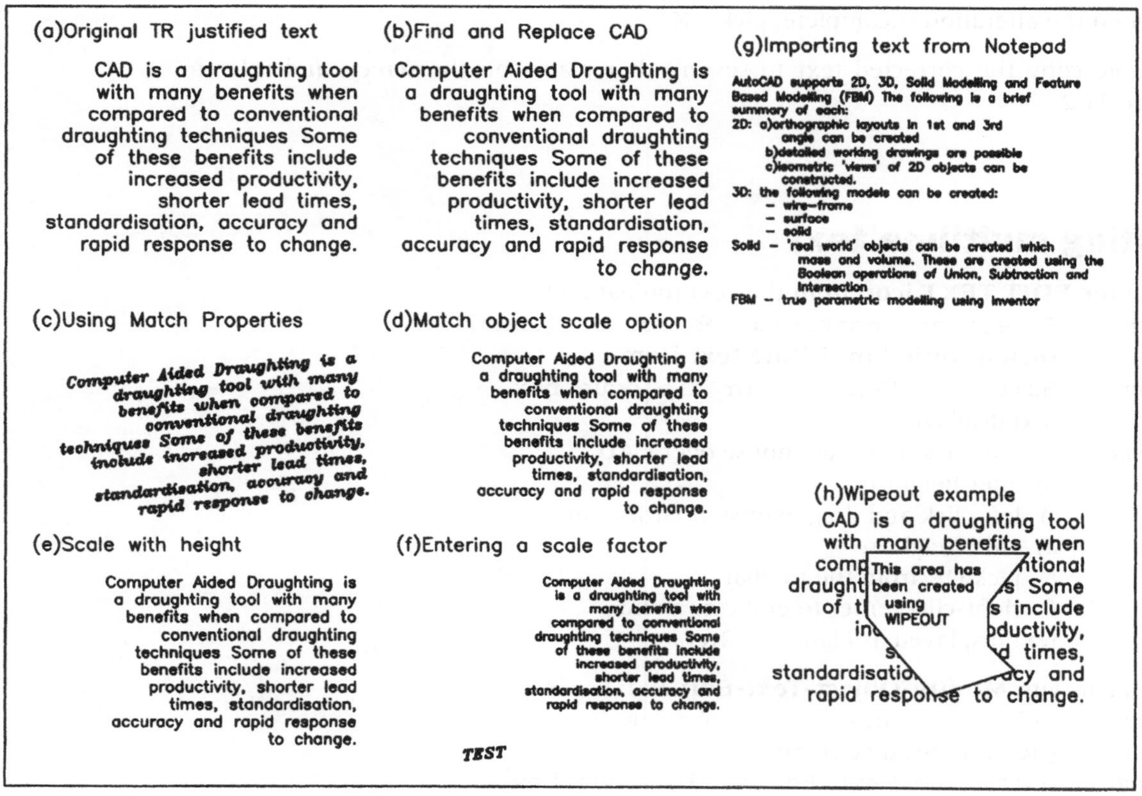

Figure 31.4 Text modifications and importing text into AutoCAD.

1   Select the **Find and Replace icon** from the Text toolbar and:
    prompt   Find and Replace dialogue box
    respond   a) Find text string: enter CAD
              b) Replace with: Computer Aided Draughting
              c) pick Select objects icon
    prompt   Select objects
    respond   **pick the multiline text item then right-click**
    prompt   Find and Replace dialogue box
    with   Search in: Current selection
    respond   **pick Find**
    prompt   Find and Replace dialogue box as Figure 31.5
    respond   **pick Replace**
    prompt   Find and Replace dialogue box
    with   No more occurrences found
    respond   **pick Close**.

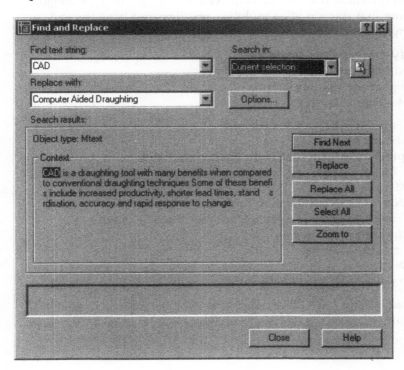

Figure 31.5 The Find and Replace dialogue box for CAD.

2   The text item will be displayed with CAD replaced by Computer Aided Draughting – Figure 31.4(b).

3   Multiple copy this replaced multiline text item to five other parts of the screen.

4   With the single line text icon from the text toolbar, create an item of text using:
    a) style: ST2
    b) start point: pick to suit
    c) height: 3.75
    d) rotation angle: 5
    e) text: TEST.

5   Select the **Match Properties icon** and:
    prompt   Select source objects
    respond   **pick the TEST text item**

      prompt   `Select destination object`
      respond   **pick the next multiline text then right-click-enter**
      and   the multiline text will be displayed with the same text 'settings' as the TEST text – Figure 31.4(c).

6. Select the **Scale Text icon** from the Text toolbar and:
      prompt   `Select objects`
      respond   **pick a copied multiline text item then right-click**
      prompt   `Enter a base point option for scaling [Existing/Left/Center/Middle...`
      respond   **<RETURN>** – i.e. accept the existing default
      prompt   `Specify new height or [Match object/Scale factor]`
      enter   **M <R>** – the match object option
      prompt   `Select a text object with the desired height`
      respond   **pick TEST item of text**
      and   the selected multiline text will be displayed at a height of 3.75 – Figure 31.4(d). Note that this modified multiline text does not have the style or rotation angle as the TEST item, only the height.

7. Menu bar with **Modify-Object-Text-Scale** and:
    a) object: pick a copied multiline text item then right-click
    b) base point for scaling: accept existing default
    c) new height: enter 4 – Figure 31.4(e).

8. At the command line enter **SCALETEXT <R>** and:
      prompt   `Select objects`
      respond   **pick a copied multiline text item then right-click**
      prompt   `Enter a base point option for scaling [Existing/Left/Center/Middle...`
      respond   **<RETURN>** – i.e. accept the existing default
      prompt   `Specify new height or [Match object/Scale factor]`
      enter   **S <R>** – the scale factor option
      prompt   `Specify scale factor`
      enter   **0.625 <R>**
      and   selected text item scaled as Figure 31.4(f).

9. As the original text had a height of 5, this scaled text item should have a height of 3.125. Use the list command to check the height of this text. It may be 3.13 – why?

Importing text files into AutoCAD

Text files can be imported into AutoCAD from other application packages using the Multiline Text Editor dialogue box. To demonstrate the concept we will use a text editor and write a new item of text, save it and then import it into our existing drawing.

1. Save the existing layout as a precaution.

2. Select Start from the Windows taskbar, then select **Programs-Applications (or Accessories)-Notepad** and:
      prompt   `Blank Notepad screen displayed`
      enter   the following lines of text with a <R> at end of each line:
                        AutoCAD supports 2D, 3D, Solid Modelling and Feature Based Modelling (FBM) The following is a brief summary of each:
                        2D: *a)* **orthographic layouts in 1st and 3rd angle can be created**
                                b) **detailed working drawings are possible**
                                c) **isometric 'views' of 2D objects can be constructed.**

3D: the following models can be created:
- **wire-frame**
- **surface**
- **solid**

Solid – 'real world' objects can be created which have mass and volume. These are created using the Boolean operations of Union, Subtraction and Intersection

FBM – true parametric modelling using Inventor.

3. When the text has been entered as above, menu bar with **File-Save As** and:
   prompt   Save As dialogue box
   respond a) scroll at Save in and pick C:\BEGIN
              b) enter File name as MYTEST
              c) note type: Text Document (*.txt)
              d) pick Save
   then   Minimise Notepad (left button from title bar) to return to the AutoCAD screen.

4. Layer TEXT still current?

5. Activate the Multiline Text command with the first corner at 270,270 and the opposite corner at 410,50 (or pick two suitable points of your own at right of drawing area) and:
   prompt   Multiline Text Editor dialogue box
   respond **right-click in blank text area**
   prompt   shortcut menu
   respond **pick Import text**
   prompt   Select File dialogue box
   respond a) scroll at Look in and pick C:\BEGIN
              b) pick MYTEST
              c) pick Open
   prompt   Multiline Text Editor dialogue box
   with   imported text displayed, but it may not appear as you would expect
   respond a) highlight all the imported text
              b) scroll at the text styles and select ST1
              c) change text height to 3
              d) right-click in text window and set a top left justification
              e) pick OK from the Text Formatting toolbar.

6. The imported text will be displayed with the ST1 text style at a height of 3. It will also be top left justified as Figure 31.4(g).

7. This part of exercise is now complete and can be saved, but remember that Notepad may still be open.

Wipeout

Wipeout is a draw command which can be used to cover existing objects with a blank polyline area to:
a) allow notes to be added
b) mask out existing objects.

To demonstrate the command:

1. Copy an item of text to a blank area of the screen.

2. Menu bar with **Draw-Wipeout** and:
   prompt   Specify first point or [Frames/Polyline]
   respond **pick a suitable point in the copied text item**

prompt   Specify next point
respond   **pick other points then right-click-enter**.

3   The added polyline area can be used as required by the user as Figure 31.4(h).

Text tables

1   A table is an object containing data in rows and columns, the user:
     a) creating and positioning the empty table
     b) adding the content to the table cells.

2   All tables can:
     a) contain three types of cell – title, header and data
     b) be created downwards or upwards
     c) be customised with colour to user requirements.

3   Figures 31.6(a) and (b) display the basic table terminology.

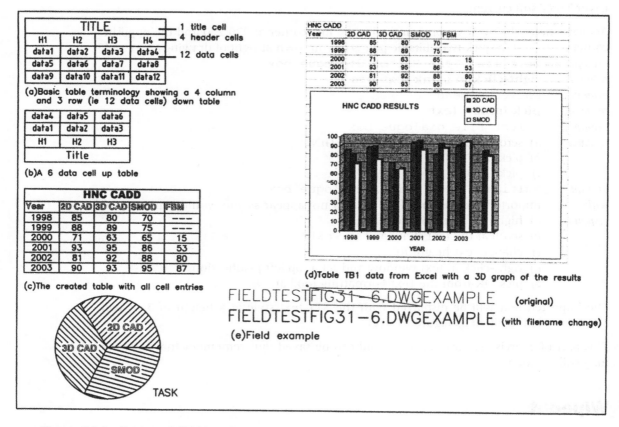

Figure 31.6   Table and field exercise.

4   *Note*: In the demonstration which follows, the procedures and dialogue boxes are very similar to those displayed when we created new dimension styles.

5   The exercise used to demonstrate how a table is created and how text is inserted is a set of results from a college which offers an HNC Computer Aided Draughting and Design course. Four topics offered in this course are 2D draughting, 3D draughting, solid modelling and feature-based modelling and it is the 'pass rates' for these topics over a six-year-period which have to be inserted into the table.

6   Open the A3SHEET and create two new text styles:
    a) ST2 with Arial Black font
    b) ST3 with Swis721 BdOul BT font.

7   Make layer TEXT current and select from the menu bar **Draw-Table** and:
    prompt   Insert Table dialogue
    with   three sections:
                   a) table style settings
                   b) insertion behaviour
                   c) column & row settings
    respond   **pick ...** at the Table Style name
    prompt   Table Style dialogue box with Standard style listed
    respond   **pick New**
    prompt   Create New Table Style dialogue box
    respond   *a*) New Style Name: TB1
                b) Start With: Standard – Figure 31.7
                c) pick Continue

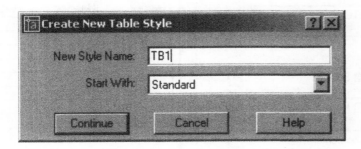

Figure 31.7 The Create New Table Style dialogue box.

prompt   New Table Style: TB1 dialogue box
with   three tab selections – Data, Column Heads, Title
respond   **select the appropriate tab and alter as described**
            a) Data tab active
                1. Cell properties
                    a) Text style: scroll and pick ST1
                    b) Text height: 4
                    c) Colours: investigate and decide yourself
                    d) Alignment: scroll and pick Middle Center
                2. Border properties: leave as given
                3. General
                    a) Table direction: Down
                4. Cell margins: both 1
            b) Column Heads tab active
                1. Cell properties
                    a) Include Header row: active – i.e. tick
                    b) Text style: scroll and pick ST3
                    c) Text height: 4.5
                    d) Colours: select to suit
                    e) Alignment: scroll and pick Top Left
                2. Border properties: leave as given
                3. General: leave as given
                4. Cell margins: leave as given

c) Title tab active
   1. Cell properties
      a) Include Title row active
      b) Text style: scroll and pick ST2
      c) Text height: 5
      d) Colours: to suit
      e) Alignment: scroll and pick Top Center
   2. Border, General, Margins: leave as given – Figure 31.8

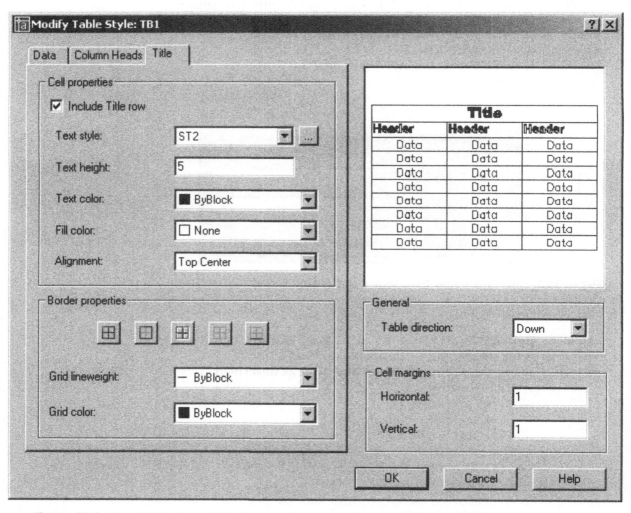

Figure 31.8 The Modify Table Style dialogue box for the new style TB1 with the Title tab active.

then	**pick OK**
prompt	Table Style dialogue box
with	a) TB1 added to the Styles list
	b) a display of the new TB1 table
respond	**pick Close**
prompt	Insert Table dialogue box

with	Table Style name: TB1
respond	*a)* Insertion Behaviour: Specify insertion point active
	b) Column & Row Settings
	1. Columns 5 with column width 12
	2. Data Rows 6 with Row Height 1 – Figure 31.9
	c) pick OK

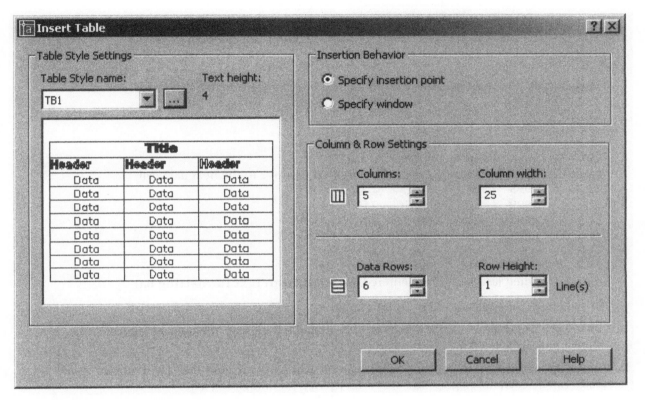

Figure 31.9 The Insert Table dialogue box for TB1.

prompt	Specify insertion point
respond	**pick to suit**
and	*a)* blank created table TB1 inserted
	b) Text Formatting bar displayed with:
	1. ST2 displayed as the text style
	2. Arial Black displayed as the font
	3. 5 displayed as the height
	c) the title bar cell highlighted with flashing cursor
respond	*a)* enter HNC CADD
	b) press the TAB key
prompt	*a)* cursor 'jumps' to next cell – the first header cell
	b) cell is highlighted
	c) Text Formatting bar displays: ST3, Swis721 BdOul BT, 4.5 – Figure 31.10
respond	*a)* enter YEAR
	b) press the TAB key and cursor jumps to next cell
	c) enter 2D CAD then TAB
	d) enter 3D CAD then TAB
	e) enter SMOD then TAB
	f) enter FBM then TAB

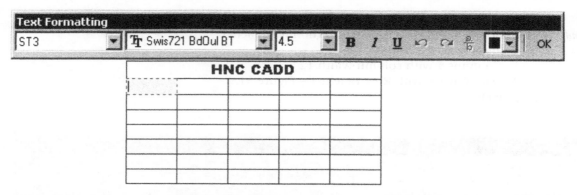

Figure 31.10 The inserted table TB1 with the Text Formatting bar for the first Header cell.

    prompt   cursor jumps to the first data cell
    respond   enter the following data, pressing TAB to move to the next data cell

1998	85	80	70	–
1999	88	89	75	–
2000	71	63	65	15
2001	93	95	86	53
2002	81	92	88	80
2003	90	93	95	87

    then   when last data (87) is entered, **pick OK from text formatting bar**.

8 The table will be displayed as Figure 31.6(c).

9 Data entered in a table can be edited with a double left-click in the appropriate cell and using the displayed text formatting bar.

10 Tables can be used for many different purposes, e.g. parts list, and the fact that the tables can be customised gives the user a very powerful draughting aid.

11 *Task*: Refer to Figure 31.6 and create a pie chart of the six-year averages for the 2D CAD, 3D CAD and SMOD results.

Exporting table data

While inserting a customised table into a drawing is useful to the user, the data contained in the table can be exported from the drawing and imported into other application software packages, e.g. Excel. While this type of operation is beyond the scope of this book, the process is very straightforward and the following exercise will demonstrate how this can be achieved

1 At the command line enter **TABLEEXPORT <R>** and:
    prompt   Select a table
    respond   **pick any point on table TB1**
    prompt   Export Data dialogue box
    respond   a) Save in: your named folder
                b) File name: TB1
                c) File type: Comma Delimited (*.csv)
                d) pick Save
    and   command prompt returned.

2   From the Windows taskbar select **Start-Programs-Excel** (or your own selection) to display a blank spreadsheet.

3   Menu bar with **File-Open** and:
    prompt   Open dialogue box
    respond   *a*) Look in: scroll and pick your named folder
                b) File types: scroll and pick Text Files
                c) pick TB1 then Open.

4   The saved table will be displayed in the spreadsheet without the text style display.

5   The table data can be used as any normal Excel data, e.g. charts, averages, etc.

6   Figure 31.6(d) displays the TB1 data from Excel 'pasted' back into an AutoCAD drawing with an associated chart created from the spreadsheet data.

    This completes the table exercise.

Fields

1   A field is text that contains instructions to display data expected to change during the 'life' of a drawing.

2   When a field is updated, the latest data is displayed.

3   Fields can be inserted in any kind of text (except tolerances), including table text and attributes.

4   As a simple demonstration, menu bar with **Draw-Text-Single Line Text** and:
    prompt   Specify start point of text or [Justify/Style]
    enter   **S <R> then ST1 <R>** – to set the text style
    prompt   Specify start point of text or [Justify/Style]
    respond   **pick a point to suit**
    prompt   Specify height and enter **8 <R>**
    prompt   Specify rotation angle of text and enter **0 <R>**
    prompt   Enter text
    enter   **FIELDTEST**
    then   **right-click**
    prompt   shortcut menu
    respond   **pick Insert Field**
    prompt   Field dialogue box
    respond   *a*) Field names: scroll and pick Filename
                b) Uppercase active
                c) Filename only active
                d) Display file extension active – Figure 31.11
                e) pick OK
    and   filename and extension added to text item
    respond   continue with text entry and enter **EXAMPLE then <R> and <R>**.

5   The text item will be displayed as FIELDTESTDrawing file nameEXAMPLE as Figure 31.6(e) original.

6   When this file name is altered, this item of text will also be altered.

7   The user can insert other field options into items of text for further practice.

    This short exercise is now complete.

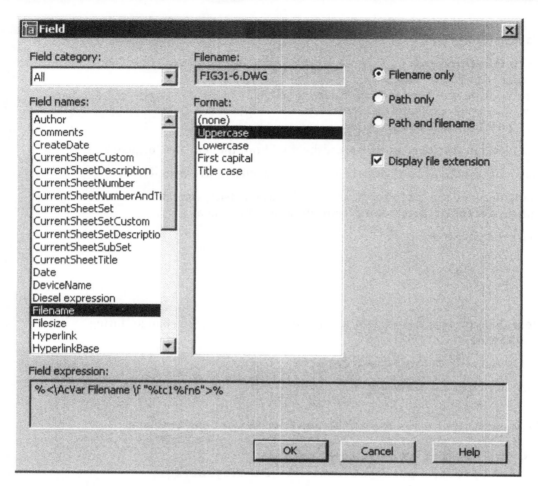

Figure 31.11 The Field dialogue box with Filename selected.

Summary

1 Multiline text is also called paragraph text.

2 The text is entered using the Multiline Text Editor dialogue box.

3 The command can be activated by icon, from the menu bar or by entering MTEXT at the command line.

4 Multiline text has powerful editing facilities.

5 Text files can be imported into AutoCAD from other application packages with the Multiline Text Editor.

6 Tables can be inserted into drawings, customised to user specifications.

7 Text for tables must be entered by the user cell-by-cell.

8 Fields can be added/inserted into text items.

Chapter 32
The array command

Array is a command which allows multiple copying of objects in either a rectangular or circular (polar) pattern. It is one of the most powerful and useful of the commands available, yet is one of the easiest to use. To demonstrate the command:

1 Open the A3SHEET standard sheet with layer OUT current and the toolbars Draw, Modify and Object snap.

2 Refer to Figure 32.1 and draw the rectangular shape using the given sizes. Do NOT ADD the dimensions.

3 Multiple copy the rectangular shape from the mid-point indicated to the points A(25,175); B(290,235); C(205,110); D(300,20) and E(35,130). The donuts in Figure 32.1 are for reference only.

4 Draw two circles:
   a) centre at 290,190 with radius 30
   b) centre at 205,75 with radius 15.

5 Move the original shape to a 'safe place' on the screen.

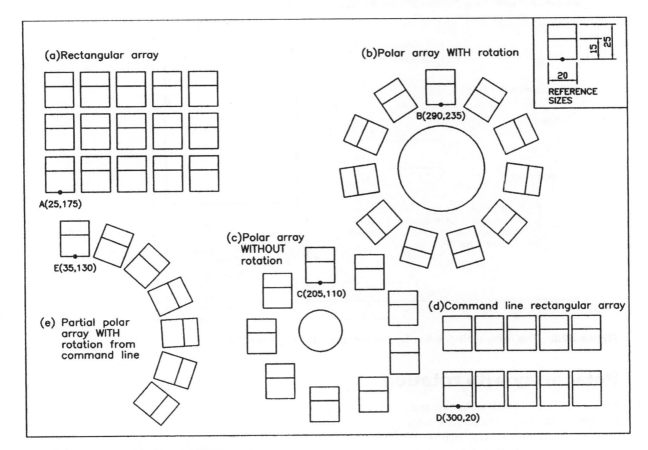

Figure 32.1 Using the ARRAY command.

Rectangular array

1 Select the **ARRAY icon** from the Draw toolbar and:
   prompt   Array dialogue box
   with   a) options: Rectangular Array or Polar Array
              b) rectangular array probably active
              c) information for creating a rectangular array
              d) a preview layout
   respond a) ensure rectangular array active
              b) enter Rows: 3 and Columns: 5
              c) alter Row offset: 30
              d) alter Column offset: 25
              e) ensure Angle of array: 0
              f) pick Select objects
   prompt   Select objects at the command line
   respond **window the shape at A then right-click**
   prompt   Array dialogue box
   with   5 objects selected and a preview display – Figure 32.2
   prompt   pick OK.

2 The shape at A will be copied 14 times into a three-row and five-column matrix pattern as Figure 32.1(a).

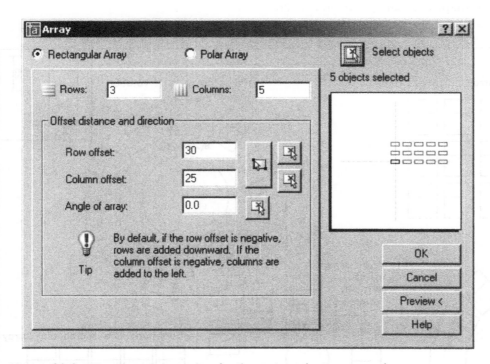

Figure 32.2 The Array dialogue box for the rectangular array exercise.

Polar array with rotation

1 Menu bar with **Modify-Array** and:
   prompt   Array dialogue box
   respond a) Polar Array active
              b) pick Pick Center point
   prompt   Select center point of array at the command line

respond	**snap to Center icon and pick larger circle**
prompt	`Array dialogue box`
with	Selected centre point co-ordinates
respond	*a*) Method: Total number of items & Angle to fill
	b) Total number of items: 11
	c) Angle to Fill: 360
	d) Rotate items as copied: active – i.e. tick
	e) pick Select objects
prompt	`Select objects at the command line`
respond	**window the shape at B then right-click**
prompt	`Array dialogue box as Figure 32.3(a)`
respond	**pick Preview<**
and	*a*) preview of the entered polar array values
	b) Array message – Figure 32.3(b)
respond	**pick Accept** if the array is correct, or **Modify** to return to the dialogue box for alterations.

2   The shape at B is copied in a circular pattern about the selected centre point. The objects are 'rotated' about this point as they are copied – Figure 32.1(b).

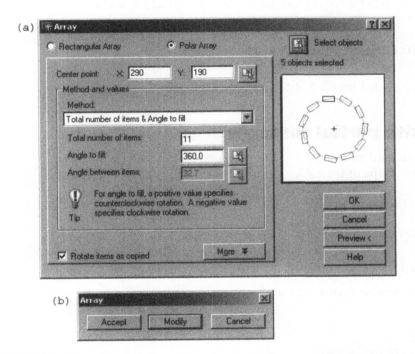

Figure 32.3 (a) The Array dialogue box for the polar array exercise and (b) The Array message box.

Polar array without rotation

1   At the command line enter **ARRAY <R>** and use the Array dialogue box with the following data:
   a) Polar Array
   b) Center point: pick and snap to center icon of smaller circle
   c) Method: Total number of items & Angle to fill
   d) Total number of items: 9
   e) Angle to fill: 360
   f) Rotate items as copied: Not active – i.e. no tick

g) Select objects: window shape at C then right-click
h) preview then accept.

2 The original shape is copied about the selected centre point but is not 'rotated' as it is copied – Figure 32.1(c).

Rectangular array from command line

The previous three examples have used the Array dialogue box to enter the details required for the various arrays. The command can also be activated from the command line, so:

1 At the command line enter **–ARRAY <R>** and:
   prompt Select objects
   respond **Window the shape at D then right-click**
   prompt Enter the type of array [Rectangular or Polar]
   enter   **R <R>** – the rectangular option
   prompt Enter the number of rows (---)<1> and enter: **2 <R>**
   prompt Enter the number of columns (||||)<1> and enter: **5 <R>**
   prompt Enter the distance between rows or specify Unit cell (---)
   enter   **40 <R>**
   prompt Specify the distance between columns (||||)
   enter   **22.5 <R>**.

2 The shape at D will be copied into a 2 × 5 column matrix pattern as Figure 32.1(d).

Polar array with partial fill angle

Activate the array command and:
a) objects: window the shape at D
b) type of array: P
c) center point: enter the point 35,70
d) number of items: 7
e) angle to fill: −130
f) rotate objects: Y
g) the result is Figure 32.1(e).

Your drawing should resemble Figure 32.1 and can now be saved if required.

Array options

Both the rectangular and polar array commands have variations to those displayed in Figure 32.1, these being:
a) rectangular   allows the angle of the array to be altered
b) polar   allows for different method items and fill angle.

To demonstrate these options:

1 Erase the array exercises from the screen (save first?) to leave the original shape which was moved to a safe place. Refer to Figure 32.4.

2 Multiple copy the original shape from the point indicated to the points P(45,175), Q(150,175), R(105,115), S(220,115), X(290,275), Y(290,185) and Z(290,25).

3 *a)* rotate the shapes at P and Q by 20 and the shapes at R and S by −10
   b) scale the shapes at X, Y and Z by 0.5
   c) in each case pick the indicated donut point as the base.

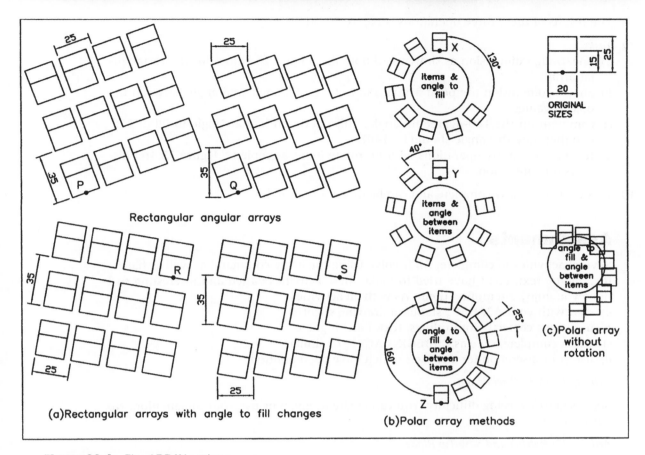

Figure 32.4 The ARRAY options.

4 Draw three circles, radius 20 with centres at the points: 290,250; 290,160; 290,65.

5 Using the Array dialogue box four times, pick the rectangular type and set:

	1st	2nd	3rd	4th
objects	shape P	shape Q	shape R	shape S
rows, columns	3,4	3,4	3,4	3,4
row offset	35	35	−35	−35
column offset	25	25	−25	−25
angle of array	20	0	−10	0

6 The result of the four rectangular operations is Figure 32.4(a).

7 Select the ARRAY command and with the polar array type from the dialogue box, enter the following three sets of data:

	1st	2nd	3rd
objects	shape X	shape Y	shape Z
method	items & angle to fill	items & angle between	angle to fill & angle between
center point	circle X	centre Y	centre Z
items	8	8	–
angle to fill	230	–	220
angle between	–	40	25
rotate items	yes	yes	yes

8 The results of these three operations is Figure 32.4(b).

9 *Tasks*
   a) linear/align dimension as shown and note the 35 and 25 values for the rectangular array
   b) angular dimension the polar arrays and note the fill angle and angle between the objects values
   c) comment on the 160-degree fill angle. This operation had an angle to fill of 220, and therefore this angle should be 140?
   d) try the polar array operations with no rotate items as copied – Figure 32.4(c) displays one operation.

10 This exercise is now complete and can be saved if required.

Assignments

Several activities to complete. All involve using the array command as well as hatching, adding text, etc. I have tried to make these activities varied and interesting so I hope you enjoy attempting them. As with all activities:
a) start with your A3SHEET standard drawing sheet
b) use layers correctly for outlines, text, hatching, etc.
c) when complete, save as C:\BEGIN\:ACT?? or similar
d) use your discretion for any sizes which have been omitted.

Activity 31: Car Wheel design

Six designs of varying difficulty. You can create as given or use your imagination. No sizes needed.

Activity 32: Ratchet and Saw Tip Blade

Two typical examples of a use for the array command
a) The ratchet tooth shape is fairly straightforward and the saw blade tooth is more difficult than you would think. Or is it?
b) Draw each tooth in a clear area of the screen, then copy them to the appropriate circle points. The trim command is extensively used.

Activity 33: Fish design

a) Refer to Figure 32.5 and create the fish design using the sizes given. Save this drawing as **C:\BEGIN\FISH** for future work.
b) When the 'fish' has been completed and saved, create the following array patterns:
   1. A 5 × 3 angular rectangular array at 10 degrees with a 0.2 scale. The distances between the rows and columns are at your discretion.
   2. Three polar arrays with:
      a) scale 0.25 with radius of 90 and 9 items
      b) scale 0.2 with radius of 60 and 7 items
      c) scale 0.15 with radius of 35 and 7 items.

Activity 34: The light bulb

a) This was one of my first array exercises and has proved very popular in previous AutoCAD books. It is harder than you would think, especially the R10 arc.
b) The basic bulb is copied and scaled three times. The polar array centre is at your discretion.
c) The position of the smaller polar bulb is relative to the larger polar array, but how is it positioned? This problem is for you to decide. No hints are given on how to draw the basic bulb.

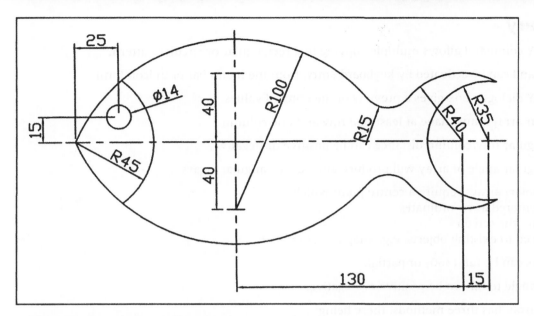

Figure 32.5 C:\BEGIN\FISH information for Activity 33.

Activity 35: Bracket and Gauge

Two engineering type applications of the array command.
a) The bracket is fairly easy, the hexagonal rectangular array to be positioned to suit by yourself.
b) The gauge drawing requires some thought with the polar arrays. I drew a vertical line and arrayed 26 items twice, the fill angles being +150 and −150. Think about this!! The other gauge fill angles are +140 and −140. What about the longer line in the gauge? The pointer position is at your discretion. Another method is to array lines and trim to circles.

Activity 36: Pinion gear wheel

Another typical engineering application of the array command.
The design details are given for you and the basic tooth shape is not too difficult. Copy and rotate the second gear wheel – but at what angle? Think about the number of teeth. I realise that the gears are not 'touching', another problem for you to think about.

Activity 37: Propeller blade

a) Draw the outline of the blade using the sizes given. Trimmed circles are useful.
b) When the blade has been drawn, use the COPY, SCALE and polar ARRAY commands to produce the following propeller designs:
   1. a 2-bladed at a scale of 0.75
   2. a 3-bladed at a scale of 0.65
   3. a 4-bladed at a scale of 0.5
   4. a 5-bladed at a scale of 0.35.
c) Note that the 4- and 5-bladed designs require some 'tidying up'.

Activity 38: Leaf design

Draw the leaf using the reference sizes given. Mirror effect is useful for the outline then 2D solid or hatch.
The arrays are with a one-fourth size object. The angular array requires some thought for the angle and the row/column offsets.

Summary

1. The ARRAY command allows multiple copying in a rectangular or circular pattern.
2. The command can be activated by keyboard entry, from the menu bar or in icon form.
3. The ARRAY dialogue box gives a preview of the entered values.
4. Rectangular arrays must have at least one row and one column.
5. The rectangular row/column distance can be positive or negative.
6. The rectangular angle of array will produce angular rectangular arrays.
7. Circular (polar) arrays require a centre point which:
   a) can be entered as co-ordinates
   b) picked on the screen
   c) referenced to existing objects, e.g. Snap to Center icon.
8. Polar arrays can be full (360) or partial.
9. The polar angle to fill can be positive or negative.
10. The polar array has three methods, these being:
    a) number of items and angle to fill
    b) number of items and angle between items
    c) angle to fill and angle between items.
11. Polar array objects can be rotated/not rotated about the centre point.

Chapter 33
Changing properties

All objects have properties, e.g. linetype, colour, layer, position, etc. Text has also properties such as height, style, width factor, obliquing angle, etc. Object properties can be changed using the command line CHANGE or the Properties palette and this chapter will demonstrate how to use these two options with a series of simple exercises.

1. Open A3SHEET with layer OUT current, toolbars to suit and refer to Figure 33.1.
2. Draw a 40-unit square and copy it to six other parts of the screen.
3. *a)* At the command line enter **LTSCALE <R>** and:
   prompt   Enter new linetype scale factor and enter: **0.6 <R>**.
   b) At the command line enter **LWDISPLAY <R>** and:
   prompt   Enter new value for LWDISPLAY and enter: **ON <R>**.

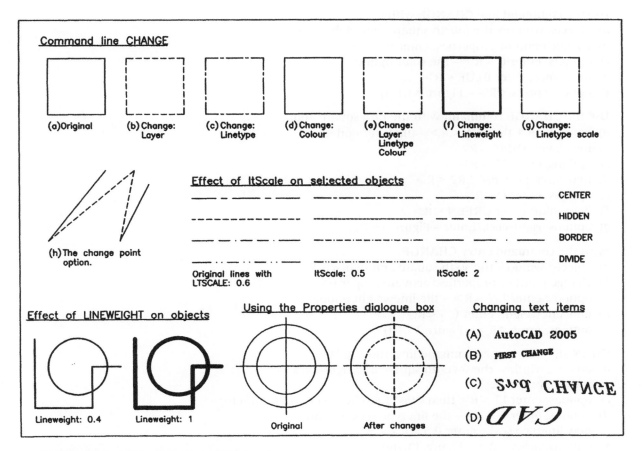

Figure 33.1 Changing properties exercise.

Command line CHANGE

1. At the command line enter **CHANGE <R>** and:
   - *prompt*   Select objects
   - *respond*   **window the second square then right-click**
   - *prompt*   Specify change point or [Properties]
   - *enter*   **P <R>** – the properties option
   - *prompt*   Enter property to change [Color/Elev/LAyer/LType/ltScale/LWeight/Thickness]
   - *enter*   **LA <R>** – the layer option
   - *prompt*   Enter new layer name<OUT>
   - *enter*   **HID <R>**
   - *prompt*   Enter property to change – i.e. any more property changes
   - *respond*   **right-click-Enter**.

2. The square will be displayed as brown hidden lines – Figure 33.1(b).

3. Repeat the CHANGE command line entry and:
   - a) objects: window the third square then right-click
   - b) change point or properties: enter P <R> for properties
   - c) options: enter LT <R> for linetype
   - d) new linetype: enter CENTER2 <R> for center linetypes
   - e) options: right-click/Enter to end command.

4. The square will be displayed with red center lines – Figure 33.1(c).

5. Use the command line CHANGE with:
   - a) objects: window the fourth square then right-click
   - b) change point or properties: enter P <R>
   - c) options: enter C <R> – the color option
   - d) new color: enter BLUE <R>
   - e) options: press <R> – Figure 33.1(d).

6. Use the CHANGE command with the fifth square and:
   - a) enter P <R> then LA <R> – the layer option
   - b) new layer: DIMS <R>
   - c) options: enter LT <R>
   - d) new linetype: CENTER2 <R>
   - e) options: enter C <R>
   - f) new color: enter GREEN <R>
   - g) options: right-click/Enter – Figure 33.1(e).

7. With the command entry CHANGE:
   - a) objects: window the sixth square and right-click
   - b) change point or properties: activate properties
   - c) options: enter LW <R> – the lineweight option
   - d) new lineweight: enter 0.5 <R>
   - e) options: press <R> – Figure 33.1(f).

8. CHANGE <R> at the command line and:
   - a) objects: window the seventh square and right-click
   - b) activate properties
   - c) options: enter LT <R> then CENTER2<R> – i.e. center linetype
   - d) options: enter S <R> – the linetype scale option
   - e) new linetype scale: enter 0.5 <R>
   - f) options: press <R> – Figure 33.1(g).

9. Compare the center linetype appearance of Figure 33.1(c) and Figure 33.1(g).

10 Menu bar with **Format-Layer** and:
    a) make layer 0 current
    b) freeze layer OUT then OK
    c) only the hidden line square and a green center line square displayed? – Figure 33.1(b) and Figure 33.1(e)
    d) thaw layer OUT and make it current.

11 *Notes*
    a) This exercise with the CHANGE command has resulted in:

Figure 33.1	square appearance	layer
a	red continuous lines	OUT
b	brown hidden lines	HID
c	red center lines	OUT
d	blue continuous lines	OUT
e	green center lines	DIMS

    b) I would suggest to you that only Figure 33.1(a) and Figure 33.1(b) are 'ideally' correct, i.e. the correct colour and linetype for the layer being used. The other squares demonstrate that it is possible to have different colours and linetypes on named layers. This can become confusing.

The change point option

The above exercise has only used the Properties option of the command line CHANGE command, but there is another option – change point.

1 Draw two lines:
   a) start point: 20,140   next point: 60,190
   b) start point: 70,140   next point: 90,180.

2 Activate the CHANGE command and:
   prompt   Select objects
   respond **pick the two lines then right-click**
   prompt   Specify change point or [Properties]
   respond **pick about the point 80,190 on the screen**.

3 The two lines are redrawn to this point – Figure 33.1(h).

4 This could be a very useful drawing aid.

LTSCALE and ltScale

One of the options available with the CHANGE command is ltScale. This system variable allows individual objects to have their linetype appearance changed. The final appearance of the object depends on the value entered and the value assigned to LTSCALE. The LTSCALE system variable is **GLOBAL**, i.e. if it is altered, all objects having center lines, hidden lines, etc. will alter in appearance and this may not be to the user's requirements. Hence the use of the ltScale option of CHANGE. As stated, the final appearance of the object depends on both the LTSCALE value and the entered value for ltScale. Thus if LTSCALE is globally set to 0.6, and the value for ltScale is entered as 0.5, the selected objects will be displayed with an effective value of 0.3. If LTSCALE is 0.6 and 2 is entered for ltScale, the effective value for selected objects is 1.2. This effect is shown in Figure 33.1 using the Center, Hidden, Border and Divide linetypes.

Lineweight

LINEWEIGHT allows selected objects to be displayed at varying width. Although this may seem to be similar to the polyline-width type of object, it is not. Objects can be drawn directly with varying lineweights or the LWeight option of the CHANGE command can be used. It is necessary to toggle the LWDISPLAY system variable to ON before lineweight can be used. This was achieved at the start of the chapter.

1   Right-click on LWT in the status bar, pick (left-click) Settings and:
    prompt   Lineweight Settings dialogue box
    respond   a) ensure Millimeters active
                b) scroll at Lineweights and pick 0.4 mm
                c) dialogue box as Figure 33.2
                d) note the default value then pick OK.

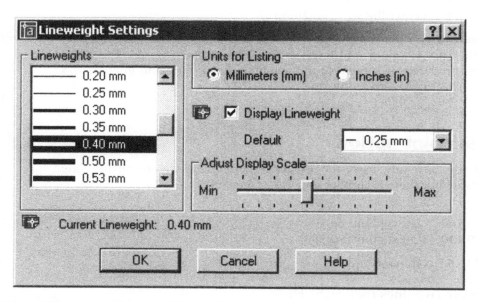

Figure 33.2   The Lineweight Settings dialogue box.

2   Now draw some lines and circles.

3   From the Properties toolbar, scroll at lineweight and set to 1.0 then draw some other lines and circles.

4   Finally set the lineweight back to the default value – 0.25?

5   Figure 33.1 displays some objects at the two set lineweights.

Pickfirst

When objects require to be modified, the normal procedure is to activate the command then select the objects. This procedure can be reversed with the PICKFIRST system variable. To demonstrate how this is achieved:

1   Draw a line and circle anywhere on the screen.

2   Menu bar with **Modify-Erase** and select the two objects then <R>.

3   The objects are erased.

4   Draw another line and circle.

5   At the command line enter **PICKFIRST <R>** and:
    prompt   Enter new value for PICKFIRST<0>
    enter   1 <R>.

6   A pickbox will be displayed on the cursor crosshairs. Do not confuse this pickbox with the grips box. They are different.

7   Pick the line and circle with the pickbox, then menu bar with Modify-Erase. The objects are erased from the screen.

8   Thus PICKFIRST allows the user to alter the selection process and:
    PICKFIRST: 0 – activate the command then select the objects
    PICKFIRST: 1 – select the objects then activate the command.

9   We will use PICKFIRST set to 1 to activate the Properties dialogue box.

Changing properties using the Properties palette

1   With layer OUT current, draw:
    a) two concentric circles
    b) two lines to represent centre lines.

2   Ensure PICKFIRST is set to 1.

3   With the pick box, select the horizontal line then the **PROPERTIES icon** from the Standard toolbar and:
    prompt   Properties palette
    with   two sets of information about the object selected:
               a) General: layer, color, linetype, etc.
               b) Geometry: start and end point of line, etc.

4   With the palette active:
    a) pick Layer line and scroll arrow appears
    b) scroll and pick layer CL
    c) alter linetype scale value to 0.5 – Figure 33.3(a)
    d) cancel the dialogue box – top left (X) box
    e) press ESC.

5   The selected line will be displayed as a green center line.

6   Pick the vertical line then the Properties icon and from the Properties palette alter:
    a) colour: Magenta
    b) linetype: Hidden
    c) lineweight: 0.3
    d) cancel dialogue box then ESC.

7   Finally pick the smaller circle then the Properties icon and alter:
    a) layer: HID
    b) linetype scale: 0.75
    c) cancel and ESC.

8   Figure 33.1 displays the 'before and after' effects of using the Properties dialogue box to change objects.

9   *Notes*
    a) The Properties palette gives useful information about the objects which are selected:
        1. lines   start and end point, delta values, length, angle
        2. circles   centre point, radius, diameter, area, circumference.
    b) These values can be altered from the palette, and the object will be modified accordingly. You can try this for yourself.

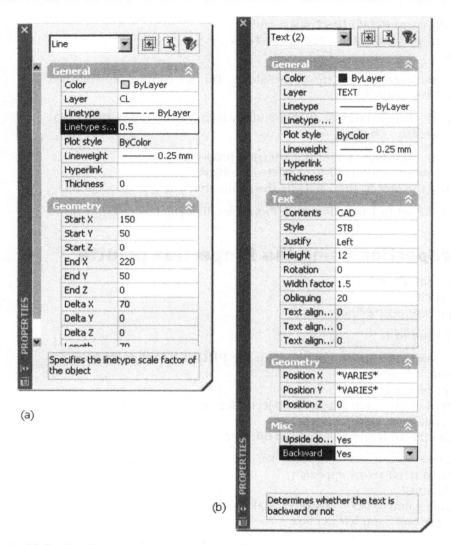

Figure 33.3 The Properties palette for (a) a selected line and (b) the selected item of text.

Changing text

Text has several properties which other objects do not, e.g. style, height, width factor, etc. as well as layer, linetype and colour. These properties can be altered with the command line CHANGE (pickfirst: 0) or the Properties palette (pickfirst: 1).

To demonstrate how text can be modified using both methods:

1 Create the following text styles:

Name	STA	STB
font	romant	italict
height	0	12
width	1	1
oblique	0	0
backwards	N	N
upside-down	N	Y
vertical	N	N

2 With layer TEXT current and STA the current style, enter the text item AutoCAD 2005 at height 5 and rotation 0 at a suitable part of the screen – Figure 33.1(A).

3 Multiple copy this item of text to three other parts of the screen.

4 At the command line enter **CHANGE <R>** and:
   prompt   Select objects
   respond   **pick the second text item then right-click**
   prompt   Specify change point or [Properties]
   respond   **<RETURN>**
   prompt   Specify new text insertion point <no change>
   respond   **<RETURN>** – no change for text start point
   prompt   Enter new text style <STA>
   respond   **<RETURN>** – no change to text style
   prompt   Specify new height and enter: **3.5 <R>**
   prompt   Specify new rotation angle and enter: **5 <R>**
   prompt   Enter new text <AutoCAD 2005>
   enter   **FIRST CHANGE <R>** – Figure 33.1(B).

5 With CHANGE <R> at the command line, pick the third item of text and:
   a) change point: <R>
   b) new text insertion point: <R>
   c) new text style: enter STB <R>
   d) new rotation angle: −2
   e) new text: enter 2nd CHANGE – Figure 33.1(C)
   f) *Question*: Why no height prompt?

6 With PICKFIRST on (i.e. set to 1) pick the fourth item of text then the Properties icon and the Properties palette will display the following details for the text item: General; Text; Geometry and Misc.
   a) Using the Properties palette alter:
      1. contents: CAD
      2. text style: STB
      3. height: 12
      4. width factor: 1.5
      5. obliquing: 20
      6. Backwards: Yes
   b) dialogue box as Figure 33.3(b)
   c) cancel the dialogue box then ESC
   d) these changes are displayed in Figure 33.1(D).

7 This exercise is now complete and can be saved if required.

Combining ARRAY and CHANGE

Combining the array command with the properties command can give interesting results. To demonstrate the effect:

1 Open your standard A3SHEET sheet and refer to Figure 33.4.

2 Draw the two arc segments as trimmed circles using the information given in Figure 33.4(a).

3 Draw the polyline and 0 text item using the reference data.

4 With the ARRAY command, polar array (twice) the polyline and text item using an arc centre as the array centre point:
   a) for 4 items, angle to fill +30 degrees with rotation
   b) for 7 items, angle to fill −60 degrees with rotation – Figure 33.4(b).

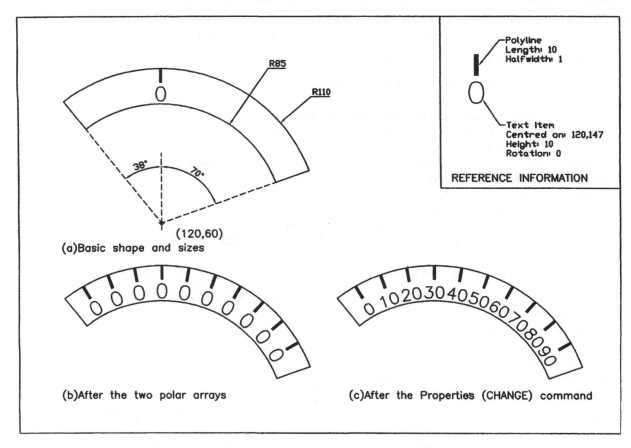

Figure 33.4 The combined ARRAY and CHANGE PROPERTIES commands.

5 Using the command line CHANGE or the Properties icon, pick each text item and alter:
   a) the text values to 10, 20, 30, etc. to 90
   b) the text height to 8.
6 The final result should be as Figure 33.4(c).
7 Save if required as the exercise is complete.

Note

The first exercise demonstrated that it was possible to change the properties of objects independent of the current layer. This means that if layer OUT (red, continuous) is current, objects can be created on this layer as green center lines, blue hidden lines, etc. This is a practice **I would not recommend** until you are proficient at using the AutoCAD draughting package. If green center lines have to be created, use the correct layer, or make a new layer if required. Try not to 'mix' different types of linetype and different colours on the one layer. Remember that this is only a recommendation – the choice is always left to the user.

Summary

1. Objects have properties such as colour, layer, linetype, etc.

2. An object's properties can be changed with:
   a) the command line CHANGE
   b) the Properties icon.

3. The icon method gives the Properties palette.

4. To use the Properties palette, the PICKFIRST variable must be set ON – i.e. value 1.

5. With PICKFIRST:
   a) value 0: select the command then the objects
   b) value 1: select the objects then the command.

6. The properties command is very useful when text items are arrayed.

7. Individual objects can have specific linetype scale factors.

Chapter 34

User exercise 3

This exercise will involve creating different arrays with already created text styles.

1. Open C:\BEGIN\STYLEX to display a blank screen with the twelve created and saved text styles (ST1–ST12) from Chapter 32. Refer to Figure 34.1.

2. With layer OUT current, draw a 30-unit square at A(10,10).

3. Read the Notes in step 8 before the next operation.

4. Multiple copy the square from point A to the points B(85,220), C(200,70), D(280,15) and E(20,255).

5. Array the squares using the following information:
    A: rectangular with 2 rows and 5 columns, both distances 35
    B: polar for 10 items, full circle with rotation, the centre point of the array being 100,170
    C: rectangular with 5 rows and 2 columns and 35 distances
    D: angular rectangular (angle of array −10) with 5 rows and 2 columns, the distances both 35. Remember to rotate first!!
    E: rectangular for 1 row and 10 columns, the distance 35.

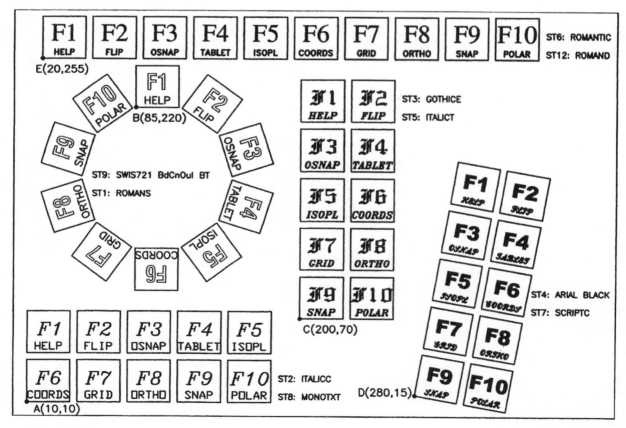

Figure 34.1 User exercise 3.

6   Add/or change the text items using the style names listed.

7   Save the completed layout as **C:\BEGIN\USEREX3**.

8   *Notes*
    a) Text can be added after the squares are arrayed, or before the first square is multiple copied.
    b) After array: the text items are added using single line text, the style, rotation angle and start point being entered by the user. This is fairly straightforward with the exception of the polar and angular arrays, i.e. what are the rotation angles?
    c) Before the multiple copy: two text items are added to the original square and it is then multiple copied and arrayed. The added text items can then be altered with the change properties command.
    d) It is the user's preference as to which method is used.
    e) I added the text to the original square, copied, arrayed and then changed properties.

Assignments

Two activities have been added which require the array and change properties commands to be used. Hopefully these activities will give you some relaxation?

Activity 39: Telephone dials

a) The 'old-fashioned' type has circles arrayed for a fill angle of ??. Is the text arrayed or just added to the drawing as text? You have to decide!
b) The modern type has a polyline 'button' with middled text. The array distances are at your discretion as is the outline shape.

Activity 40: Flow Gauge and Dartboard

a) Flow Gauge: A nice simple drawing to complete, but it takes some time! The text is added during the array operation.
b) Dartboard: Draw the circles then array the 'spokes'. The filled sections are trimmed donuts. The text is middled, height 10 and ROMANT. Array then change properties?

Chapter **35**

Dimension styles 2

In Chapter 20 we investigated how dimensions could be customised to user requirements by setting and saving a dimension style. The dimension style A3DIM was created and saved with our A3SHEET standard sheet.

In this chapter we will:
a) create and use several new dimension styles
b) add tolerance dimensions
c) investigate geometric tolerance.

The process for creating new dimension styles involves altering the values for specific variables which control how the dimensions are displayed on the screen.

Getting started

1   Open your A3SHEET standard drawing sheet.

2   Create the following text styles, all with height 0:

Name	Font
ST1	romans.shx
ST2	scriptc.shx
ST3	italict.shx
ST4	Arial Black

3   Menu bar with **File-Save** to update the A3SHEET standard sheet with the four created text styles. These are then available for future use if required.

4   Now continue with the exercise.

Creating the new dimension styles

1   Is the A3SHEET standard sheet still on the screen?

2   Menu bar with **Dimension-Style** and:
   prompt   `Dimension Style Manager dialogue box with style A3DIM current`
   respond **pick New**
   prompt   `Create New Dimension Style dialogue box`
   with   a) New Style Name: Copy of A3DIM
              b) Start with: A3DIM
              c) Use for: All dimensions
   respond a) alter New Style Name to DIMST1
              b) pick Continue
   prompt   `New Dimension Style: DIMST1 dialogue box with tab selections`
   respond **pick OK at present**
   prompt   `Dimension Style Manager dialogue box`
   with   Styles: A3DIM and DIMST1

respond	**pick A3DIM then NEW**	step A
prompt	Create New Dimension Style dialogue box as before	step B
respond	a) alter New Style Name to DIMST2	step C
	b) pick Continue	step D
prompt	New Dimension Style: DIMST2 dialogue box	step E
respond	**pick OK at present**	step F
prompt	Dimension Style Manager dialogue box	
with	Styles: A3DIM, DIMST1 and DIMST2	
respond	using steps A, B, C, D, E and F, pick A3DIM then New and create new dimension styles DIMST3, DIMST4, DIMST5 and DIMST6	
then	when DIMST6 has been created	
prompt	Dimension Style Manager dialogue box	
with	Styles: A3DIM, DIMST1–DIMST6 as Figure 35.1	
respond	a) pick A3DIM then Set Current	
	b) pick Close.	

3   We have now created six new dimension styles (DIMST1–DIMST6) and these all have the same 'settings' as the A3DIM dimension style.

4   These six new dimension styles now have to be individually modified to meet our requirements.

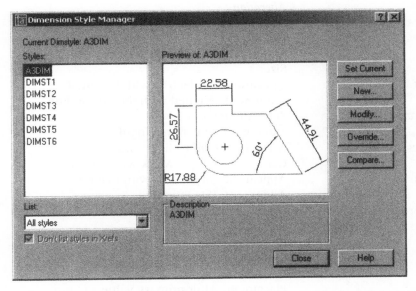

Figure 35.1 The Dimension Style Manager dialogue box with the original A3DIM style and the newly created DIMST1–DIMST6 styles.

Modifying the new styles

In the exercise which follows, we will only alter certain dimension variables. The rest of the dimension variables will have the same 'settings' as A3DIM. We will start by modifying DIMST1, so:

1   Menu bar with **Dimension-Style** and:
    prompt   Dimension Style Manager dialogue box
    respond   a) pick DIMST1
                 b) pick Modify

prompt		Modify Dimension Style: DIMST1 dialogue box with tab selections
respond		**pick the Text tab**
alter	*a*)	Draw frame around text active – i.e. tick
	b)	Text Placement: Vertical: Centered
		Horizontal: Centered
		Offset from dim line: −1.5
	c)	Text Alignment: Aligned with dimension line
	d)	Text Appearance: Text style: ST1 with height: 4
	e)	pick OK
prompt		Dimension Style Manager dialogue box
respond		**pick Close**.

2   The above process is the basic procedure for 'customising' dimension styles, the steps being:
   a) pick **Dimension-Style** or the Dimension Style icon
   b) pick required style name, e.g. DIMST1
   c) pick Modify
   d) pick required tab and alter as required
   e) pick OK then Close.

3   Using the procedure described, alter the named dimension styles to include the following:

Style	*Tab*	*Alteration*
DIMST2	Lines and Arrows	Arrowheads: all Box filled
		Arrow size: 4
		Center Mark Type: Line
		Size: 3
	Text	Appearance Style: ST1 with height: 4
		Alignment: Aligned with dimension line
	Primary Units	Zero suppression
		Trailing off: linear and angular
DIMST3	Lines and Arrows	Arrowheads: all Oblique, size: 4
		Center Mark Type: Mark with size: 4
	Text	Appearance Style: ST2 with height: 4
		Alignment: ISO Standard
	Alternative Units	Display alternative units active
		Unit format: Decimal
		Precision: 0.00
		Multiplier: 0.03937 (default?)
		Round distances to: 0
		Placement: After primary value
DIMST4	Lines and Arrows	Arrowheads: all Dot Small, height: 8
	Text	Appearance Style: ST1, height: 4
		Alignment: Horizontal
	Primary Units	Measurement scale factor: 2
DIMST5	Text	Appearance Style: ST3 with height: 5
		Alignment: ISO Standard
		Offset from dimension line: 0
DIMST6	Text	Appearance Style: ST4 with height: 6
		Offset from dimension line: 3
		Alignment: ISO Standard

4   When the six dimension styles have been altered, return to the Dimension Style Manager dialogue box, pick A3DIM-Set Current then Close.

Using the created dimension styles

1. Using Figure 35.2 for reference, draw seven horizontal and vertical lines, seven circles and seven angled lines – the sizes are not of any significance but try and 'fit' all the objects into your drawing 'frame'.

2. Dimension one set of objects, i.e. a horizontal and vertical line, a circle diameter and an angled line. The dimensions added will be with the A3DIM style as Figure 35.2(a).

3. Menu bar with **Dimension-Style** and:
   prompt   Dimension Style Manager dialogue box
   respond   a) pick DIMST1
                 b) pick Set Current
                 c) pick Close.

4. Using the set dimension style (DIMST1) dimension another set of four objects.

5. Display the dimension toolbar (**View-Toolbars**) and note that DIMST1 is displayed, i.e. it is the current dimension style.
   a) pick the scroll arrow to display the other styles
   b) pick DIMST2 to make it current
   c) cancel or reposition the dimension toolbar
   d) dimension a set of objects with the DIMST2 style.

6. At the command line enter **–DIMSTYLE <R>** and:
   prompt   Current dimension style: DIMST2
   then   Enter a dimension style option [Save/Restore . . .

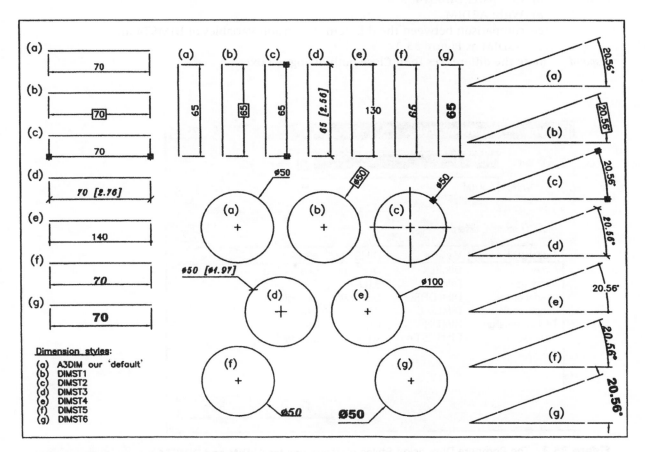

Figure 35.2 The dimension styles exercise.

enter	**R <R>** – the restore option	
prompt	Enter a dimension style name, [?] or <select dimension>	
enter	**DIMST3 <R>**	
prompt	Current dimension style: DIMST3	
and	DIMST3 displayed in Dimension toolbar if displayed	
note	if the text window appears, cancel it.	

7   Dimension a set of objects with the DIMST3 style.

8   Using the menu bar selection **Dimension-Style**, the Dimension toolbar or the command line **–DIMSTYLE** entry:
    a) make each created dimension style current
    b) dimension a set of objects with each style.

9   Figure 35.2 displays the result of the exercise which can be saved if required.

Comparing dimension styles

It is possible to compare dimension styles with each other and note any changes between them. To demonstrate this comparison, select from the menu bar **Dimension-Style** and:

prompt	Dimension Style Manager dialogue box
respond	*a)* ensure A3DIM is the current style
	b) pick DIMST4 then Compare
prompt	Compare Dimension Styles dialogue box
with	*a)* Compare: DIMST4
	b) With: A3DIM
	c) comparison between the different dimension variables of DIMST4 and A3DIM as Figure 35.3
respond	note the differences then Close both dialogue boxes.

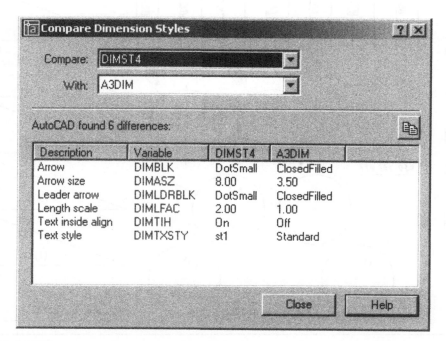

Figure 35.3 The Compare Dimension Styles dialogue box for A3DIM and DIMST4.

Dimension variables

The above comparison between the two dimension styles displays some of the AutoCAD dimension variables. All dimension variables (dimvars) can be altered by keyboard entry and as a demonstration:

1. Make DIMST1 the current dimension style and note the dimensions which used this style. These are the second set of dimensioned objects as Figure 35.2(b).

2. At the command line enter **DIMASZ <R>** and:
   prompt   Enter new value for DIMASZ <3.50>
   enter   **8 <R>**.

3. Pick the **Dimension Style icon** from the Dimension toolbar and:
   prompt   Dimension Style Manager dialogue box
   with   a) Current Style: DIMST1
               b) DIMST1
                    └──<style overrides>
               c) Description: DIMST1 + Arrow size = 8.00 – Figure 35.4
                  (indicating that DIMST1 has been altered)
   respond   a) move pointer to <style overrides>
               b) right-click the mouse
   prompt   Shortcut menu displayed
   respond   a) pick Save to current style
               b) close the Dimension Style Manager dialogue box.

4. The dimensions which used the DIMST1 dimension style will be displayed with an arrow size of 8.

5. Now change the DIMASZ variable for DIMST1 back to 3.5.

6. This exercise is now complete and can be saved if required.

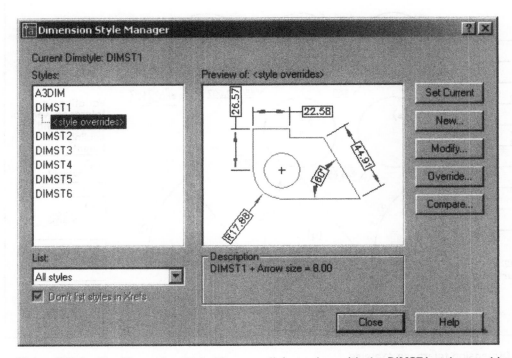

Figure 35.4 The Dimension Style Manager dialogue box with the DIMST1 style overrides.

Task

1. Open your A3SHEET standard sheet with the four saved text styles ST1–ST4.
2. Using the Dimension Style Manager dialogue box:
   a) check current style is A3DIM
   b) with Modify-Text tab, set the text style to ST1 with a height of 4
   c) pick OK then close the dialogue box.
3. Menu bar with **File-Save** to update the A3SHEET standard sheet with the modified A3DIM dimension style.

Tolerance dimensions

Dimensions can be displayed with tolerances and limits, the 'types' available being symmetrical, deviation, limits and basic. These are displayed in Figure 35.5(a) along with the standard default dimension type, i.e. no tolerances displayed. To use tolerance dimensions correctly, a dimension style should be created for each 'type' which is to be used in the drawing. As usual we will investigate the topic with an exercise, so:

1. Open your A3SHEET standard sheet and refer to Figure 35.5.
2. Draw a line, circle and angled lines and copy them to four other areas of the screen.
3. Dimension one set of objects with your standard A3DIM dimension style setting – Figure 35.5(p).

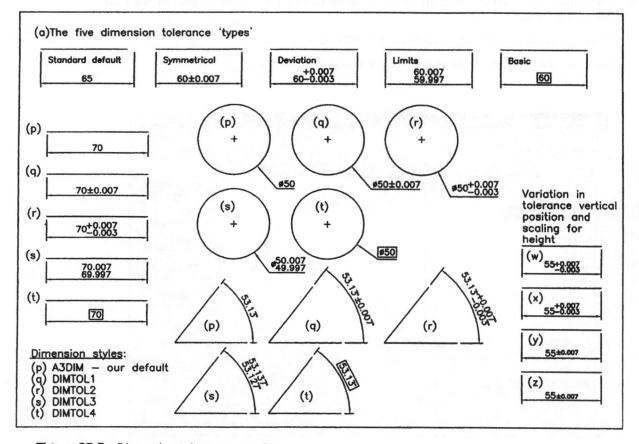

Figure 35.5 Dimension tolerance exercise.

Creating the tolerance dimension styles

1. Using the Dimension Style Manager dialogue box, create four new named dimension styles, all starting with A3DIM. The four new style names are DIMTOL1, DIMTOL2, DIMTOL3 and DIMTOL4.

2. With the Dimension Style Manager dialogue box:
   a) pick each new dimension style
   b) pick Modify then select the:
      1. Primary Units tab and alter both the Linear and Angular precision to 0.000
      2. Tolerances tab and alter as follows:

Method	DIMTOL1 Symmetrical	DIMTOL2 Deviation	DIMTOL3 Limits	DIMTOL4 Basic
Precision	0.000	0.000	0.000	–
Upper value	0.007	0.007	0.007	–
Lower value	–	0.003	0.003	–
Scaling for height	1	1	1	–
Vertical position	Middle	Middle	Middle	Middle

3. Making each new dimension style current, dimension a set of three objects. The effect is displayed in Figure 35.5 with:
   Figure 35.5(q): style DIMTOL1   Figure 35.5(r): style DIMTOL2
   Figure 35.5(s): style DIMTOL3   Figure 35.5(t): style DIMTOL4

4. *Task*
   Investigate the vertical position and the scaling for height options available in the Tolerances tab dialogue box. Figure 35.5 displays the following:

Figure 35.5	Method	Vertical pos	Scaling for height
w	deviation	top	0.8
x	deviation	bottom	0.8
y	symmetrical	top	0.7
z	symmetrical	bottom	0.7

5. This exercise is complete and can now be saved.

Geometric tolerancing

AutoCAD 2005 has geometric tolerancing facilities and to demonstrate this feature:

1. Open your A3SHEET standard sheet.

2. Menu bar with **Dimension-Tolerance** and:
   prompt Geometric Tolerance dialogue box
   with   Feature Control Frames
   respond **move pointer to box under Sym and left-click**
   prompt Symbol dialogue box as Figure 35.6
   respond a) press ESC to cancel the Symbol dialogue box
             b) study the Geometric Tolerance dialogue box
             c) pick Cancel at present.

Feature control frames

Geometric tolerances define the variations which are permitted to the form or profile of a component, its orientation and location as well as the allowable runout from the exact geometry of a feature. Geometric tolerances are added by the user in feature control frames, these being displayed in the Geometric Tolerances dialogue box.

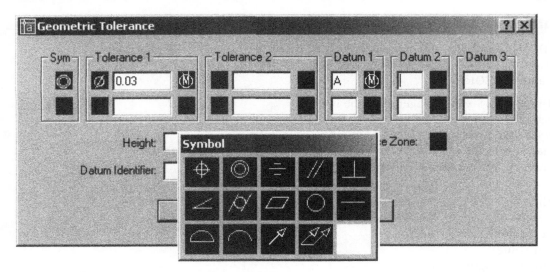

Figure 35.6 The Symbol dialogue box superimposed on the Geometric Tolerance dialogue box.

1   A feature control frame consists of two/three sections or compartments. These are displayed in Figure 35.7(a) and are:
   a) *The geometric characteristic symbol*
       1. This is a symbol for the tolerance which is to be applied.
       2. These geometric symbols can represent location, orientation, form, profile and runout. The symbols include symmetry, flatness, straightness, angularity, concentricity, etc.

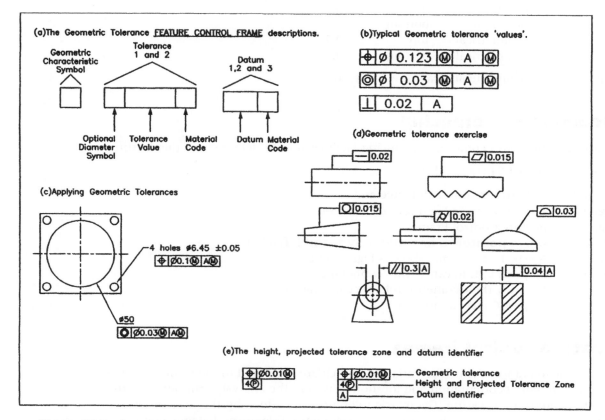

Figure 35.7 Geometric tolerance terminology and usage.

b) *The tolerance section (1 and 2)*
      1. This creates the tolerance value in the Feature Control Frame.
      2. The tolerance indicates the amount by which the geometric tolerance characteristic can deviate from a perfect form. It consist of three boxes:
         a) first box: inserts an optional diameter symbol
         b) second box: the actual tolerance value
         c) third box: displays the material code which can be:
            1. M: at maximum material condition
            2. L: at least material condition
            3. S: regardless of feature size.
   c) *The primary, secondary and tertiary datum information*
      1. This can consist of the reference letter and the material code.

2 Typical types of geometric tolerance 'values' are displayed in Figure 35.7(b).

3 *Notes*
   a) Feature control frames are added to a dimension by the user.
   b) A feature control frame contains all the tolerance information for a single dimension. The tolerance information must be entered by the user.
   c) Feature control frames can be copied, moved, erased, stretched, scaled and rotated once 'inserted' into a drawing.
   d) Feature control frames can be modified with DDEDIT.

Geometric tolerance example

1 Refer to Figure 35.7(c) and create a square with circles as shown. The size and layout is of no importance.

2 Diameter dimension the larger circle using the A3DIM dimension style (should be current) and the DIMS layer.

3 Select the **TOLERANCE icon** from the Dimension toolbar and:
   prompt   Geometric Tolerance dialogue box
   respond **pick top box under Sym**
   prompt   Symbol dialogue box
   respond **pick Concentricity symbol** (2nd top left)
   prompt   Geometric Tolerance dialogue box
   with   Concentricity symbol displayed
   respond a) at Tolerance 1, pick left box
              b) at Tolerance 1, enter 0.03
              c) at Tolerance 1, pick MC box and:
                 prompt   Material Condition dialogue box
                 respond **pick M**
              d) at Datum 1, enter A
              e) at Datum 1, pick MC box, pick M
              f) Geometric Tolerance dialogue box as Figure 35.6
   then   pick OK
   prompt   Enter tolerance location
   respond **pick 'under' the diameter dimension**.

4 Menu bar with **Dimension-Tolerance** and:
   prompt   Geometric Tolerance dialogue box
   respond **pick top Sym box**
   prompt   Symbol dialogue box
   respond **pick Position symbol** (top left)
   prompt   Geometric Tolerance dialogue box

respond *a*) at Tolerance 1 pick left box (for diameter symbol)
           b) at Tolerance 1 enter value of 0.1
           c) at Tolerance 1 pick MC box and pick M
           d) at Datum 1 enter datum A
           e) at Datum 1 pick MC box and pick M
           f) pick OK
prompt   Enter tolerance location
respond **pick to suit** – refer to Figure 35.7(c).

5 Select the **QUICK LEADER icon** from the Dimension toolbar and:
   prompt   Specify first leader point and **pick a smaller circle**
   prompt   Specify next point and **pick a point to suit then right-click**
   prompt   Specify text width and enter: **0 <R>**
   prompt   Enter first line of annotation text and right-click
   prompt   Multiline Text Formatting Editor dialogue box
   enter   **4 holes diameter 6.45 plus/minus 0.05** (use text control codes for diameter and plus/minus) – Figure 35.8
   then   pick OK.

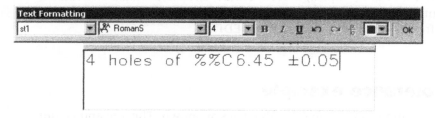

Figure 35.8 The Text Formatting dialogue box for the Quick Leader example.

Geometric tolerance exercise

1 Refer to Figure 35.7(d) and create 'shapes' as shown – size is not important.

2 Add the geometric tolerance information.

3 Investigate the height, projected tolerance zone and datum identifier options from the Geometric Tolerance dialogue box as Figure 35.7(e).

Assignments

Two assignments on dimension styles, tolerance dimensions and geometric tolerancing. Adding dimensions to a drawing is generally a tedious process and if dimension styles and geometric tolerancing are required, the process does not get any easier.

Activity 41: Tolerance dimensions

The two drawings are easy to complete. Adding the dimensions requires four created styles, all from the A3DIM default. The dimension styles use the symmetrical, limits, deviation and basic tolerance methods. You have to enter the upper and lower values.

Activity 42: Geometric tolerances

A slightly more complex drawing to complete. Several dimension styles have to be created and two geometric tolerances have to be added.

Summary

1. Dimension styles are created by the user and 'customised' for company/customer requirements.
2. Dimension styles are created with the Dimension Style Manager dialogue box using several tab selections.
3. It possible to alter dimension variables (dimvars) from the command line.
4. Tolerance dimensions can be created as dimension styles. The four 'types' are symmetrical, deviation, limits and basic.
5. Geometric tolerances can be added to drawings using feature control frames.
6. There are 14 geometric tolerance 'types' available.
7. All geometric tolerance information must be added by the user in feature control frames.
8. Feature control frames can be copied, moved, scaled, etc.

Chapter 36
Drawing with various units and paper sizes

Many new users of the AutoCAD draughting package frequently ask specific questions about drawings, three of the most common being:

1. Can I draw with imperial units?
2. Can I set a scale at the start of a drawing?
3. How do I work with very large/small components?

The answer to these three questions is **yes**:

1. You can draw in feet and inches.
2. You can set a scale at the start of a drawing.
3. You can draw large/small components on an A3-size sheet.

The most difficult concept for many CAD users is that all work should be completed full size.

In this chapter we will investigate three different types of exercise:

1. a component drawn and dimensioned in inches
2. a large-scale drawing
3. a small-scale drawing.

Each exercise will involve modifying the existing A3SHEET standard sheet and, hopefully, all users will appreciate what it means by drawing everything full size.

Drawing in inches

To draw in inches we require a new standard sheet which could be created:
a) from scratch
b) using the Wizard facility
c) altering the existing A3SHEET standard sheet which is set to metric sizes.

We will alter our existing standard sheet, the only reason being that it has already been 'customised' to our requirements, i.e. it has layers, a text style and a dimension style. This may save us some time? The conversion from a metric standard sheet to one which will allow us to draw in inches is relatively straightforward. We have to alter several parameters, e.g. units, limits, drawing aids and the dimension style variables.

1. Open your A3SHEET standard sheet and erase the black border as it is the wrong size for drawing in inches.

2. *Units*
   Menu bar with **Format-Units** and:
   prompt Drawing Units dialogue box
   respond set the following:
   　　a) *Length*
   　　　　Type: Engineering; Precision: 0'-0.00"
   　　b) *Angle*
   　　　　Type: Decimal Degrees; Precision: 0.00
   　　c) *Drag-and-drop scale*
   　　　　Units to scale drag-and-drop content: Inches
   then pick OK.

3. *Limits*
   Menu bar with **Format-Drawing Limits** and:
   prompt Specify lower left corner and enter: **0,0 <R>**
   prompt Specify upper right corner and enter: **16",12" <R>**
   Note: Shift 2 (") for the inches symbol.

4. *Drawing area*
   a) make layer 0 current
   b) menu bar with **Draw-Rectangle** and:
   　　prompt Specify first corner point and enter: **0,0 <R>**
   　　prompt Specify other corner point and enter: **15.5",11.75"**
   c) menu bar with **View-Zoom-All**.

5. *Drawing Aids*
   a) set grid spacing to 0.5
   b) set snap spacing to 0.25
   c) set global LTSCALE value to 0.025.

6. *Dimension Style*
   Menu bar with **Dimension-Style** and:
   prompt Dimension Style Manager dialogue box
   respond create a new dimension style with:
   　　a) New Style Name: STDIMP
   　　b) Start with: A3DIM
   　　c) Tab alterations as follows:
   　　　　1. *Lines and Arrows*
   　　　　　　Baseline spacing: 0.75"
   　　　　　　Extend beyond dim line: 0.175"
   　　　　　　Offset from origin: 0.175"
   　　　　　　Arrowheads: Closed Filled, Size: 0.175"
   　　　　　　Centre Marks for Circles: None.
   　　　　2. *Text*
   　　　　　　Style: ST1 with height: 0.18"
   　　　　　　Offset from dim line: 0.08"
   　　　　　　Text alignment: with dimension line.
   　　　　3. *Primary Units*
   　　　　　　Unit Format: Engineering
   　　　　　　Precision: 0'-0.00"
   　　　　　　Degrees: Decimal
   　　　　　　Precision: 0.00
   　　　　　　Zero suppression: both ON – i.e. tick
   　　　　　　Scale factor: 1.

7 Make STDIMP the current dimension style.

8 *a*) make layer OUT current
   b) save the drawing as **C:\BEGIN\STDIMP** – it may be useful.

9 Refer to Figure 36.1 and draw the component as shown. Add all text and dimensions. You should have no problems in completing the drawing or adding the dimensions. What should be the value for text height? Use the grid spacing as a guide.

10 When complete, save the drawing – plot?

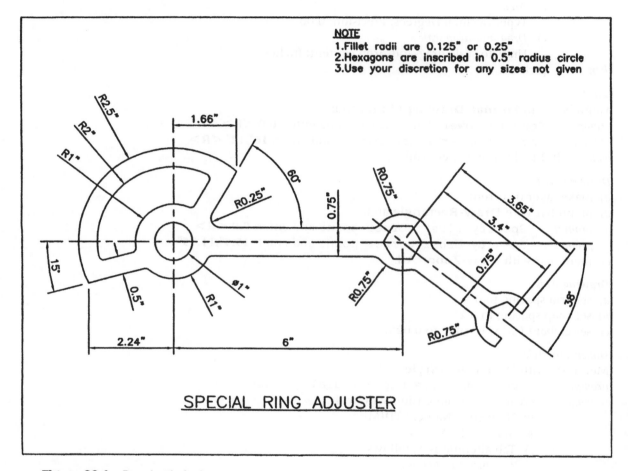

Figure 36.1 Drawing in inches.

Large-scale drawing

This exercise will require the A3SHEET standard sheet to be modified to allow a very large drawing to be drawn (full size) on A3 paper.

1 Open A3SHEET standard sheet and erase the black border – it is too small for the limits to be used.

2 Make layer 0 current.

3 At the command line enter **MVSETUP <R>** and:
   prompt Enable paper space? (No/Yes)
   enter   **N <R>**
   prompt Enter units type [Scientific/Decimal...
   enter   **M <R>** – for metric units

prompt	AutoCAD Text Window with Metric scales
and	Enter the scale factor
enter	**50 <R>** – i.e. scale of 1:50
prompt	Enter the paper width and enter: **420 <R>**
prompt	Enter the paper height and enter: **297 <R>**.

4 A polyline black border will be displayed. This is our drawing area.

5 Pan the drawing to suit the screen.

6 With the menu bar **Tools-Inquiry-ID Point**, obtain the co-ordinates of the lower left and upper right corners of the black border. These should be:
 a) lower left: X = 0.00   Y = 0.00   Z = 0.00
 b) upper right: X = 21000.00   Y = 14850.00   Z = 0.00.

7 Set the grid to 400 and the snap to 200.

8 With layer OUT current refer to Figure 36.2 and complete the factory plan layout using the sizes as given, the start point being at A(2000,2000). The wall thickness is 250.

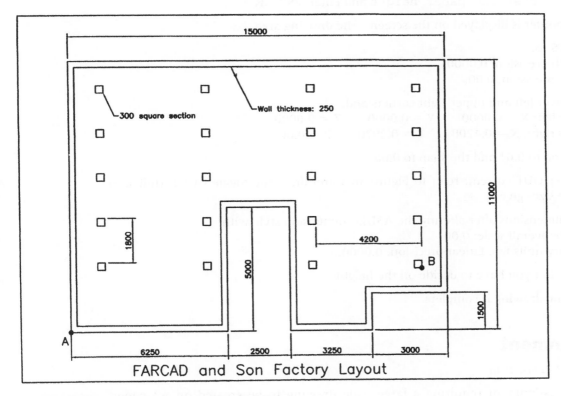

Figure 36.2 Large-scale drawing on A3 paper.

9 Create the 300 square section column at the point B(16000,4500), then rectangular array it (5 rows and 4 columns) using the information given.

10 With the **Dimension Style icon** create a new dimension style and alter as follows:
 a) New Style Name: STDLRG
 b) start with: A3DIM
 c) Fit tab: Use overall scale: alter to 50
 d) make STDLRG the current dimension style
 e) close the Dimension Style dialogue box.

11 Add the given dimensions and appropriate text – but what will be the text height?

12 Save this drawing as **C:\BEGIN\LARGESC** as it will be used to demonstrate the model/paper space concept in a later chapter.

13 If you have access to a plotter/printer, obtain a hard copy on A3 paper – the problem is the plot figures (50 is a useful number to know).

Small-scale drawing

This exercise requires the A3 standard sheet to be modified to suit a very small-scale drawing, which (as always) will be drawn full size.

1 Open A3SHEET standard sheet and erase the border.

2 Make Layer 0 current, enter **MVSETUP <R>** and:
   prompt   Enable paper space? and enter: **N <R>**
   prompt   Enter units type and enter: **M <R>** – metric
   prompt   Enter the scale factor and enter: **0.001 <R>**
   prompt   Enter the paper width and enter: **420 <R>**
   prompt   Enter the paper height and enter: **297 <R>**.

3 A black border is displayed on the screen – the drawing area.

4 Set UNITS to:
   a) Length precision: 0.0000
   b) Angle precision: 0.00.

5 ID the lower left and upper right corners and:
   a) lower left: X = 0.0000   Y = 0.0000   Z = 0.0000
   b) upper right: X = 0.4200   Y = 0.2970   Z = 0.0000.

6 Set the grid to 0.01 and the snap to 0.005.

7 With layer OUT current, refer to Figure 36.3 and draw the toggle switch (full size) using the sizes given.

8 Add all dimensions after altering the A3DIM dimension style with:
   a) Fit tab: overall scale: 0.001
   b) Primary units tab: Linear precision: 0.0000.

9 Add text, but you have to decide on the height.

10 Save as the drawing is complete.

Assignment

Activity 43: House Plan

A single assignment requiring a large-scale drawing to be created on A3 paper. The assignment requires a house plan to be drawn at a scale of 1:50. I would suggest:
   a) Open the LARGESC drawing saved earlier in the chapter as it has all the required 'settings'.
   b) Erase the factory layout.
   c) Complete the house plan using the metric sizes given.
   d) Add the text and dimensions, remembering that the drawing is 50 times larger than the A3 paper. This means that the text and dimensions should be 50 times larger than normal. The text height is easy to enter, and the dimension overall scale factor should have been 'set' to 50. As an addition, modify the dimension style primary tab to Architectural units and note the effect.

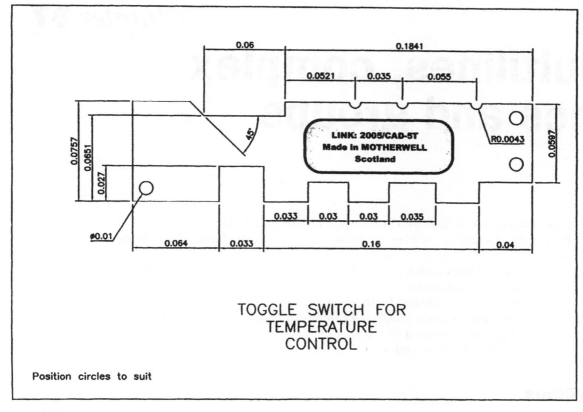

Figure 36.3 Small-scale drawing on A3 paper.

Summary

1 Drawings can be created with several different unit types.

2 Scale drawings require the MVSETUP command to be used.

3 Large- and small-scale drawings can be 'fitted' onto any size of paper.

4 Dimensions with scaled drawings are controlled with the Overall scale factor in the Fit tab of the Dimension Styles dialogue box. This is equivalent to altering the dimension variable DIMSCALE.

5 Text is scaled according to the overall scale factor.

Chapter 37

Multilines, complex lines and groups

Layers have allowed us to display continuous, centre and hidden linetypes, but AutoCAD has the facility to display multilines and complex lines, these being defined as:

Multiline
    a) parallel lines which can consist of several linear elements of differing linetype and colour
    b) they must be created by the user.

Complex
    a) lines which can be displayed containing text items and shapes
    b) they can be created by the user
    c) AutoCAD has several 'stored' complex linetypes.

Multilines

Multilines consist of between 1 and 16 parallel line elements. The ends of multilines can be capped with lines and arcs or be left uncapped. The basic multiline terminology is displayed in Figure 37.1(a). To investigate how to use multilines:

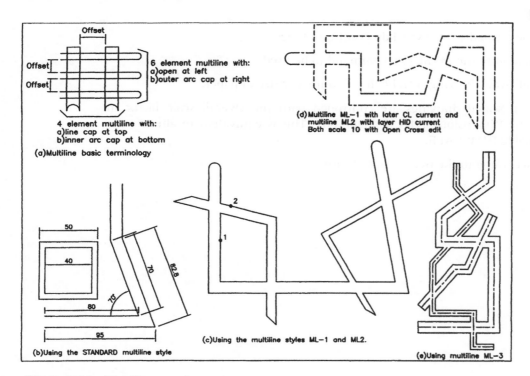

Figure 37.1 Multiline exercise.

1   Open your A3SHEET standard sheet with layer OUT current and refer to Figure 37.1.

2   From the menu bar select **Draw-Multiline** and:
    prompt   `Justification=Top,Scale=??,Style=STANDARD`
    then   `Specify start point or [Justification/Scale/STyle]`
    enter   **S <R>** – the scale option
    prompt   `Enter mline scale<??>`
    enter   **10 <R>**
    prompt   `Specify start point and enter` **20,40 <R>**
    prompt   `Specify next point and enter` **@80,0 <R>**
    prompt   `Specify next point and enter` **@70<110 <R>**
    prompt   `Specify next point and enter` **@0,50 <R>**
    prompt   `Specify next point and` **right-click-Enter**.

3   Menu bar again with **Draw-Multiline** and:
    a) set scale to 5
    b) draw a square of side 40 from the point 20,60 using the close option – Figure 37.1(b).

4   From the menu bar select **Format-Multiline Style** and:
    prompt   `Multiline Styles dialogue box`
    respond   *a)* alter Name to: ML-1
                b) enter description as: MY FIRST ATTEMPT
                c) pick Element Properties
    prompt   `Element Properties dialogue box` – Figure 37.2(a)
    respond   **note layout then pick OK**
    prompt   `Multiline Styles dialogue box`
    respond   **pick Multiline Properties**
    prompt   `Multiline Properties dialogue box`
    respond   *a)* Display Joints not active – i.e. blank
                b) Caps: Start – Outer arc
                          End – Outer arc – Figure 37.2(b)
                c) Fill not active – i.e. blank
                d) pick OK
    prompt   `Multiline Styles dialogue box`
    respond   *a)* pick Add – Figure 37.2(c)
                b) pick OK.

5   At the command line enter **MLINE <R>** and:
    prompt   `Specify start point or [Justification/Scale/STyle]`
    enter   **ST <R>** – the style option
    prompt   `Enter mline style name and enter` **ML-1 <R>**
    prompt   `Specify start point or [Justification/Scale/Style]`
    enter   **S <R>** – the scale option
    prompt   `Enter mline scale and enter` **10 <R>**
    prompt   `Specify start point and enter` **170,170 <R>**
    prompt   `Specify next point and enter` **@0,-100 <R>**
    prompt   `Specify next point and enter` **@150,0 <R>**
    prompt   `Specify next point and enter` **@120<100 <R>**
    prompt   `Specify next point and` **right-click-Enter**.

6   Menu bar with **Format-Multiline Style** and:
    prompt   `Multiline Styles dialogue box`
    respond   *a)* alter name to: ML2
                b) enter description as: MY SECOND ATTEMPT
                c) pick Multiline Elements

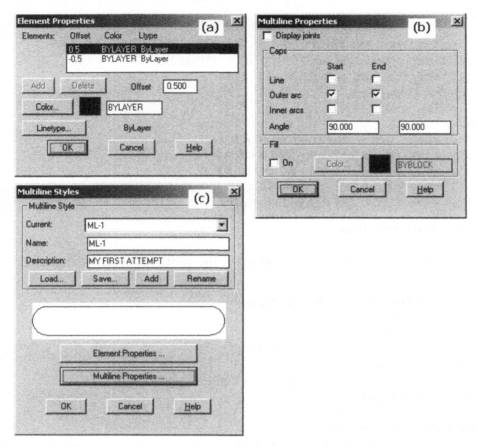

Figure 37.2 The dialogue boxes for creating the ML1 multiline: (a) Element Properties (b) Multiline Properties (c) Multiline Styles.

prompt	Element Properties dialogue box
respond	**note layout then pick OK**
prompt	Multiline Styles dialogue box
respond	**pick Multiline Properties**
prompt	Multiline Properties dialogue box
respond	a) cancel any existing end caps
	b) pick Line-Start
	c) pick Line-End
	d) alter start angle to 45 and end angle to 30
	e) pick OK
	f) Multiline Styles dialogue returned
	g) pick Add then OK.

7   Activate the MLINE command and:
    prompt   Specify start point or [Justification/Scale/STyle]
    enter   **ST <R>** – the style option
    prompt   Enter mline style name (or ?)
    enter   **? <R>** – the query option
    prompt   AutoCAD Text Window with:

Name	Description
ML-1	MY FIRST ATTEMPT
ML2	MY SECOND ATTEMPT
STANDARD	

    respond   **cancel the text window (F2)**

prompt	Enter mline style name and enter **ML2 <R>**
prompt	Specify start point or [Justification/Scale/STyle]
enter	**S <R>** then **8 <R>** – the scale option
prompt	Specify start point and enter **210,40 <R>**
prompt	Specify next point and enter **@0,80 <R>**
prompt	Specify next point and enter **@−80,20 <R>**
prompt	Specify next point and **press <RETURN>**.

8 Repeat the MLINE command and:
   a) set the scale to 5
   b) ensure ML2 is current style
   c) draw from: 350,160 to: @100<−150 to: @80<−60 to: <R>.

9 Menu bar with **Modify-Object-Multiline** and:

prompt	Multiline Edit Tools dialogue box – Figure 37.3
respond	**pick Open Cross then OK** (middle row left)
prompt	Select first mline and **pick multiline 1**
prompt	Select second mline and **pick multiline 2**
prompt	Select first mline
then	**pick as required** until all the 'crossed' multilines are opened – Figure 37.1(c).

10 *Tasks*
   a) The multilines created so far have been with layer OUT current. Make layers CL and HID current and draw multilines with the ML-1 and ML2 styles using your own scale values.
   b) Dimension the multilines in Figure 37.1(b) to determine which element has been drawn to the correct 'length'.

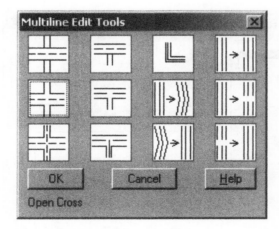

Figure 37.3 The Multiline Edit Tools dialogue box.

Creating a three-element multiline

The two multilines (ML-1 and ML2) both have been created with two continuous linetype elements. We will now investigate how to create a three-element multiline with varying linetype and colour.

1 From the menu bar select **Format-Multiline Style** and:

prompt	Multiline Styles dialogue box
respond	a) alter Name to: ML-3
	b) enter description as: 3 ELEMENT ML
	c) pick Element Properties

prompt	Element Properties dialogue box
respond	a) pick the highlighted 0.5 offset line
	b) pick Add
and	a 0.0 offset line added to the elements list
respond	**pick Color**
prompt	Select Color palette
respond	**pick blue then OK**
prompt	Element Properties dialogue box
respond	**pick Linetype**
prompt	Select Linetype dialogue box
respond	**pick CENTER2 then OK**
prompt	Element Properties dialogue box – Figure 37.4(a)
respond	**pick OK**
prompt	Multiline Styles dialogue box
respond	**pick Multiline Properties**
prompt	Multiline Properties dialogue box
respond	a) Display Joints not active – i.e. blank
	b) Caps: Start – Line
	End – Outer arc
	c) Angle: both 90
	d) Fill not active – i.e. blank
	e) pick OK
prompt	Multiline Styles dialogue box
respond	a) pick Add – Figure 37.4(b)
	b) pick OK.

2 Now use the created ML-3 multiline style to draw some line segments – Figure 37.1(e).

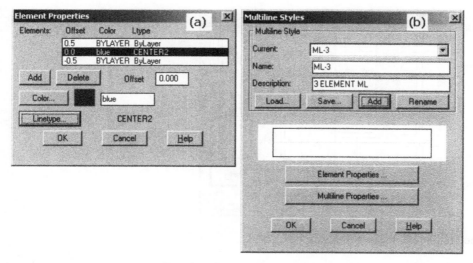

Figure 37.4 The Element Properties and Multiline Styles dialogue boxes for the ML-3 multiline.

Saving multilines

When multilines have been created, they are only available for use in the current drawing. If these multilines are required in other drawings they must be saved. We will now investigate how to save our three created multilines to our named folder, as they will be required for the assignment at the end of the chapter.

1   Menu bar with **Format-Multiline Style** and:
    prompt   Multiline Styles dialogue box
    respond a) scroll at Current and pick ML-1
               b) pick Save
    prompt   Save Multiline Style dialogue box
    respond a) Save in: scroll and pick **C:\BEGIN** or your named folder
               b) File name: alter to **MYMULTIS.MLN**
               c) pick Save
    prompt   Multiline Styles dialogue box
    respond a) scroll at Current and pick ML2
               b) save to **C:\BEGIN\MYMULTIS.MLN** file
               c) save the ML-3 multiline style to the new MLN file
    prompt   Multiline Styles dialogue box
    respond **pick OK**.

2   Open your A3SHEET drawing and menu bar with **Format-Multiline Styles** and:
    prompt   Multiline Styles dialogue box
    respond a) scroll at Current and only STANDRAD listed
               b) pick Load
    prompt   Load Multiline Styles dialogue box
    respond **pick File**
    prompt   Load Multiline Styles from File dialogue box
    respond a) Look in: scroll and select C:\BEGIN or your named folder
               b) pick MYMULTIS
               c) pick Open
    prompt   Load Multiline Styles dialogue box with ML-1 highlighted –
               Figure 37.5
    respond **pick OK**
    prompt   Multiline Styles dialogue box with ML-1 listed
    respond a) select Load, pick ML2 then OK
               b) select Load, pick ML-3 then OK
    prompt   Multiline Styles dialogue box with all three created multilines
               available
    prompt   **pick OK**.

3   Now save A3SHEET with the created multilines for future use.

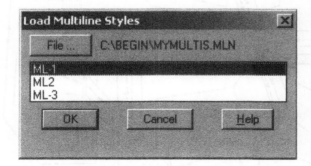

Figure 37.5 The Load Multiline Styles dialogue box after opening file C:\BEGIN\MYMULTIS.MLN.

Using complex linetypes

Complex linetypes can contain text items or shapes and can be created by the user. This is beyond the scope of this book and we will investigate using the AutoCAD stored complex linetypes to create a drawing layout. AutoCAD has seven complex linetypes stored in the .lin file so:

1. Open your A3SHEET standard drawing.

2. Menu bar with **Format-Layer** and create a new layer:
   a) Name: L1
   b) Colour: to suit
   c) Linetype: pick 'Continuous' in L1 layer line and:

prompt	Select Linetype dialogue box
respond	**pick Load**
prompt	Load or Reload Linetypes dialogue box
with	acadiso.lin file displayed
respond	**scroll and pick GAS_LINE then OK**
prompt	Select Linetype dialogue box
with	GAS_LINE displayed
respond	**pick GAS_LINE then OK**
prompt	Layer Properties Manager dialogue box
with	layer L1 displayed with GAS_LINE linetype
respond	**pick OK**.

3. Using step 2 as a guide, create another six layers and load the appropriate linetype using the following information:

Name	Colour	Linetype
L2	to suit	BATTING
L3	to suit	FENCELINE1
L4	to suit	FENCELINE2
L5	to suit	HOT_WATER_SUPPLY
L6	to suit	TRACKS
L7	to suit	ZIGZAG

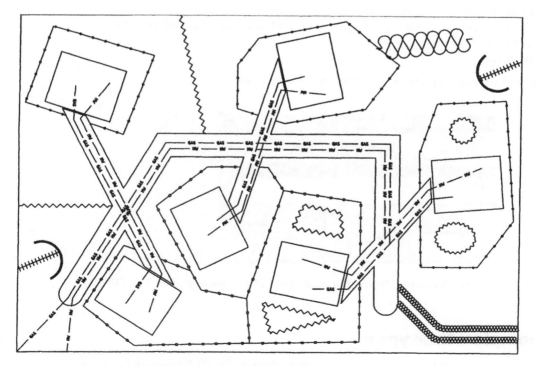

Figure 37.6 Using the AutoCAD complex linetypes.

Multilines, complex lines and groups

4   With each layer current, refer to Figure 37.6 and create a small housing estate layout using your imagination and design ability. With the CHANGE PROPERTIES command (command line CHANGE or icon), use the ltScale option to optimise the appearance of the new added linetypes.

5   The exercise is now complete and can be saved if required.

Groups

A group is a named collection of objects saved by the user within a drawing. Group definitions can be externally referenced. To demonstrate how groups are created and used:

1   Open your standard sheet, layer OUT current and refer to Figure 37.7.

2   Draw the reference shape using your discretion for any sizes not given. Ensure the lowest vertex is at the point 210,120.

3   At the command line enter **GROUP <R>** and:
    prompt   Object Grouping dialogue box
    respond   a) at Group Name enter: **GR1**
                b) at Description enter: FIRST GROUP
                c) ensure Selectable is ON – i.e. tick in box
                d) pick Create Group: New<
    prompt   Select objects for grouping – at the command line
    then   Select objects
    respond   **pick the two inclined lines and the large circle then right-click**

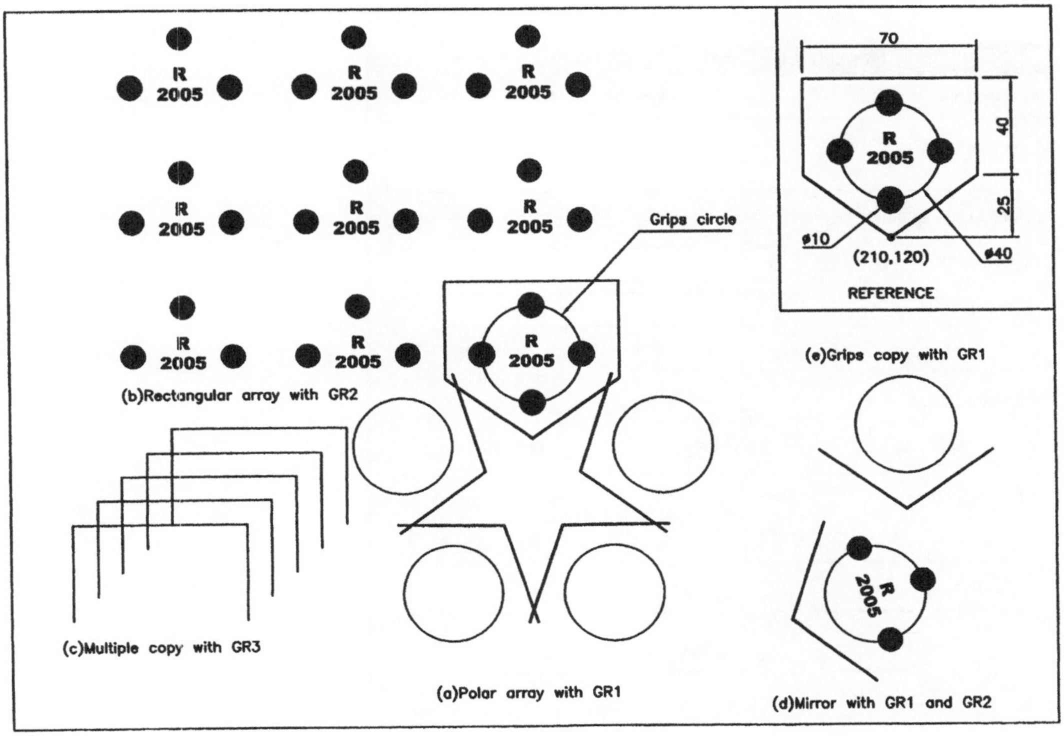

Figure 37.7 Group exercise.

prompt	Object Grouping dialogue box	
with	Group Name	Selectable
	GR1	Yes
respond	**pick OK**.	

4. At the command line enter **GROUP <R>** and:
   prompt   Object Grouping dialogue box
   respond a) at Group Name enter: **GR2**
              b) at Description enter: SECOND GROUP
              c) Selectable ON – should be
              d) pick Create Group: New<
   prompt   Select objects
   respond **pick the items of text and the top three filled circles then right-click**
   prompt   Object Grouping dialogue box with GR2 listed?
   respond **pick OK**.

5. Repeat the GROUP command and:
   a) Name: GR3
   b) Description: THIRD GROUP
   c) Create Group: New<
   d) Objects: pick the two vertical and the horizontal lines
   e) Dialogue box as Figure 37.8
   f) pick OK.

6. At the command line enter **–ARRAY <R>** and:
   prompt   Select objects
   enter   **GROUP <R>**
   prompt   Enter group name and enter: **GR1 <R>**
   prompt   3 found

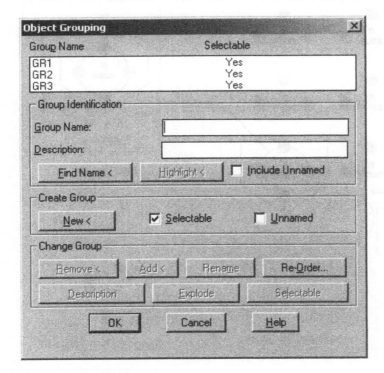

Figure 37.8 The Object Grouping dialogue box.

then	Select objects
respond	**right-click**
prompt	Enter type of array and enter **P <R>**
prompt	Specify center point of array and enter **210,100 <R>**
prompt	Enter the number of items and enter **5 <R>**
prompt	Specify the angle to fill and enter **360 <R>**
prompt	Rotate arrayed objects and enter **Y <R>**.

7   The named group (GR1) will be displayed as Figure 37.7(a).

8   Select the **ARRAY icon** and:
prompt	Array dialogue box
respond	**pick Select objects**
prompt	Select objects at command line
enter	**GROUP <R>** then **GR2 <R><R>** – why two returns?
prompt	Array dialogue box
respond	alter as follows:

   a) Rectangular array
   b) Rows: 3 and Columns: 3
   c) Offsets: Row 55 and Column 70
   d) preview then Accept – Figure 37.7(b).

9   With the COPY command:
   a) objects and enter: GROUP <R>
   b) group name and enter: GR3 <R><R>
   c) base point and enter: 175,145 <R> – why these co-ordinates?
   d) second point and enter the following co-ordinate pairs:
      @–150,–100   @–140,–90   @–130,–80   @–120,–70   @–110,–60.

10   Select the MIRROR icon and:
   a) select objects and enter: GROUP <R>
   b) group name and enter: GR1 <R>
   c) 3 found then select objects and enter: GROUP <R>
   d) group name and enter: GR2 <R>
   e) 5 found, 8 total then select objects and right-click
   f) first point of mirror line – Center icon and pick bottom right circle
   g) second point – Intersection icon and pick top right of border
   h) delete source objects and enter: N <R> – Figure 37.7(d).

11   Ensure GRIPS are on (GRIPS 1 at command line) and pick the circle indicated in Figure 37.7
prompt	grip boxes at circle and inclined lines – i.e. GR1
prompt	make the circle centre grip the base grip
prompt	STRETCH and enter **<R>**
prompt	MOVE and enter **C <R>**
prompt	Specify move point and enter **@150,–30 <R>** – Figure 37.7(e)
then	exit and cancel the grip command.

12   The exercise is complete and can be saved if required.

13   Do not quit the drawing.

14   *Task*
   a) Erase the original text item – complete group (GR2) erased?
   b) Undo this effect with the UNDO icon.
   c) Select the explode icon and select any group object and:
prompt	0 found, 1 group
	3 were not able to be exploded (or similar message).

d) At the command line enter GROUP <R> and from the dialogue box:
  1. pick GR2 line
  2. pick Explode
  3. pick OK.
e) The text item can now be erased.

Assignment

Activity 44: Rail layout

A model railway enthusiast has in his collection a old style three-rail track, the middle rail being 'live'. It is this layout that you have to create using multilines only. One problem with multilines is that there is not a facility to draw arcs segments, so how did I create the curved three-rail track? Think about using multiline ML-1 three times and changing the layer of the middle multiline. No more help.

Summary

1 Multilines can be created by the user with different linetypes and can have several elements in their definition.

2 Multilines have line or arc end caps.

3 Multilines have their own editing facility.

4 Complex linetypes can have text or shape items in their definition and can be created by the user – not considered in this book.

5 Groups are collections of objects defined by the user.

6 Once created, a group can be exploded from the Object Grouping dialogue box.

7 The GROUP command can only be entered from the keyboard.

Chapter 38

Blocks

A block is part of a drawing which is 'stored away' for future recall **within the drawing in which it was created**. The block may be a nut, a diode, a tree, a house or any part of a drawing. Blocks are used when repetitive copying of objects is required, but they have another important feature – text can be attached to them. This text addition to blocks is called **attributes** and will be considered in a later chapter.

Remember the statement '**within the drawing in which it was created**'. We will refer to this in a later chapter.

Getting started

1 Open the A3SHEET standard sheet and refer to Figure 38.1.
2 Draw the stamp shape using the reference sizes given, or your own design, but with:
   a) the outline on layer OUT
   b) the circular shape on layer OUT but green (CHANGE command?)
   c) a text item on layer TEXT

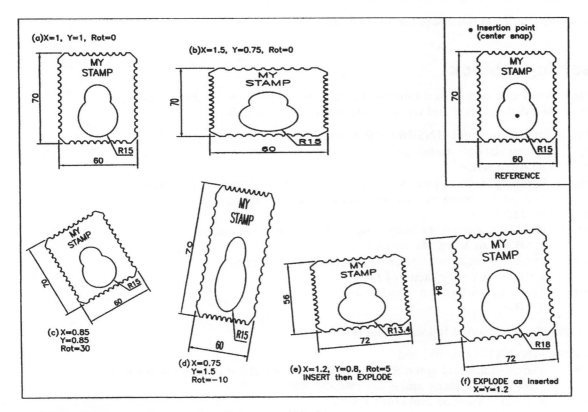

Figure 38.1 Creating and inserting a saved block.

d) three dimensions on layer DIMS
e) remember to use your discretion for the design and any size not given.

3   Make layer OUT current.

4   At this stage save as **C:\BEGIN\STAMP** for future recall.

Creating a block

Blocks can be created from both the command line and via a dialogue box. This first exercise will consider the keyboard entry method.

1   At the command line enter **–BLOCK <R>** and:
    prompt   Enter block name or [?]
    enter   **STAMP <R>**
    prompt   Specify insertion base point
    respond **Center icon and pick circle as indicated**
    prompt   Select objects
    respond **window the stamp and dimensions**
    prompt   '??' found then Select objects
    respond **right-click**.

2   The stamp shape may disappear from the screen. If it does (or does not), do not panic!

3   *Notes*
    a) the stamp shape has been 'stored' as a block within the current drawing
    b) this drawing (with the stamp block) has not yet been saved
    c) the –BLOCK keyboard entry will 'bypass' the dialogue box.

4   If the stamp shape did not disappear, erase it now.

Inserting a block

Created blocks can be inserted into the current drawing either by direct keyboard entry or via a dialogue box and we will investigate both methods.

1   At the command line enter **–INSERT <R>** and:
    prompt   Enter block name or [?]
    enter   **STAMP <R>**
    prompt   Specify insertion point or [Scale/X/Y/Z/Rotate/PScale...
    and   'ghost' image of block attached to cursor
    enter   **60,220 <R>**
    prompt   Enter X scale factor, specify opposite corner...
    enter   **1 <R>** – an X scale factor of 1
    prompt   Enter Y scale factor <use X scale factor>
    enter   **1 <R>** – a Y scale factor of 1
    prompt   Specify rotation angle<0>
    enter   **0 <R>**.

2   The stamp block is inserted full size as Figure 38.1(a).

3   Repeat the **–INSERT** command and:
    prompt   Block name and enter: **STAMP <R>** (STAMP should be default name?)
    prompt   Insertion point and enter **190,220 <R>**
    prompt   X scale factor and enter **1.5 <R>**
    prompt   Y scale factor and enter **0.75 <R>**
    prompt   Rotation angle and enter **0 <R>**
    and   Stamp block inserted as Figure 38.1(b).

4   From the Draw toolbar select the **INSERT BLOCK icon** and:
    prompt   Insert dialogue box
    with   STAMP as the Name
                (This is because we entered STAMP at the command line INSERT command)
    respond   a) ensure that the three Specify On-screen prompts are active – i.e. tick
                   in box for Insertion point, Scale and Rotation
                b) ensure that Explode is NOT active – i.e. tick
                c) pick OK
    prompt   Specify insertion point and enter **60,100 <R>**
    prompt   Enter X scale factor and enter **0.85 <R>**
    prompt   Enter Y scale factor and enter **0.85 <R>**
    prompt   Specify rotation angle and enter **30 <R>**
    and   Stamp block inserted as Figure 38.1(c).

5   Menu bar with **Insert-Block** and:
    prompt   Insert dialogue box
    with   STAMP as block name
    respond   a) deactivate the three On-screen prompts (no tick)
                b) alter Insertion Point to X: 160, Y: 90, Z: 0
                c) alter Scale to X: 0.75, Y: 1.5, Z: 1
                d) alter Rotation to Angle: −10
                e) ensure Explode not active (no tick)
                f) dialogue box as Figure 38.2
                g) pick OK – Figure 38.1(d).

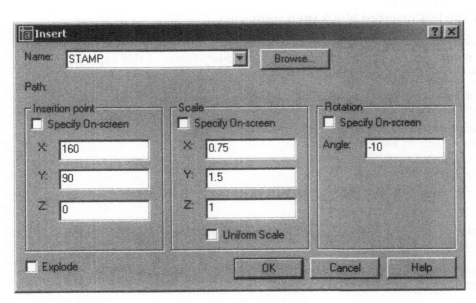

Figure 38.2   The Insert dialogue box for the STAMP block.

6   Notes
    a) An inserted block is a single object. Select the erase icon and pick any point on one
       of the inserted blocks then right-click. The complete block is erased. Undo the erase
       effect.
    b) Blocks are inserted into a drawing with layers 'as used'. Freeze the DIMS layer and
       turn off the TEXT layer. The inserted blocks will be displayed without dimensions
       or text. Now thaw the DIMS layer and turn on the TEXT layer.
    c) Blocks can be inserted at varying X and Y scale factors and at any angle of
       rotation. The default scale is X = Y = 1, i.e. the block is inserted full size. The
       default rotation angle is 0.

d) Dimensions which are attached to inserted blocks are not altered if the scale factors are changed.
e) A named block can be redefined and will be discussed later.

Exploding a block

The fact that a block is a single object may not always be suitable to the user, i.e. you may want to copy parts of a block. AutoCAD uses the **EXPLODE** command to 'convert' an inserted block back to its individual objects.

The explode option can be used:
a) after a block has been inserted
b) during the insertion process.

1   At the command line enter **–INSERT <R>** and:
    prompt   Block name and enter **STAMP**
    prompt   Insertion point and enter **260,70**
    prompt   X scale factor and enter **1.2**
    prompt   Y scale factor and enter **0.8**
    prompt   Rotation angle and enter **5**.

2   Note the dimensions of this inserted block.

3   Select the **EXPLODE icon** from the Modify toolbar and:
    prompt   Select objects
    respond **pick the last inserted block then right-click**.

4   The exploded block is restored to its individual objects and the dimensions are scaled to the factors entered in step 1 – Figure 38.1(e). The individual objects of this exploded block can now be modified if required.

5   Select the **INSERT BLOCK icon** and from the Insert dialogue box:
    a) Name: STAMP
    b) Deactivate the three Specify On-screen prompts – i.e. no tick
    c) Insertion Point X: 325; Y: 35; Z: 0
    d) Scale X: 1.2; Y: 0.8; Z: 1
    e) Rotation angle: 5
    f) Explode: ON – i.e. tick and *note the scale factor values*
    g) pick OK.

6   The block is exploded as it is inserted at a scale of X = Y = 1.2 and the dimensions display this scale effect as Figure 38.1(f).

7   Compare the dimensions of Figure 38.1(e) with those of Figure 38.1(f) and note the dimension 'orientation' of the two exploded blocks.

8   *Notes*
    a) a block exploded after insertion will retain the original X and Y inserted scale factors
    b) a block exploded as it is inserted has X = Y scale factors.

Block exercise

1   Open the A3SHEET standard sheet with layer OUT current.

2   Refer to Figure 38.3 and draw the four lines using the sizes given. Do not add the dimensions.

3   Menu bar with **Draw-Block-Make** and:
    prompt   Block Definition dialogue box
    respond   *a)* Name: enter SEAT
                b) Base point: select Pick point
    prompt   Specify insertion base point at command line
    respond   **midpoint icon and pick line indicated in Figure 38.3**

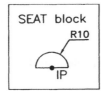

Figure 38.3 Seat block details.

   prompt   Block Definition dialogue box
   with   Base point: X and Y co-ordinates of selected point
   respond   Objects: pick Select objects
   prompt   Select objects at command line
   respond   **window the objects then right-click**
   prompt   Block Definition dialogue box
   with   preview icon displayed
   respond   *a)* Objects: ensure Delete is active (black dot)
               b) Create icon from block geometry: active
               c) Drag-and-drop units: scroll and pick Millimeters
               d) at Description enter: basic SEAT block
               e) pick OK.

4   The seat shape should disappear.

5   *Notes*
    The Objects part of the Block dialogue box allows the user one of three options, these being:
    a) Retain   will keep the selected objects as individual objects on the screen after the block has been made
    b) Convert to block   will display the selected objects on the screen as a block after the block has been made
    c) Delete   will remove the selected objects from the screen after the block has been made – our selection.
    It is user preference as to which of these options is used. I generally select delete, although retain is also useful.

6   Draw the following three objects (the centre points being important):
    a) pentagon, centred on 80,150, inscribed in a 30-radius circle
    b) hexagon, centred on 210,150 and inscribed in a 30-radius circle
    c) circle, centre 340,150 and radius 30.

7   Select the INSERT BLOCK icon from the Draw toolbar and use the Insert dialogue box to insert the block SEAT three times using the following information:

Insertion point	Scale	Rotation	Explode
80,120	X = Y = 1	180	not active
210,120	X = Y = 1	180	not active
340,185	X = Y = 1	0	not active

8 Using the ARRAY dialogue box, array the three inserted seat blocks with the following information:

	First array	*Second array*	*Third array*
objects	pick left SEAT block	pick middle SEAT block	pick right SEAT block
type	Polar	Polar	Polar
centre	80,150	210,150	340,150
method	items and angle	items and angle	items and angle
items	5	6	7
angle	360	360	360
rotate	Y	Y	Y

9 Select the **MAKE BLOCK icon** from the Draw toolbar and:
prompt   Block Definition dialogue box
respond a) enter Name: TABLE1
           b) enter Base point as X: 80; Y: 150; Z: 0
           c) at Objects: pick Select objects
prompt   Select objects at command line
respond **window the 5 seat arrangement then right-click**
prompt   Block Definition dialogue box with icon preview display
respond a) Objects: ensure that Delete is active
           b) Create icon from block geometry: active
           c) enter Description: 5 SEATER TABLE
           d) Drag-and-drop units: Millimeters
           e) dialogue box as Figure 38.4

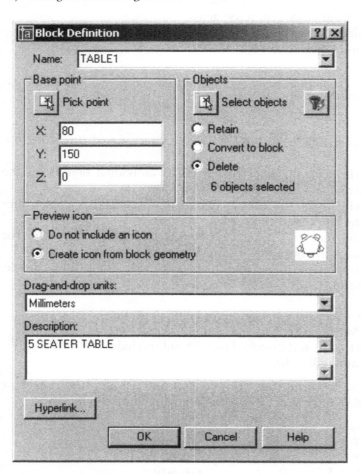

Figure 38.4 The Block Definition dialogue box for the 5 SEATER TABLE.

	f) pick OK
and	the 5 seater table and chairs will disappear.

10 Repeat the Make Block icon selection and create blocks of the 6 and 7 seat table layouts with the following data:

	Second block	*Third block*
name	TABLE2	TABLE3
base point	210,150	340,150
select objects	window 6 seats and table	window 7 seats and table
objects	delete active	delete active
units	millimetres	millimeters
description	6 SEATER TABLE	7 SEATER TABLE

11 Now erase any objects from the screen.

12 With layer OUT current, menu bar with **Insert-Block** and using the Insert dialogue box, insert each table arrangement using the following data:

Name	Insertion point	Scale	Rotation	Explode
TABLE1	60,250	0.75	0	not active
TABLE2	60,150	1	0	not active
TABLE3	60,60	0.75	0	not active

13 Menu bar with **Modify-Array** and use the Array dialogue box with the inserted tables with:

Objects	Type	Rows, Columns	Rows, Column Offsets	Angle of Array
TABLE1	Rect	1, 4	0, 90	−2
TABLE2	Rect	1, 3	0, 140	0
TABLE3	Rect	1, 5	0, 80	3

14 The final layout should be as Figure 38.5 and can be saved with a suitable name.

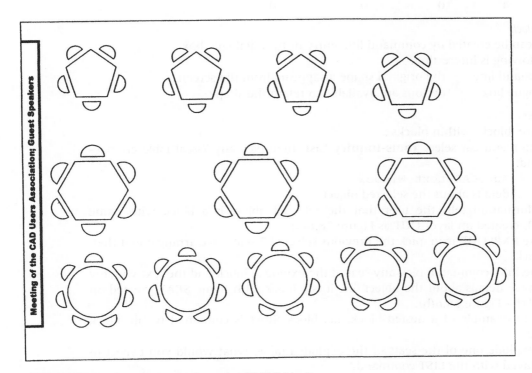

Figure 38.5   Block exercise: Inserting TABLE blocks.

Notes on blocks

Several points are worth discussing about blocks:

1. *The insertion point*
   a) When a block is being inserted using the Insert dialogue box, the user has the option to decide whether the parameters are entered via the dialogue box or at the command line.
   b) The Specify On-screen selection allows this option.

2. *Exploding a block*
   a) I would recommend that blocks are inserted before they are exploded.
   b) This maintains the original X and Y scale factors.
   c) Remember that blocks do not need to be exploded.

3. *Command line*
   a) Entering –BLOCK and –INSERT will 'bypass' the dialogue boxes.
   b) Parameters can then be entered from the keyboard.

4. *The ? option*
   a) Both the command line –BLOCK and –INSERT commands have a query (?) option which will list all the blocks in the current drawing.
   b) At the command line enter **–BLOCK <R>** and:

prompt	Enter block name (or ?) and enter ? <R>
prompt	Enter block(s) to list<*> and enter * <R>
prompt	AutoCAD Text Window
with	Defined blocks
	"SEAT"
	"TABLE1"
	"TABLE2"
	"TABLE3"

User Blocks	External Blocks	Dependent Blocks	Unnamed Blocks
4	0	0	0

5. *Making a block*
   a) Blocks can be created by command line entry or by a dialogue box.
   b) The following is interesting:
      1. command line — the original shape disappears from the screen
      2. dialogue box — options are available to retain the shape.

6. *Nested blocks*
   a) These are 'blocks within blocks'.
   b) With the menu bar select **Tools-Inquiry-List**, then pick any 7-seat table arrangement and:

prompt	AutoCAD text Window
with	details about the selected object.

   c) This information tells the user that the selected object is a block with name TABLE3, created on layer OUT as Figure 38.6(a).
   d) With the EXPLODE icon, pick the previous selected 7 seat table arrangement then right-click.
   e) With the LIST command, pick any seat of the exploded block and the text window will display details about the object, i.e. it is a block with name SEAT, created on layer OUT – Figure 38.6(b).
   f) This is an example of a nested block, i.e. block SEAT is contained within block TABLE2.
   g) If you explode one of the seats of the exploded table, what would you expect to be displayed with the LIST command?

```
        (a)     BLOCK REFERENCE   Layer: "OUT"
                        Space: Model space
                Handle = 422
                "TABLE3"
             at point, X=   139.89   Y=      64.19   Z=      0.00
                X scale factor       0.75
                Y scale factor       0.75
         rotation angle      0.0
                Z scale factor       0.75

        (b)     BLOCK REFERENCE   Layer: "OUT"
                        Space: Model space
                Handle = 434
                "SEAT"
             at point, X=   139.89   Y=      90.44   Z=      0.00
                X scale factor       0.75
                Y scale factor       0.75
         rotation angle      0.0
                Z scale factor       0.75
```

Figure 38.6 The AutoCAD text window for the NESTED blocks: (a) the inserted TABLE block (b) the SEAT block.

Using blocks

1. Open your A3SHEET standard sheet with layer OUT current.

2. Refer to Figure 38.7 and draw a 20-sided square with diagonals and then make a block with:
   a) name: BL1
   b) insertion point: diagonal intersection
   c) objects: window the shape – no dimensions.

3. Draw an inclined line, a circle, an arc and a polyline shape of line and arc segments – discretion for sizes, but use Figure 38.7 as a guide for the layout.

4. *The DIVIDE command*
   Menu bar with **Draw-Point-Divide** and:
   prompt   Select object to divide
   respond **pick the line**
   prompt   Enter the number of segments or [Block]
   enter   **B <R>** – the block option
   prompt   Enter name of block to insert
   enter   **BL1 <R>**
   prompt   Align block with object? [Yes/No] <Y>
   enter   **N <R>**
   prompt   Enter the number of segments
   enter   **4 <R>** – Figure 38.7(a).

5. *The MEASURE command*
   Menu bar with **Draw-Point-Measure** and:
   prompt   Select object to measure
   respond **pick the circle**
   prompt   Specify length of segment or [Block] and enter **B <R>**
   prompt   Enter name of block to insert and enter **BL1 <R>**
   prompt   Align block with object and enter **Y <R>**
   prompt   Specify length of segment and enter **27 <R>** – Figure 38.7(b).

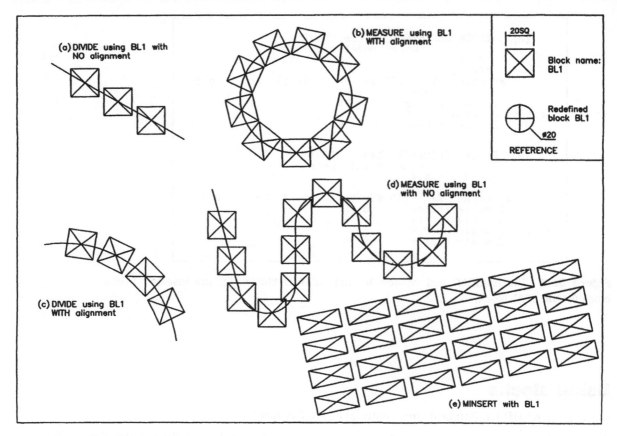

Figure 38.7 Using blocks exercise.

6 Use the *DIVIDE* and *MEASURE* commands and:
   a) divide the arc with block BL1, with alignment and with five segments – Figure 38.7(c)
   b) measure the polyline with block BL1, no alignment and with a segment length of 27 – Figure 38.7(d).

7 *The MINSERT or multiple insert command*
   At the command line enter **MINSERT <R>** and:
   prompt   Enter block name and enter **BL1 <R>**
   prompt   Specify insertion point and **pick a suitable point on screen**
   prompt   Enter X scale factor and enter **1.5 <R>**
   prompt   Enter Y scale factor and enter **0.65 <R>**
   prompt   Specify rotation angle and enter **12 <R>**
   prompt   Enter number of rows and enter **4 <R>**
   prompt   Enter number of columns and enter **6 <R>**
   prompt   Enter distance between rows and enter **20 <R>**
   prompt   Specify distance between columns and enter **35 <R>**.

8 The block BL1 is arrayed in a 4 × 6 rectangular matrix – Figure 38.7(e).

9 Using the *EXPLODE* icon, pick any of the minsert blocks and:
   prompt   1 was minserted
   i.e. **you cannot explode minserted blocks**.

10 *Redefining a block*
   a) Draw a circle of radius 10 and add the vertical and horizontal diagonals.
   b) Change the colour of the circle and lines to blue.

c) At the command line enter **–BLOCK <R>** and:
   prompt   Enter block name and enter: **BL1 <R>**
   prompt   Block "BL1" already exists. Redefine it? [Yes/No]<N>
   enter   Y <R>
   prompt   Specify insertion base point
   respond   **pick the centre of the circle**
   prompt   Select objects
   respond   **window the circle and lines then right-click**.
d) Interesting result? – all the red squares should be replaced by blue circles with the same alignment, scales, etc.

11 This exercise is now complete – save?

Layer 0 and blocks

Blocks have been created with the objects drawn on their 'correct' layers. Layer 0 is the AutoCAD default layer and can be used for block creation with interesting results. We will use the command line entry to demonstrate this example.

1 Open A3SHEET and draw:
   a) a 50-unit square on layer 0 – black
   b) a 20-radius circle inside the square on layer OUT – red
   c) two centre lines on layer CL – green.

2 With the command line entry –BLOCK, make a block of the complete shape with block name TRY, using the circle centre as the insertion point.

3 Make layer OUT current and with the command line –INSERT, insert block TRY at 55,210, full size with no rotation. The square is inserted with red continuous lines.

4 Make layer CL current and –INSERT block TRY at 135,210, full size and with no rotation. The square has green centre lines.

5 With layer HID current, –INSERT the block at 215,210, full size with no rotation and the square will be displayed with coloured hidden lines.

6 Make layer 0 current and –INSERT at 295,210 – square is black.

7 With layer 0 still current, freeze layer OUT and:
   a) no red circles
   b) no red square from first insertion – when layer OUT was current.

8 Thaw layer OUT and, still with layer 0 current, insert block TRY using the Insert dialogue box with:
   a) explode option active – i.e. tick in box
   b) insertion point: 175,115
   c) full size with 0 rotation
   d) square is black – it is on layer 0.

9 Explode the first three inserted blocks, and the square should be black, i.e. it has been 'transferred' to layer 0 with the explode command.

10 Finally make layer OUT current and freeze layer 0 and:
   a) no black squares
   b) no objects from fourth insertion – why?
   c) the screen displays four red circles with green centre lines.

11 This completes this block exercise – no save necessary.

Assignments

Two activities requiring blocks to be created have been included for you to attempt.

Activity 45: In-Line Cam and Roller Follower

1. Draw the CAM and FOL shapes using the reference sizes given (use your discretion for any omitted size).
2. Make a block of each shape with the block names CAM and FOL. Use the given insertion points.
3. Insert the CAM block using the information given on the activity drawing.
4. Insert the FOL block:
   a) insertion point: 25,90
   b) X and Y scale: 0.5 with 0 rotation.
5. Rectangular array the inserted FOL block:
   a) for 1 row and 10 columns
   b) column offset: 40.
6. The followers have to be moved vertically downwards until they 'touch' the cams. This sounds easier than it may seem.
7. Complete the tasks detailed in the activity drawing.

Activity 46: Modifying an existing drawing layout

1. Open the LARGSC drawing created in Chapter 36 and:
   a) erase the square columns and any dimensions
   b) modify the lower right corner
   c) position the circular outline at the midpoint of the right vertical wall.
2. Draw the BEAM shape using the sizes given at any suitable part of the screen then make a block of the shape with name BEAM, using the insertion point indicated.
3. Insert block BEAM twice using the following:

	First	*Second*
insertion point	4000,3750	16750,5000
scale	full size	full size
rotation	45	0

4. Produce the rectangular and polar arrays using the inserted blocks as displayed in the drawing and erase any wrongly positioned beams.
5. Save the modified layout as C:\BEGIN\LARGESCMOD.

Summary

1. A block is a single object 'stored' for recall in the drawing in which it was created. We will return to this statement in a later chapter.
2. Blocks are used for the insertion of frequently used 'shapes'.
3. An inserted block is a single object.
4. Blocks can be inserted full size (X = Y) or with differing X and Y scale factors and varying rotation angles.
5. The explode command 'restores' an inserted block to its original objects.

6 The explode command can be used after insertion or during insertion.
7 Blocks are inserted with 'layers intact'.
8 Blocks can be used with the divide and measure commands.
9 Multiple block inserts are permissible with the MINSERT command. The insertion gives a rectangular array pattern.
10 Existing blocks can be redefined. The current drawing will be 'updated' to display the new block definition.
11 Unused blocks can be purged from a drawing.

Chapter 39

Wblocks

Blocks are useful when shapes/objects are required for repetitive insertion in a drawing, but they are drawing specific, i.e. they can only be used within the drawing in which they were created (for now). There are however, blocks which can be created and accessed by all AutoCAD users, i.e. they are global. These are called **WBLOCKS** (**write blocks**) and they are created in a similar manner to 'ordinary' blocks. Wblocks are stored and recalled from a 'named folder/directory' which we will assume to be C:\BEGIN.

Creating Wblocks

1 Open your A3SHEET standard sheet with layer OUT current.

2 Refer to Figure 39.1 and draw a rectangular polyline shape using the information in Figure 39.1(a):
   a) the start point is to be 5,5
   b) the polyline has to have a constant width of 5
   c) close the shape with the close option
   d) the rectangle sizes will become apparent during insertion.

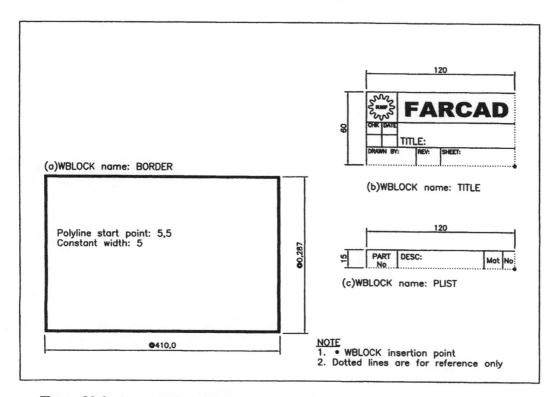

Figure 39.1 Layout and sizes for creating the WBLOCKS.

3   At the command line enter **WBLOCK <R>** and:
    prompt   Write Block dialogue box (similar to Block dialogue box)
    respond a) Source: ensure Objects active (dot)
               b) Base point: alter to X: 2.5; Y: 2.5; Z: 0
               c) Objects: pick Select objects
    prompt   Select objects at the command line
    respond **pick any point on the polyline then right-click**
    prompt   Write Block dialogue box
    respond a) Objects: Delete from drawing active
               b) Destination
                  1. File name and path: **C:\BEGIN\BORDER**
                  2. Insert units: millimeters – Figure 39.2
               c) pick OK
    and   as the file is created, the WBLOCK preview is displayed at the top left of drawing screen.

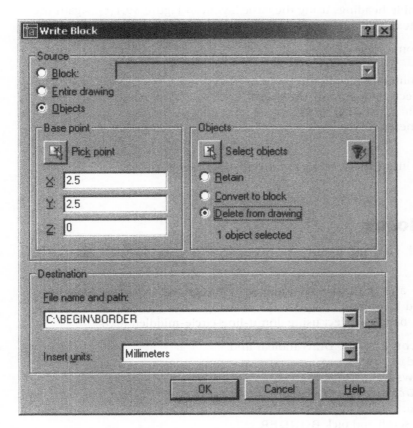

Figure 39.2 The Write Block dialogue box for BORDER.

4   *Note*: As the delete from the drawing option was active, the border will disappear from the current drawing. Entering OOPS at the command line will restore it.

5   Construct the title box from the information in Figure 39.1(b), the actual detail being to your own design, e.g. text style, company name and logo, etc. The only requirement is to use the given sizes – the donut and dotted lines are for 'guidance' only and should not be drawn. Use layers correctly for lines, text, etc.

6 Command line with **WBLOCK <R>** and:
  prompt   Write Block dialogue box
  respond a) Source: Objects active
             b) Base point: select Pick point
  prompt   Specify insertion base point at command line
  respond **pick point indicated by donut**
  prompt   Write Block dialogue box
  respond Objects: pick Select objects
  prompt   Select objects at command line
  respond **window the title box then right-click**
  prompt   Write Block dialogue box
  respond a) Objects: Delete from drawing active
             b) File name and path: C:\BEGIN\TITLE
             c) Insert units: millimeters
             d) pick OK
  and   WBLOCK preview of saved file.

7 Create the parts list table headings using the basic layout in Figure 39.1(c). Again use your own design for text, etc.

8 With the WBLOCK command:
  a) Source: Objects
  b) Base point: Pick point and pick donut point indicated
  c) Objects: Select objects and window the shape
  d) File name and path: C:\BEGIN\PLIST
  e) Insert units: millimeters
  f) pick OK.

9 We are now ready to insert these Wblocks into existing/new drawings.

Inserting Wblocks

1 Close A3SHEET and pick no to any save changes prompt (why?) then open A3SHEET again.

2 Refer to Figure 39.3 and draw the sectional pulley assembly. No sizes have been given, but you should be able to complete the drawing – use the snap to help. Add the hatching and the 'balloon' effect using donut-line-circle-middled text.

3 Select the **INSERT icon** from the Draw toolbar and:
  prompt   Insert dialogue box
  respond **pick Browse**
  prompt   Select Drawing File dialogue box
  respond a) Look in: scroll and pick **C:\BEGIN**
             b) Name: scroll and pick **BORDER**
             c) pick Open
  prompt   Insert dialogue box with BORDER listed
  respond a) ensure Insertion point, Scale and Rotation are Specify On-screen – i.e. tick in three boxes
             b) Explode: active or not – your decision
             c) pick OK
  prompt   Specify insertion point and enter **2.5,2.5 <R>**
  prompt   Enter X scale and enter **1 <R>**
  prompt   Enter Y scale and enter **1 <R>**
  prompt   Specify rotation angle and enter **0 <R>**.

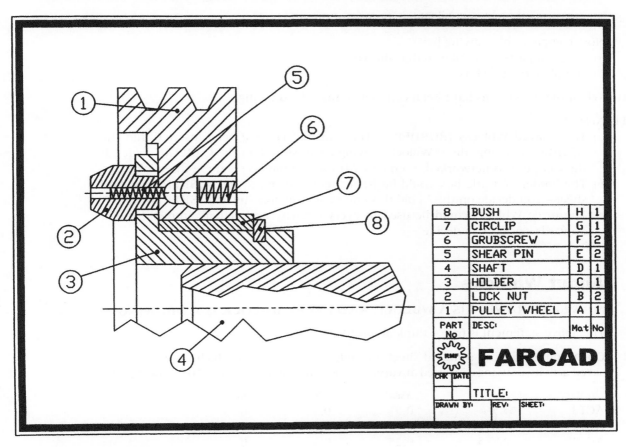

Figure 39.3 Inserting Wblocks exercise.

4 The polyline border should be inserted within the 'drawing frame' of the A3SHEET standard sheet.

5 Menu bar with **Insert-Block** and:
   prompt   Insert dialogue box
   respond   **pick Browse**
   prompt   Select Drawing File dialogue box
   respond   a) Look in: scroll and pick C:\BEGIN
               b) Name: scroll and pick **TITLE**
               c) pick Open
   prompt   Insert dialogue box
   respond   a) cancel the three on-screen prompts – no tick
               b) alter insertion point to X: 412.5; Y: 7.5; Z: 0
               c) uniform scale of X = 1
               d) rotation angle to be 0
               e) pick OK.

6 The Wblock TITLE will be inserted into the lower right corner of the BORDER Wblock.

7 At the command line enter **–INSERT <R>**
   prompt   Enter block name and enter **PLIST <R>**
   prompt   Specify insertion point and enter **412.5,67.5 <R>**
   prompt   Enter X scale and enter **1 <R>**
   prompt   Enter Y scale and enter **1 <R>**
   prompt   Specify rotation angle and enter **0 <R>**.

8   The parts list Wblock will be added 'on top' of the title box.

9   Now complete the drawing by:
    a) adding a parts list similar to that shown
    b) complete the title box.

10 When these additions have been completed, save the drawing.

11 *Notes*
    a) Three saved Wblocks (BORDER, TITLE and PLIST) have been inserted into the A3SHEET drawing, these Wblocks having been 'stored' in the C:\BEGIN folder. If the computer is networked, then all CAD users could access these three Wblocks.
    b) The border and title box could be permanently added to the A3SHEET standard sheet – you decide on this? I did this and saved the drawing as A3SHEET-1.
    c) The three Wblocks could be used in every exercise and activity from this point onwards – again you decide.

About Wblocks

EVERY DRAWING IS A WBLOCK AND EVERY WBLOCK IS A DRAWING

The above statement is true – think about it!

1   Open your A3SHEET standard sheet and refer to Figure 39.4 which displays several saved activity drawings, inserted at varying scales and rotation angles. The drawings are:

Drawing	IP	Xscale	Yscale	Rot
ACT8	5,5	0.5	0.5	0
ACT23	25,155	0.4	0.3	5
ACT33	220,5	0.25	0.5	0
ACT37	330,5	0.15	0.75	0
ACT39	220,170	0.2	0.3	−5
ACT44	10,260	0.5	0.1	0
ACT46	320,230	0.004	0.004	0

2   Note that certain layers have frozen layers TEXT and DIMS for clarity.

3   Insert some other previously saved drawings.

4   *WBLOCK options*
    When a Wblock is being created, the user has three options for selecting the source. These are:
    a) Block   allows a previously created block to be 'converted' to a Wblock
    b) Entire drawing   the user can enter a new file name for the existing drawing
    c) Objects   allows parts of a drawing to be saved as a Wblock, i.e. as a drawing file. This is probably the most common source selection method.

Exploding Wblocks

1   Open A3SHEET and draw anywhere on the screen a circle of radius 25 and a square of side 50 – do not use the rectangle command.

2   Make WBLOCKS of the circle and square using:
    a) circle: Base point at the circle centre and file name C:\BEGIN\CIR
    b) square: Base point at any corner of the square and file name C:\BEGIN\SQ.

3   Now close the drawing (no save) and re-open A3SHEET.

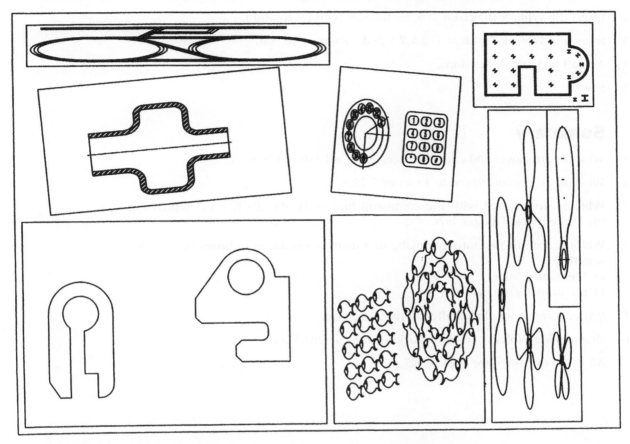

Figure 39.4 Inserting 'WBLOCK' activity drawings into the A3SHEET drawing.

4 Menu bar with **Insert-Block** and:
  a) browse and select CIR from C:\BEGIN
  b) insert at a suitable point, full size with 0 rotation
  c) Explode not active.

5 Menu bar with **Insert-Block** and:
  a) browse and pick SQ from your named folder
  b) Explode active from the dialogue box
  c) insert at any suitable point with zero rotation.

6 With **LIST <R>** at the command line:
  a) pick the circle and BLOCK REFERENCE listed for the object selected
  b) pick any line of square and LINE listed for the selected object.

7 This exercise has demonstrated that unexploded Wblocks become blocks within the current drawing.

8 This exercise is complete. Do not save.

Assignment

A single activity for you to attempt.

Activity 47: Coupling arrangement

1 Complete the drawing using your discretion for sizes not given. I have deliberately not given all sizes in this drawing.

2   Insert the Wblock BORDER at 5,5 – full size with no rotation.

3   Insert the Wblock TITLE at 412.5,7.5 and customise to suit.

4   Add all text and dimensions.

5   Save?

Summary

1   Wblocks are global and could be accessed by all AutoCAD users.

2   Wblocks are usually saved to a named folder.

3   Wblocks are created with the command line entry WBLOCK <R> which displays the Write Block dialogue box.

4   Wblocks are inserted into a drawing in a manner similar to ordinary blocks, the user selecting:
    a) the named folder
    b) the drawing file name.

5   Wblocks can be exploded after/during insertion.

6   Unexploded Wblocks become blocks in the current drawing.

7   All saved drawings are WBLOCKS.

Chapter 40

Attributes

An attribute is an item of text attached to a block or a Wblock and allows the user to add repetitive type text to frequently used blocks when they are inserted into a drawing. The text could be:
a) weld symbols containing appropriate information
b) electrical components with values
c) parts lists containing codes, number off, material, etc.

Attributes used as text items are useful, but their main advantage is that attribute data can be extracted from a drawing and stored in an attribute extraction file. This data could then be used as input to other computer packages, e.g. databases, spreadsheets, etc. for creating a Bill-of-Material or an Inventory.

This chapter is only a 'taster' as the topic will not be investigated fully. The editing and extraction features of attributes are beyond the scope of this book. The purpose of this chapter is to introduce the user to:
a) attaching attributes to a block
b) inserting an attribute block into a drawing.

Getting started

The attribute example for demonstration is a fisherman's trophy and will use a previously created and saved drawing. The fish 'symbol' will represent the block for adding the attributes. The added attributes will give information about the type of fish, the year it was caught and the river.

1   Open the **C:\BEGIN\FISH** drawing created during Chapter 32 and refer to Figure 40.1.

2   Erase any centre lines, text and dimensions.

3   a) Move the complete shape from 'the nose' to the point 50,50
    b) Scale the shape about the point (50,50) by 0.5 – Figure 40.1(a)
    c) *Note*: The 50,50 point is essential for positioning the text items during the attribute exercise.

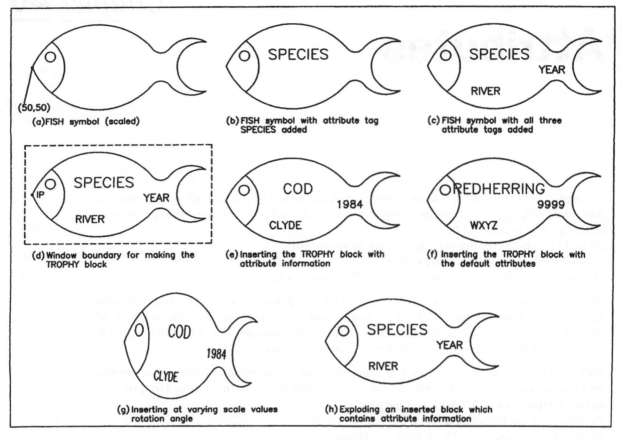

Figure 40.1 Making and using attributes with block TROPHY.

Defining the attributes

Before a block containing attributes can be inserted into a drawing, the attributes must be defined, so:

1 Make layer TEXT current.

2 Menu bar with **Draw-Block-Define Attributes** and:
   prompt   Attribute Definition dialogue box
   with   options for Mode, Attribute, Insertion Point and Text
   respond   with the following:
        a) *At Mode*: leave all four options un-selected – no tick
        b) *At Attribute enter*:
           1. Tag: SPECIES
           2. Prompt: What displayed?
           3. Value: REDHERRING
        c) *At Insertion point enter*:
           1. X: 100
           2. Y: 55
           3. Z: 0
        d) *At Text options alter*:
           1. Justification: scroll and pick Center
           2. Text Style: ST1
           3. Height: 7
           4. Rotation: 0 – dialogue box as Figure 40.2
        e) pick OK.

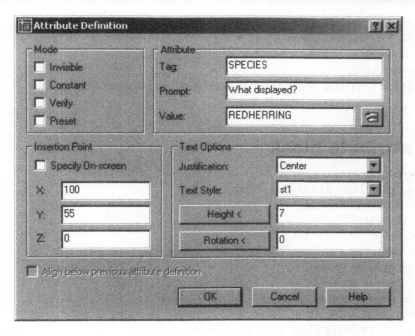

Figure 40.2 The Attribute Definition dialogue box for the SPECIES tag.

3   The attribute tag SPECIES will be displayed in the fish symbol as Figure 40.1(b).

4   Activate the Attribute Definition command two more times and enter the following attribute information in the Attribute Definition dialogue box using the same procedure as step 2.

	First entry	*Second entry*
Attribute modes	blank	blank
Attribute tag	RIVER	YEAR
Attribute prompt	Where caught?	What year?
Default value	WXYZ	9999
Insertion Pt X	80	145
Insertion Pt Y	30	45
Insertion Pt Z	0	0
Justification	Left	Right
Text style	ST1	ST1
Height	5	5
Rotation	0	0

5   When all the attribute information has been entered, the fish symbol will display the three tags – Figure 40.1(c).

6   *Notes*
   a) When attributes are used for the first time, the words Tag, Prompt and Value can cause confusion. The following descriptions may help to overcome this confusion:
      1. tag   is the actual attribute 'label' which is attached to the drawing at the specified text start point. This tag item can have any text style, height and rotation
      2. prompt   is an aid to the user when the attribute data is being entered with the inserted block
      3. value   is an artificial name/number for the attribute being entered. It can have any alpha-numeric value.
   b) The Insertion point in the Attribute Definition dialogue box refers to the attribute text tag and not to a block.

7   In our first attribute definition sequence, we created the following attribute information:
    a) Tag: SPECIES
    b) Prompt: What displayed?
    c) Default value: REDHERRING
    d) Text insertion point for SPECIES, centred on 100,55 with height 7 and 0 rotation angle.

Creating the attribute block

1   Menu bar with **Draw-Block-Make** and:
    prompt   Block Definition dialogue box
    respond   enter/activate the following:
        a) Name: TROPHY
        b) Base point: X: 50   Y: 50   Z: 0
        c) Objects: Select and window symbol and attributes as Figure 40.1(d) then right-click
        d) Objects: Delete active
        e) Preview: Create icon from Block geometry active
        f) Drag-and-drop units: Millimeters
        g) Description: TROPHY block with 3 attributes
    and   dialogue box as Figure 40.3
    then   pick OK.

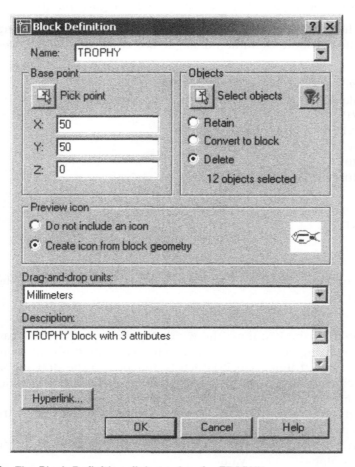

Figure 40.3   The Block Definition dialogue box for TROPHY.

2   The symbol and attributes have been made into a block and should disappear from the screen as we activate this option from the dialogue box.

Testing the created block with attributes

Now that the block with attributes has been created, we want to 'test' the attribute information it contains. This requires the block to be inserted into the drawing, and this will be achieved with both command line and dialogue box entries.

1   Make layer OUT current.

2   At the command line enter **ATTDIA <R>** and:
    prompt   Enter new value for ATTDIA<?>
    enter   **0 <R>**.

3   ATTDIA is a system variable, and when set to 0 will only allow attribute values to be entered from the keyboard.

4   At the command line enter **–INSERT <R>** and:
    prompt   Enter block name and enter **TROPHY <R>**
    prompt   Specify insertion point and **pick any point to suit**
    prompt   Enter X scale and enter **1 <R>**
    prompt   Enter Y scale and enter **1 <R>**
    prompt   Specify rotation angle and enter **0 <R>**
    prompt   What year? <9999> and enter **1984 <R>**
    prompt   Where caught? <WXYZ> and enter **CLYDE <R>**
    prompt   What displayed? <REDHERRING> and enter **COD <R>**.

5   The fish trophy symbol will be displayed with the attribute information as Figure 40.1(e).

6   *Notes*
    a) the prompt and default values are displayed as entered in the Attribute Definition dialogue box
    b) the order of the last three prompt lines (i.e. displayed, caught and year) may not be in the same order as listed in step 4. Do not worry if they are not the same.

7   Now insert the trophy block twice more with –INSERT from the command line using:
    a) at any suitable point, full size with 0 rotation and accept the default values, i.e. right-click or <R> at each attribute prompt line – Figure 40.1(f)
    b) at another point on the screen with the X scale factor as 0.75, the Y scale factor as 1.25 and the rotation angle −5. Use the same attribute entries as step 4, i.e. COD, CLYDE and 1984. The result should be as Figure 40.1(g).

8   Explode any inserted block which contains attribute information and the tags will be displayed – Figure 40.1(h).

9   We are now ready to insert the 'real' attribute data.

Attribute information

The fisherman's trophy cabinet contains five prime examples of what he has caught over the past few years, and each catch is represented in the trophy cabinet by the block symbol containing the appropriate attribute information. The attribute data to be displayed is:

Species	River	Year
SALMON	SPEY	1992
TROUT	TAY	1994
PIKE	DART	2000
CARP	NENE	2002
EELS	DERWENT	2004

Attribute information can be added to an inserted block:
a) from the keyboard – as previous example
b) via a dialogue box which will now be discussed.

1   Erase all objects from the screen and make layer OUT current.

2   At the command line enter **ATTDIA <R>** and:
prompt   Enter new value for ATTDIA<0>
enter   1 <R>.

3   Menu bar with **Insert-Block** and:
prompt   Insert dialogue box
with   Block name: TROPHY – from previous insertion
respond   *a*) ensure all on-screen prompts not active – i.e. no tick
          b) insertion point at X: 40; Y: 230; Z: 0
          c) explode not active
          d) scale X: 1.2; Y: 1.2; Z: 1
          e) rotation angle 0
          f) pick OK
prompt   Edit Attributes dialogue box
with   Prompts and default values from the attribute definition
        sequence as Figure 40.4(a)
respond   *a*) alter What displayed to SALMON
          b) alter Where caught to SPEY
          c) alter What year to 1992
          d) dialogue box as Figure 40.4(b)
          e) pick OK.

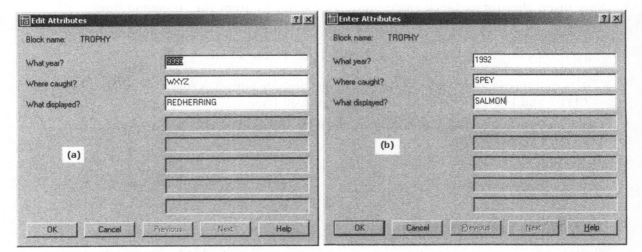

Figure 40.4 The Edit Attributes dialogue box.

4 The trophy block will be inserted with the attribute information displayed.
5 Using step 3 as a guide with the attribute data of species, river and year, refer to Figure 40.5 and insert the TROPHY block to complete the cabinet – use your imagination with the scales, then complete the cabinet and save?
6 This completes our brief 'taster' into attributes.

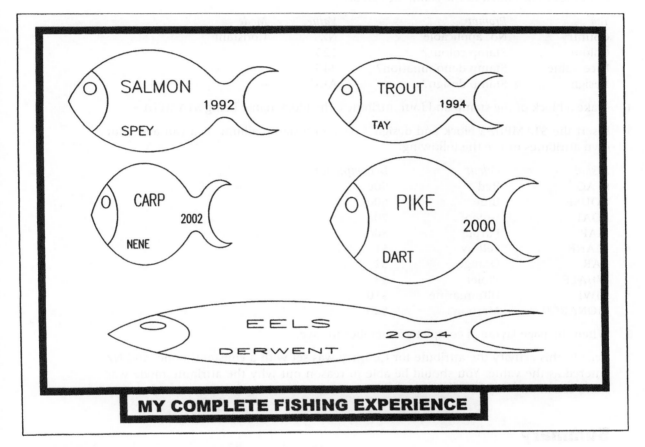

Figure 40.5 The fisherman's display trophy cabinet.

Point of interest?

In the previous chapter we created a title box as a Wblock. This title box had text items attached to it, e.g. the drawing name, date, revision, etc. These text items could have been made as attributes. When the title box was inserted into a drawing, the various text items (attributes) could have been entered to drawing requirements. Think about this application of attributes!

Assignment

Activity 48: Stamp Page layout

A single activity involving making attributes using a previously saved drawing. The activity involves the stamp enthusiast from Chapter 38 and you have to design a page of the stamp album.

1. Open the saved STAMP drawing.

2. Refer to the activity drawing which displays:
   a) the original stamp design
   b) the modified stamp
   c) the stamp with attribute tags.

3. Add four attribute definitions using your own text style, height and layout. The attributes to be added to the stamp design are:

Tag	Prompt	Value	Mode
Country	Not applicable	NZ	Constant
Colour	Stamp colour?	123	–
Face Value	Stamp denomination?	345	–
Design	Stamp design?	ABC	–

4. Make a block of the stamp and four attributes, the block name being STAMPDES.

5. Insert the STAMPDES block and design a page of a stamp album. You can add your own attributes or use the following:

Design	Colour	Denomination
FLAG	Red	50c
HOUSE	Blue	60c
BOAT	Green	70c
MAP	Purple	80c
PLANE	Yellow	$1
CAR	Orange	$2
WHALE	Violet	$3
KIWI	Ultramarine	$10
MONARCH	Maroon	$20

6. When the page layout is complete remember to save.

7. *Note*: In this activity the attribute for Country was set with a constant mode, and NZ entered as the value. You should be able to reason out why the attribute mode was set to constant.

Summary

1. Attributes are text items added to BLOCKS or WBLOCKS.

2. Attribute must be defined by the user.

3. Attribute data is added to a block when it is inserted into a drawing.

4. Attributes can be edited and extracted from a drawing, but these topics are beyond the scope of this book.

Chapter **41**

External references

Wblocks contain information about objects, colour, layers, linetypes, dimension styles, etc. and all this information is inserted into the drawing with the Wblock. All the information may not be required by the user, and it also takes time and uses memory space. Wblocks have another disadvantage, this being that drawings which contain several Wblocks are not automatically updated if one of the original Wblocks is altered.

External references (or Xrefs) are similar to Wblocks in that they are created by the user and can be inserted into a drawing, but they have one major advantage over the Wblock. Drawings which contain external references are automatically updated if the original external reference 'Wblock' is modified.

A worked example will be used to demonstrate external references. The procedure may seem rather involved as it requires the user to save and open several drawings, but the final result is well worth the effort. For the demonstration we will:
a) create a Wblock
b) use the Wblock as an Xref to create two drawing layouts
c) modify the original Wblock
d) view the two drawing layouts
e) use the existing C:\BEGIN folder.

Getting started

1 Open your A3SHEET standard drawing sheet and refer to Figure 41.1.

2 Make a new current layer with:
name: XREF; colour: red; linetype: continuous.

3 Draw:
a) a circle of radius 21
b) a item of text, middled on the circle centre with height 5 and rotation angle 0. The item of text is to be AutoCAD and is to be blue – reference Figure 41.1(a).

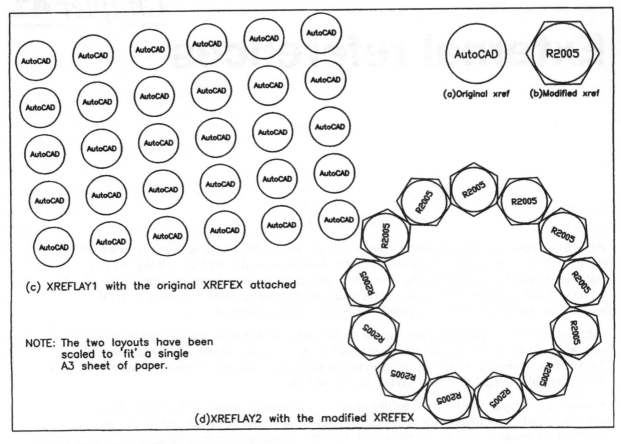

Figure 41.1 External reference example.

Creating the Xref (a Wblock)

1   At the command line enter **WBLOCK <R>** and:
    prompt   Write Block dialogue box
    respond   a) Source: Objects
                b) Base point: Pick point and pick circle centre point
                c) Objects: Select objects, pick circle and text then right-click
                d) Objects: Delete from drawing active
                e) File name and path: **C:\BEGIN\XREFEX**
                f) Insert units: Millimeters
                g) pick OK.

2   A preview of the Wblock will be displayed and a blank screen returned, due to the delete from drawing option being active.

Inserting the Xref (drawing layout 1)

1   Menu bar with **File-Close** (no to any save changes) then menu bar with **File-Open** and select your A3SHEET standard sheet again with layer OUT current.

2   Menu bar with **Insert-External Reference** and:
    prompt   Select Reference File dialogue box (looks familiar?)
    respond   a) Look in: scroll and pick C:\BEGIN
                b) scroll and pick XREFEX
                c) pick Open
    prompt   External Reference dialogue box

with	Name: XREFEX
	Found in: C:\BEGIN\XREFEX.dwg
	Saved path: C:\BEGIN\XREFEX.dwg
respond	a) Reference Type: Attachment active (black dot)
	b) Path type: Full path active
	c) all on-screen options active – i.e. tick
	d) dialogue box similar to Figure 41.2
	e) pick OK
prompt	Attach Xref "XREFEX": C:\BEGIN\XREFEX.dwg
and	"XREFEX" loaded
then	Specify insertion point and enter **50,50 \<R\>**
prompt	Enter X scale and enter **1 \<R\>**
prompt	Enter Y scale and enter **1 \<R\>**
prompt	Specify rotation angle and enter **0 \<R\>**.

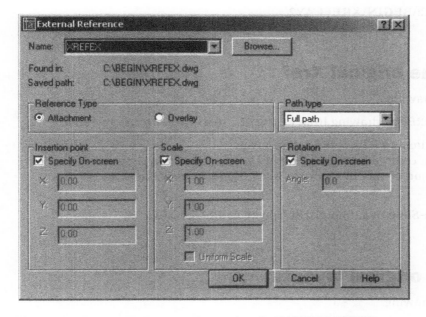

Figure 41.2 The External Reference dialogue box for C:\BEGIN\XREFEX.

3 The named external reference (XREFEX) will be displayed at the insertion point entered. Does the complete process seem similar to inserting a Wblock?

4 Now rectangular array the inserted attached Xref for:
   a) 5 rows with row offset: 50
   b) 6 columns with column offset: 60
   c) Angle of array: 5.

5 Save the layout as **C:\BEGIN\XREFLAY1** – Figure 41.1(c).

Inserting the Xref (drawing layout 2)

1 Close the existing drawing then re-open A3SHEET.

2 At the command line enter **XREF \<R\>** and:
   prompt   Xref Manager dialogue box
   respond **pick Attach**

prompt	Select Reference File dialogue box
respond	*a)* scroll and pick XREFEX from your C:\BEGIN folder
	b) pick Open
prompt	External Reference dialogue box
respond	*a)* Reference Type: Attachment
	b) De-activate the three on-screen prompts (no tick)
	c) Insertion point at X = 200, Y = 250, Z = 0
	d) X,Y,Z scales: 1
	e) Rotation angle: 30
	f) pick OK.

3   Now polar array the inserted attached Xref with:
    a) centre point: 200,150
    b) number of items: 13
    c) angle to fill: 360
    d) rotate items as copied active.

4   Save the layout as **C:\BEGIN\XREFLAY2**.

Modifying the original Xref

1   Close the current drawing.

2   Open the original XREFEX drawing from C:\BEGIN. This is the original Wblock.

3   *a)* Change the text item to R2005 and colour green
    b) Draw a hexagon, centred on the circle and circumscribed in a circle of radius 22. Change the colour of this hexagon to blue. These modifications are shown in Figure 41.1(b).

4   Menu bar with **File-Save** to automatically update C:\BEGIN\XREFEX.

Viewing the original layouts

1   Close the existing drawing.

2   Menu bar with **File-Open** and:
    a) pick XREFLAY1
    b) note the Preview then pick Open
    c) interesting result?

3   Menu bar with **File-Open** and:
    a) pick XREFLAY2
    b) note the Preview then pick Open
    c) again an interesting result? – Figure 41.1(d).

4   The layout drawings should display the modified XREFEX without any 'help' from us. This is the power of external references. Surely this is a very useful (and dangerous) concept?

5   Menu bar with **Format-Layer** and note the layer: XREFEX|XREF. This indicates that an external reference (Xrefex) has been attached to layer Xref. The (|) is a pipe symbol indicating an attached external reference.

6   This completes our simple investigation into Xrefs.

Assignment

A single external reference activity requires the creation of several Wblocks, attaching these as Xrefs to a drawing and then modifying the original Wblocks. There is quite a bit of work involved, but the end result is worth the effort.

Activity 49: Weather map

1. Refer to Figure 49A and using the grid information given, create:
   a) an outline map of a land mass and save as MAP
   b) the seven ORIGINAL weather symbols.
2. Make Wblocks of the seven weather symbols using the names given and your own insertion point.
3. Close all drawings then open the saved MAP drawing.
4. Refer to Figure 49B and attach the seven weather symbols as Xrefs, creating a weather map of your own design. When the weather map is complete, save with a suitable name. Remember that you can copy attached Xrefs.
5. Open each saved weather symbol and modify as Figure 49A then re-save.
6. Open the saved weather map to display the modified weather symbol map.
7. To bind or not to bind, that is the question.
8. Note that you could also create the seven weather symbols as blocks then re-define these blocks, but would that defeat the purpose of the activity?

Summary

1. External references are Wblocks Which can be attached to drawings.
2. When the original Xref Wblock is altered, all drawing files which have the external reference attached are automatically updated to include the modifications to the original Wblock.

Chapter 42
Pictorial drawings

Pictorial drawings allow two-dimensional objects to be displayed in 3D. This is a very useful concept as it can convey additional information about a component which is not always apparent with the traditional orthographic views. The reader should always be aware that all pictorial drawings are themselves two-dimensional. I am constantly surprised that there are still 'draughtspersons' and teachers/lecturers who refer to an isometric as a 3D drawing. Nothing could be far from the truth. An isometric is a 2D representation of a 3D drawing and although it appears to be displayed in 3D, the user should never forget that it is a 'flat 2D' drawing without any 'depth'.

In this chapter we will investigate three pictorial visualisations, these being isometric, oblique and planometric.

Isometric drawings

Isometric drawings are created with the X and Y axes at 30 degrees and accuracy is achieved with polar co-ordinate entry. Isometric drawings are created 'full size'. AutoCAD has the facility to display an isometric grid as a drawing aid. To discuss isometric 'concepts', open your A3SHEET standard sheet.

Setting the isometric grid

There are two methods for setting the isometric grid, these being by using a dialogue box and via the keyboard.

1 To demonstrate the dialogue box method, from the menu bar select **Tools-Drafting Settings** and:
    prompt   Drafting Settings dialogue box with three tab selections:
        a) Snap and Grid
        b) Polar Tracking
        c) Object Snap
    respond   **pick the Snap and Grid tab** and:
        a) Snap On (F9) active – tick
        b) Grid On (F7) active
        c) Snap type & Style with:
            1. Grid snap active – black dot
            2. Isometric snap active
        d) set Grid Y Spacing: 10
        e) set Snap Y Spacing: 5
        f) Angle, X base, Y base: all 0 – Figure 42.1
        g) pick OK.

2 The screen will display an isometric grid of 10 spacing, with the on-screen cursor 'aligned' to this grid with a snap of 5.

3 Use the Drafting Settings dialogue box to 'turn the isometric grid off', i.e. pick the Rectangular snap and set the grid spacing to 10 and the snap spacing to 5.

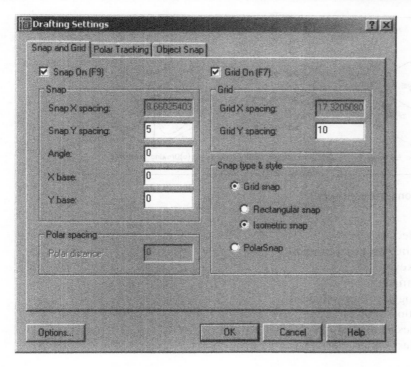

Figure 42.1 The Drafting Settings dialogue box with the Snap and Grid tab active.

4   The screen will display the standard grid pattern.

5   For the keyboard entry method, at the command line enter **SNAP <R>** and:
    prompt   Specify snap spacing or [ON/OFF/Aspect/Rotate/Style/Type]
    enter   **S <R>** – the style option
    prompt   Enter snap grid style [Standard/Isometric]
    enter   **I <R>** – the isometric option
    prompt   Specify vertical spacing<5.00>
    enter   **10 <R>**.

6   The screen will again display the isometric grid pattern with the cursor 'snapped to the grid points'.

7   Leave this isometric grid on the screen.

Isoplanes

1   AutoCAD uses three 'planes' called isoplanes when creating an isometric drawing, these being named top, right and left. The three planes are designated by two of the X, Y and Z axes as shown in Figure 42.2(a) and are:
    a) isoplane top: XZ axes
    b) isoplane right: XY axes
    c) isoplane left: YZ axes.

2   When an isoplane is 'set' or 'current', the on-screen cursor is aligned to the axes of the isoplane axis as displayed in Figure 42.2(b).

3   The isoplane can be set by one of three methods:
    a) At the command line enter **ISOPLANE <R>** and:
        prompt   Enter isometric plane setting [Left/Top/Right]
        enter   **R <R>** – right plane.

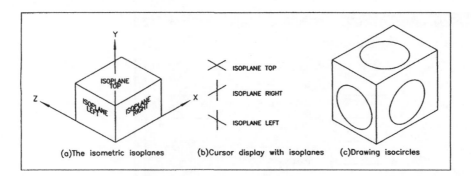

Figure 42.2 Three isometric concepts.

   b) Using a 'toggle' effect by:
     1. holding down the Ctrl key (control)
     2. pressing the E key
     3. toggles to isoplane left
     4. Ctrl E again – toggles to isoplane top.
   c) Using the F5 function key:
     1. press F5 – toggles isoplane right
     2. press F5 – toggles isoplane left, etc.

4 *Notes*
It is user preference as to what method is used to set the isometric grid and isoplane, but my recommendation is:
   a) set the isometric grid ON from the Drafting Settings dialogue box with a grid spacing of 10 and a snap spacing of 5
   b) toggle to the required isoplane with Ctrl E or F5
   c) isoplanes are necessary when creating 'circles' in an isometric 'view'.

Isometric circles

Circles in isometric are often called **isocircles** and are created using the ellipse command and the correct isoplane MUST be set. Try the following exercise:

1 Set the isometric grid on with spacing of 10 and toggle to isoplane top.

2 Using the isometric grid as a guide and with the snap on, draw a cuboid shape as Figure 42.2(c). The size of the shape is not important at this stage – only the basic shape.

3 Select the **ELLIPSE icon** from the Draw toolbar and:
   prompt   `Specify axis endpoint or [Arc/Center/Isocircle]`
   enter   **I <R>** – the isocircle option
   prompt   `Specify center of isocircle`
   respond   **pick any point on top 'surface'**
   prompt   `Specify radius of isocircle`
   respond   **drag and pick as required**.

4 Toggle to isoplane right with Ctrl E.

5 At the command line enter **ELLIPSE <R>** and:
   prompt   `Specify axis endpoint or [Arc/Center/Isocircle]`
   enter   **I <R>**
   prompt   `Specify center of isocircle` and pick a point on 'right side'
   prompt   `Specify radius of isocircle` and drag/pick to suit.

6 Toggle to isoplane left, and draw an isocircle on the left side of the cuboid.

7 The cuboid now has an isometric circle on the three 'sides'.

8 Now continue with the example which follows.

Isometric example

1 Erase any objects from the screen or re-open your A3SHEET standard sheet and refer to Figure 42.3.

2 Set the isometric grid on, with a grid spacing of 10 and a snap spacing of 5.

3 With the **LINE icon** draw:
First point: pick towards lower centre of the screen
Next point and enter @80<30 <R>
Next point and enter @100<150 <R>
Next point and enter @80<-150 <R>
Next point and enter @100<-30 <R>
Next point and enter @50<90 <R>
Next point and enter @80<30 <R>
Next point and enter @50<-90 <R><R> – Figure 42.3(a).

4 With the **COPY icon**:
   a) objects: pick lines D1, D2 and D3 then right-click
   b) base point: pick intersection of pt 1
   c) displacement: pick intersection of pt 2.

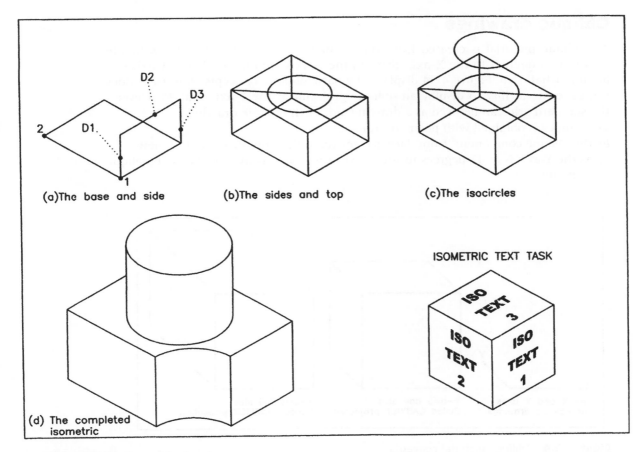

Figure 42.3 Isometric exercise.

5 Draw the two lines (endpoint–endpoint) to complete the sides and top then draw a top diagonal line as Figure 42.3(b).

6 Toggle to isoplane top.

7 With the **ELLIPSE-Isocircle** command:
   a) pick midpoint of diagonal as the centre
   b) enter a radius of 30.

8 Copy the isocircle:
   a) from the diagonal midpoint
   b) by @50<90 – Figure 42.3(c).

9 Erase the diagonal.

10 Draw in the two 'cylinder' sides using the quadrant snap and picking the top and bottom isometric circles.

11 Trim objects to these lines and erase unwanted objects to give the complete isometric.

12 Now draw two additional 30-radius isometric circles and modify to give the completed isometric as Figure 42.3(d).

13 *Task*: Text with Isometrics
Adding text to an isometric drawing requires the user to set text styles to suit the appropriate 'face' of the component. During the text style chapter, we created two styles which would suit isometric text. Refer to the task drawing in Figure 42.3 and create three text styles and add text to the three faces on an isometric cube.

Oblique drawings

An oblique pictorial is created from the normal X-axis orientation and with the Y-axis at 45 degrees to the X-axis. Sizes in the X-axis are full, while the Y-axis can be full or half size. Figure 42.4 displays the basic oblique concepts. AutoCAD does not have the facility to display an oblique grid, but as the Y-axis is at 45 degrees, the standard grid can be used as a drawing aid. With oblique drawings:
a) accuracy is obtained with polar co-ordinates
b) the oblique component 'front' face is identical to the front orthographic view
c) as the Y-axis is at 45 degrees to the X-axis, there are always two possible oblique solutions.

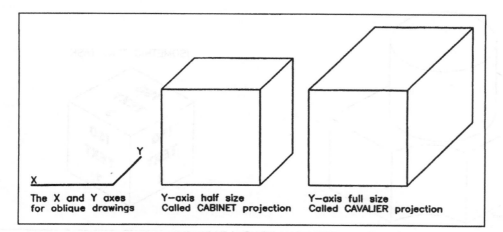

Figure 42.4 Oblique pictorial concepts.

A single example will demonstrate how an oblique pictorial is obtained.

1. Open you A3SHEET and set a standard 10 grid with 5 snap.
2. Refer to Figure 42.5 and using the sizes in Figure 42.5(a), create the front 'face' of the component – Figure 42.5(b).

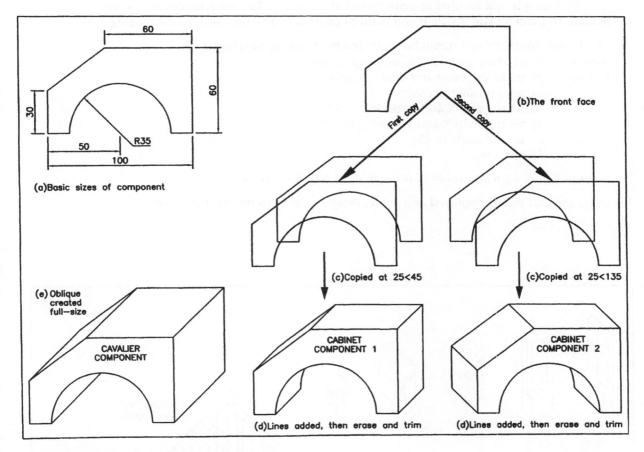

Figure 42.5 Creating an oblique pictorial.

3. Copy this front face to another part of the screen.
4. The component is 50 'wide', so copy one complete front face from an endpoint by @25<45, and the other from an endpoint by 25<135 – Figure 42.5(c).
5. Draw in the edge lines then erase and trim unwanted objects to display the complete component – Figure 42.5(d).
6. Create another oblique using the 50-width size – Figure 42.5(e).
7. Decide for yourself which is better – the cabinet projection (half-size) or the cavalier (full size).
8. The exercise is now complete.
9. *Note*: I have not discussed how to draw an oblique circle in the Y-axis. This is not as easy as creating isometric circles.

Planometric drawings

Planometric projections have recently become popular with the introduction of customised CAD systems in kitchen and bedroom design. This type of projection allows a 'birds-eye' view of a complete room, floor layout, etc. to the customer. Planometric drawings are created with both the X and Y axes at 45 degrees to the horizontal and this can be achieved in AutoCAD by modifying the standard grid.

A single example will be used to demonstrate the concept. This exercise can be rather tedious, so I will let you decide if you want to attempt it. The final result is interesting.

1   Open your A3SHEET and menu bar with **Tools-Drafting Settings** and:
    prompt   Drafting Settings dialogue box
    respond   a) make the Snap and Grid tab active
                b) set a rectangular snap
                c) set the grid X and Y spacing to 5
                d) set the snap X and Y spacing to 2.5
                e) set the angle to 45
                f) pick OK.

2   Refer to Figure 42.6 and zoom in on a small area of the screen.

3   Using the grid points displayed as a guide, design some planometric furniture.

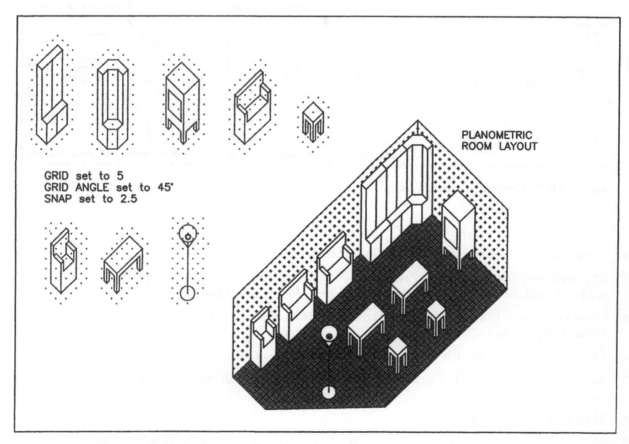

Figure 42.6 Planometric exercise.

4 Design a planometric room layout using your furniture and add hatching to the floor and walls.

5 Save if required.

Assignments

Three pictorial assignments have been included and I have tried to make these interesting for the user.

Activity 50: Bicycle spanner

Draw the two orthographic views, then create both an isometric and an oblique pictorial using the sizes given. The isometric hexagon is interesting to complete.

Activity 51: Garden wall

Design an isometric garden wall block using the basic 100 square × 40 sizes displayed. Scale the wall block by 0.25 then create the wall. I have added some isometric text for effect.

Activity 52: Plug

Using the sizes given, draw the two views of the component, and an isometric and oblique pictorial. The larger square is interesting to create, especially as it is not dimensioned. You need to think.

Summary

1 Pictorial drawings are flat 2D drawings with 3D visualisation.

2 The three types of pictorial drawings are isometric, oblique and planometric.

3 The pictorial drawing axes are:
   a) isometric: both X and Y axes at 30 degrees to the horizontal
   b) oblique: X-axis is horizontal, Y-axis is at 45 degrees to the horizontal
   c) planometric: both X and Y axes at 45 degrees to the horizontal.

4 Oblique pictorials can be CABINET (half Y-axis size) or CAVALIER (full Y-axis size).

5 An isometric and planometric grid can be used as a drafting aid. There is no oblique grid aid.

6 Pictorial drawings are generally constructed with polar co-ordinates for accuracy.

7 With isometrics, circles are drawn with the ellipse-isocircle command and the correct isoplane must be set. The isocircle option of the ellipse command is only available when the isometric grid is 'set on'. Try this for yourself with a normal rectangular grid.

8 The objects snaps (endpoint, midpoint, etc.) are available with pictorial drawings.

9 The modify commands (e.g., copy, trim, etc.) are available with pictorials.

10 The OFFSET command does not give the effect that the user would expect, and should not be used.

Chapter 43

Model space and paper space

AutoCAD has two drawing environments:
a) model space: used to draw the component
b) paper space: used to layout the drawing paper for plotting.

The two drawing environments are independent of each other and while the concept is particularly applicable to 3D modelling, we will demonstrate its use with a previously saved 2D drawing.

Until now all work has been completed in model space

Notes

1. The user must realise that this chapter is an introduction to the model/paper space concept. Paper space is particularly targeted for plotting and as I have no idea what type of plotter the reader has access to, I have assumed that no plotter is available. This does not affect the exercise.

2. When the paper space environment is entered, a new icon will be displayed. The model space and paper space icons are shown in Figure 43.1. The icon display is dependent on whether the UCS icon style has been set to 2D or 3D. I generally display the 2D icon style.

3. Before paper space can be used, the paper space environment must be 'entered'. This can be achieved by:
   a) picking MODEL from the status bar
   b) picking one of the layout tabs from the drawing screen
   c) entering TILEMODE then 0 at the command line.

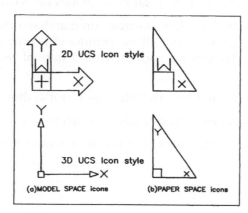

Figure 43.1 The model space and paper space icons.

Getting started

1. Open your A3SHEET standard drawing sheet and:
   a) erase the rectangular border
   b) make two new layers, VP and SHEET both with your own colour selection and with continuous linetype
   c) make layer VP current.

2. AutoCAD generally defaults with a Model tab, a Layout1 tab and a Layout2 tab. While the layout tabs could be used for this exercise, we will create a new tab.

3. Menu bar with **Tools-Wizards-Create Layout** and:
   prompt   Create Layout - Begin dialogue box
   respond **alter Layout name to FACTORY then pick Next** – Figure 43.2
   prompt   Create Layout - Printer dialogue box
   respond **pick None then Next**
   prompt   Create Layout - Paper Size dialogue box
   respond a) Drawing units: Millimeters
              b) scroll at paper size and pick ISO A3 (420.00 × 297.00 mm)
              c) pick Next
   prompt   Create Layout - Orientation dialogue box
   respond **pick Landscape then Next**
   prompt   Create Layout - Title Block dialogue box
   respond **pick None then Next**
   prompt   Create Layout - Define Viewports dialogue box
   respond a) Viewport setup: Single
              b) Viewport scale: Scaled to Fit
              c) pick Next
   prompt   Create Layout - Pick Location dialogue box
   respond **pick Select location <**
   prompt   Drawing screen returned
   prompt   Specify first corner at the command prompt
   enter   **10,10 <R>**
   prompt   Specify opposite corner

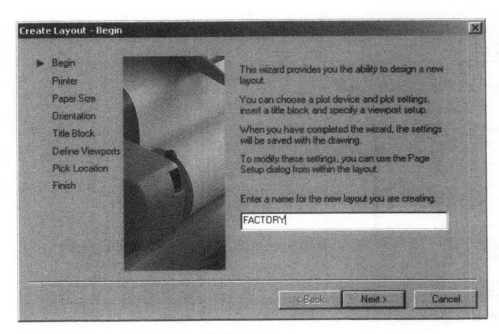

Figure 43.2 The Create Layout - Begin dialogue box.

enter **210,145 <R>**
prompt Create Layout - Finish dialogue box
respond **pick Finish**.

4 The drawing screen will be returned with:
   a) a white area – the A3-size drawing paper
   b) a dotted line area – the plottable area
   c) a coloured rectangle – the created viewport
   d) a new tab – FACTORY
   e) the paper space icon displayed (relative to your UCS style icon)
   f) PAPER is displayed in the status bar.

5 As the paper space icon is displayed, the user is 'in paper space'.

The sheet layout

Our A3 drawing paper now has a paper space viewport and we want to create another three viewports. Refer to Figure 43.3 and:

1 Still in paper space with layer VP current, menu bar with **View-Viewports-1 Viewport** and:
   prompt Specify corner of viewport and enter **230,75 <R>**
   prompt Specify opposite corner and enter **380,170 <R>**.

2 Repeat the menu bar **View-Viewports-1 Viewport** selection and create another viewport from 190,180 to 280,250.

3 Menu bar with **View-Viewports-Polygonal Viewport** and:
   prompt Specify start point and enter **50,160 <R>**
   prompt Specify next point and enter **@50<0 <R>**

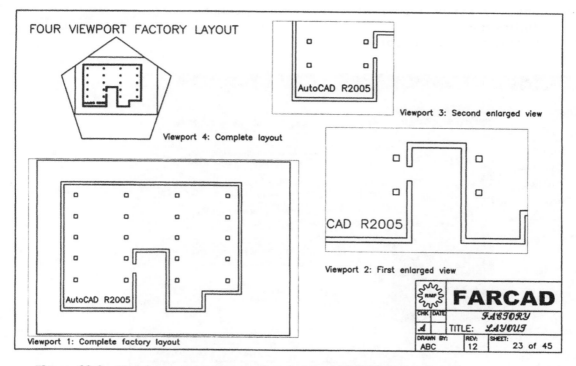

Figure 43.3 Model–paper space example using the LARGESC factory layout.

prompt	Specify next point and enter **@50<72 <R>**
prompt	Specify next point and enter **@50<144 <R>**
prompt	Specify next point and enter **@50<-144 <R>**
prompt	Specify next point and enter **C <R>**.

4 We have now created three rectangular paper space viewports and one pentagonal paper space viewport.

The model space – example 1

Rather than draw a new component we will insert an already completed and saved (I hope) drawing into our created viewports. This drawing is the large-scale factory layout (LARGESC) from Chapter 36.

1 Enter model space with a left-click on PAPER in the status bar.

2 The model space environment will be entered and the traditional model space icon will be displayed in the rectangular viewports.

3 Make the first created viewport (lower left) active by:
   a) moving the pointer into the rectangular area
   b) left-click
   c) cursor cross-hairs displayed and the viewport border appears highlighted.

4 Make layer OUT current then menu bar with **Insert-Block** and:

prompt	Insert dialogue box
respond	*a*) pick Browse
	b) scroll and pick C:\BEGIN folder
	c) scroll and pick LARGESC (or your entered name)
	d) pick Open
prompt	Insert dialogue box with Name: LARGESC
respond	*a*) ensure Specify On-screen active – i.e. tick
	b) ensure Explode active – i.e. tick
	c) pick OK
prompt	Specify insertion point for block and enter **0,0 <R>**
prompt	Specify scale factor for XYZ axes and enter **1 <R>**
prompt	Specify rotation angle and enter **0 <R>**.

5 Now select from the menu bar **View-Zoom-All** and the active viewport will display the complete factory layout while the other three viewports may display some lines.

6 With the large rectangular viewport still active:
   a) erase all text
   b) freeze layer DIMS.

7 Make each viewport active in turn by moving the pointing arrow into the viewport and left-click, then **View-Zoom-All** from the menu bar and the complete drawing (with border) will be displayed in each viewport.

Using the viewports

1 With the lower right viewport active select **View-Zoom-Window** from the menu bar and:

prompt	Specify first corner and enter **5000,1500 <R>**
prompt	Specify opposite corner and enter **13000,8000 <R>**.

2 With the top right viewport active, zoom a window from 1000,1000 to 9000,8000.

3   With the pentagonal viewport active, menu bar with **View-Zoom-Scale** and:
    prompt   Enter a scale factor
    enter   **0.015 <R>**.

4   Make layer TEXT current and the lower left viewport active and select from menu bar **Draw-Text-Single Line Text** and:
    prompt   Specify start point of text
    enter   **C <R>** – the center option
    prompt   Enter center point of text
    enter   **5000,2850 <R>**
    prompt   Specify height and enter **450 <R>**
    prompt   Specify rotation angle and enter **0 <R>**
    prompt   Enter text and enter **AutoCAD R2005 <R><R>**.

5   The entered item of text will be displayed in all viewports.

6   Make layer OUT current with the lower right viewport active.

7   Draw two lines:
    a) first point at 7250,5250, next point at 8750,5250
    b) first point at 7250,6000, next point @1500,0.

8   Use the TRIM command to trim these lines to give an opening into the factory – displayed in all viewports. You may have to zoom in on the wall area to complete the trim, but remember to zoom previous.

Using the paper space environment

1   Enter paper space with a left-click on MODEL in the status bar and try and erase any object from a viewport. You cannot as they were created in model space.

2   Make layer TEXT current and add the following text items to the layout:

Start	Ht	Rot	Text item
a) 10,4	4	0	Viewport 1: Complete factory layout
b) 230,60	4	0	Viewport 2: First enlarged view
c) 285,180	4	0	Viewport 3: Second enlarged view
d) 110,160	4	0	Viewport 4: Complete layout
e) 10,240	6	0	FOUR VIEWPORT FACTORY LAYOUT

3   Now enter model space with a left-click on PAPER in the status bar and try to erase any of the added text items – you cannot as they were created in paper space.

Completing the layout

1   Enter paper space and make layer SHEET current.

2   With the RECTANGLE command draw a rectangle with first corner point at 0,0 and the other corner point at 405,257.

3   Menu bar with **Insert-Block** and:
    a) select Browse
    b) pick your C:\BEGIN folder
    c) scroll and pick TITLE then Open
    d) activate Explode
    e) insertion point: X – 405; Y – 0; Z – 0
    f) scale: uniform at 0.9

g) rotation: 0
h) pick OK.

4 Modify the title box as required and the factory layout created in paper space is now complete and can be saved.

5 *Task*
If you have access to a printer/plotter:
a) Plot from model space with any viewport active.
b) Plot from paper space.
c) In model space, investigate the co-ordinates of the top right corner of the left viewport – about 22,000,15,000?
d) In paper space investigate the co-ordinates of:
    1. top right corner of our sheet: 405,257?
    2. top right corner of white area: 410,275?
e) *Question*: How can a drawing area on 22,000,15,000 be displayed on an A3 sheet of paper? This is the 'power' of model/paper space.

Model/Paper space – example 2

The previous exercise created the viewports prior to 'inserting' a drawing into the layout. In this exercise, we will open a previously created drawing, and then adapt the paper space layout. We will also investigate model and paper space dimensioning.

1 Open WORKDRG (which has not been used for some time) and erase any text and dimensions to leave the original red outline, two red circles and four green centre lines. Erase the black border.

2 Make a new layer, named VP with continuous linetype, colour to suit and current.

3 Left-click on the status bar **Layout1 tab** and:
    prompt   Page Setup Manager – Layout1 dialogue box
    respond   **pick Close**.

4 The screen will be returned and will display:
    a) a paper space layout with the paper space icon
    b) a white area – the drawing paper
    c) a dotted area – the plottable area
    d) a coloured outline – an active viewport
    e) the WORKDRG and the black border within the coloured viewport.

5 Erase the coloured viewport and WORKDRG 'disappears'.

6 Layer VP still current and still in paper space.

7 Menu bar with **View-Viewports-New Viewports** and:
    prompt   Viewports dialogue box
    with   two tabs:
                a) New Viewports
                b) Named Viewports
    respond *a*) New Viewports tab active
               b) pick Three: Left
               c) Viewport Spacing: 1
               d) Setup: 2D
               e) Preview: displays *Current* – dialogue box as Figure 43.4
               f) pick OK
    prompt   Specify first corner and enter **5,5 <R>**
    prompt   Specify opposite corner and enter **250,190 <R>**.

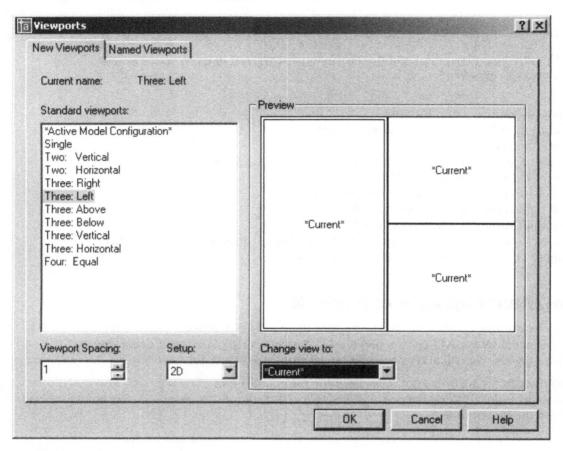

Figure 43.4 The Viewports dialogue box for the WORKDRG example.

8 The drawing screen will be returned (in paper space) with a three-viewport configuration, each viewport displaying WORKDRG.

9 Enter model space with one of the following methods then refer to Figure 43.5:
   a) pick PAPER from the status bar
   b) enter MS <R> at the command prompt.

10 Note the model space icon in all three viewports.

11 To make a viewport 'active', move the cursor into the required viewport and left-click.

12 a) with the lower right viewport active, zoom extents and the WORKDRG will 'fill the viewport'
   b) now zoom to a scale of 1.5 in this viewport.

13 a) make the top right viewport active
   b) Zoom window from 60,90 to 140,160 and the circle and centre lines 'fill the viewport'.

14 Still with the top right viewport active:
   a) make layer OUT current
   b) draw a hexagon, centred on 100,140 and inscribed in a 5-radius circle
   c) add an item of text, middled on 100,140 with height 3 and 0 rotation, the text item being CAD
   d) Polar array the hexagon and text, about the large circle centre, for 7 items with 360 angle to fill and rotated as copied.

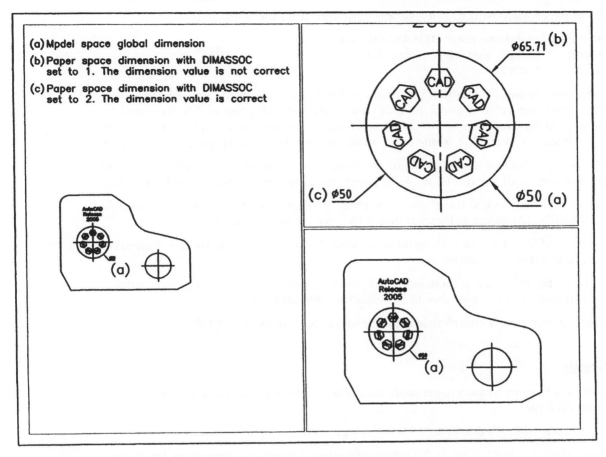

Figure 43.5 The model–paper space example with WORKDRG.

15 Make the lower right viewport active and layer TEXT current.

16 Menu bar with **Draw-Text-Single Line Text** and:
   a) start point: enter C <R> – the centre text option
   b) centre point of text at 100,175
   c) height of 5 and rotation angle of 0
   d) text: AutoCAD <R>, then Release <R> then 2005 <R><R>.

17 With layer 0 current, draw a rectangle from 0,0 to 257,195.

Dimensioning in model space and paper space

When a multi-viewport layout has been created in paper space, many users are unsure whether dimensions should be added in model space or paper space. We will investigate this concept with our screen layout.

1 Ensure model space and Layout1 tab still active.

2 Make the top right viewport active and layer DIMS current.

3 Modify the current dimension style to suit your own requirements if necessary.

4 Diameter dimension the circle, illustrated by dimension (a) in Figure 43.5.

5 This diameter dimension is displayed in all three viewports, and is one of the main 'drawbacks' of model space dimensioning with a multi-viewport layout.

6 Enter paper space with PS <R> or select MODEL from the status bar.

7 At the command line enter **DIMASSOC <R>** and:
   prompt   Enter new value for DIMASSOC
   enter   1 <R>.

8 Select the diameter dimension command and pick the circle in the top right viewport. The dimension value is displayed only at the selected pick point and is illustrated by dimension (b). The value of this dimension is 65.71 and is obviously wrong as the circle has a diameter of 50. This is one of the main 'drawbacks' of paper space dimensioning.

9 We seem to have a slight problem. Dimensioning in model space will display the dimensions in all viewports, while in paper space the wrong dimension is obtained.

10 AutoCAD overcomes these two apparent problems with the system variable DIMASSOC (Associate Dimensioning) which we set to 1 in step 7.

11 Set DIMASSOC to 2 and diameter dimension the circle and the correct dimension will be displayed as dimension (c).

12 Correct paper space dimensioning is thus obtained with the system variable DIMASSOC set to 2. This should be the actual default value.

13 This exercise is now complete and can be saved, but not as WORKDRG.

Task

You now have to create a paper space layout and add dimensions using the procedures from example 2, so:

1 Open your USEREX drawing and refer to Figure 43.6.

2 Using steps 2–7 from example 2, create a two-vertical viewports paper space layout and:
   a) erase the black border
   b) zoom to a factor of 2 in each viewport.

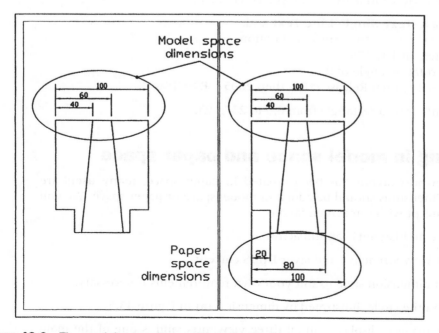

Figure 43.6 The model–paper space dimension task.

3   Add model space baseline dimensions (three) to the top edge of the component.
4   Add paper space baseline dimensions to the bottom edge and an angular dimension.
5   Remember that DIMASSOC should be set to 2.
6   The task is complete and can be saved.

This completes the discussion on model space and paper space layouts but before leaving the chapter, refer to Figure 43.7 which displays a multi-viewport configuration of a solid model flange-pipe arrangement. The viewports have been 'set-up' to display a 3D view of the model as well as a top, front and end view, these with hidden detail added. The flange-pipe composite was created in model space and the four-viewport configuration was created in paper space. This is one of the main uses of the model–paper space drawing environments.

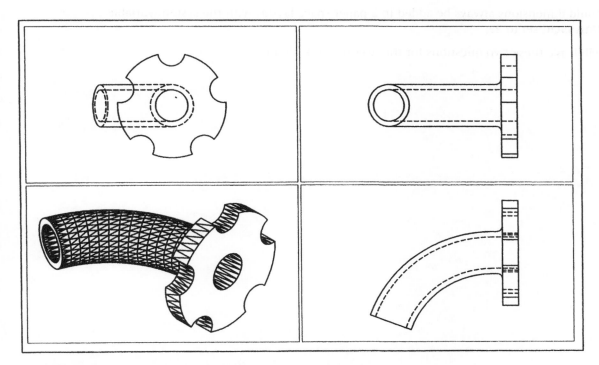

Figure 43.7 A solid model in a multi-viewport configuration.

Summary

1   Model space and paper space are two drawing 'environments' which 'exist together'.
2   Model space is used for traditional draughting purposes.
3   Paper space is used to lay out the paper to user specifications.
4   While the model/paper space concept is especially suitable for 3D and Solid modelling, the techniques can also be used in 2D.
5   Viewports are created in paper space and all drawing work completed in active model space viewports.
6   Viewports can be created to user requirements.
7   Layout tabs can be used to 'set' viewport configurations and can be renamed.

8 Dimensions can be added in model space using viewport-specific layers – not considered in this book.

9 AutoCAD 2005 has true model and paper space associativity.

10 DIMASSOC is a system variable which determines the 'type' of associativity available and:
    set to 0: dimensions have no associativity
    set to 1: model space associativity and dimensions are global
    set to 2: paper space associativity.

Final thoughts

1 Should paper space layouts be used at all times?

2 Should dimensions always be added to a paper space layout with the system variable DIMASSOC set to 2?

I will leave these two questions for the user to think about!

Chapter 44

Templates

This topic has been left close to the end of the book, when it should probably have been included nearer the beginning. The reasons for this were to allow the user to:
a) become proficient at draughting with AutoCAD
b) understand attributes
c) understand the concepts of paper space layouts.

What is a template?

1. A template is a prototype drawing, i.e. it is similar to our A3SHEET standard sheet which has been used when every new exercise/activity was started. The term 'prototype/standard drawing' refers to a drawing which has various default settings, e.g. layers, text styles, dimension styles, etc. The drawing has thus been customised to company/user requirements.

2. With AutoCAD, all drawings are saved with the file extension **.dwg** while template files have the extension **.dwt**. Any drawing can be saved with the .DWG or .DWT extensions.

3. Template drawings (files) are used to 'safeguard' the prototype drawing being mistakenly overwritten – have you ever saved work on your A3SHEET standard sheet by mistake? Templates help overcome this problem.

4. AutoCAD has templates which conform to several drawing conventions, including ANSI, DIN, Gb, ISO and JIS. These templates are available in several paper sizes.

5. Other templates are also available.

6. Template files can be opened from the Startup dialogue box.

    In this chapter we investigate:
    a) completing and saving a drawing using an opened template file
    b) creating our own template file.

Opening an AutoCAD template file

1. Close any existing drawing.
2. Menu bar with **File-New** and:
    prompt   Create New Drawing dialogue box
    respond   **pick Use a Template**
    prompt   Create New Drawing (Use a Template) dialogue box

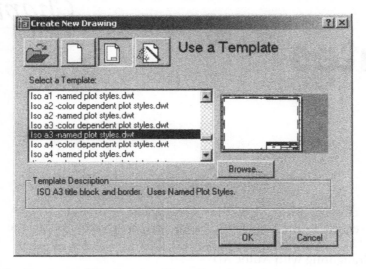

Figure 44.1 The Create New Drawing (Use a Template) dialogue box.

    respond   *a*) scroll at Select a Template
                  b) pick **Iso a3-named plot styles.dwt**
                  c) note the preview and description – Figure 44.1
                  d) pick OK.

3 The screen will display:
    a) a layout with a title box
    b) the paper space icon
    c) a new tab – ISO A3 Title Block.

4 Menu bar with **Format-Layer** and note the layers:
    a) layer 0: the AutoCAD default layer
    b) FRAME layers: 025, 050, 070 and 200 with various colours
    c) Title Block layer
    d) Viewport layer.

5 The new layers all have continuous linetypes and the colours match those in the title box.

6 Menu bar with **Dimension-Style** and note that the current style is **ISO-25**?

7 To investigate the title block information, enter **ATTEDIT <R>** at the command line and:
    prompt   Select block reference
    respond   **pick any X in the title block**
    prompt   Edit Attributes dialogue box
    with   11 attributes which can be edited by the user (need to pick Next), e.g.:
        a) attribute 1: File name, default XXX
        b) attribute 2: Drawing number, default X
        c) attribute 10: Edition, default 0
        d) attribute 11: Sheet, default 1/1
    respond   study the dialogue box (including Next)
    then   **pick Cancel**.

8 Figure 44.2 displays the Edit Attributes dialogue box (original and next) and matches the actual attributes from the dialogue box to the title block (sorry about this diagram being a bit messy).

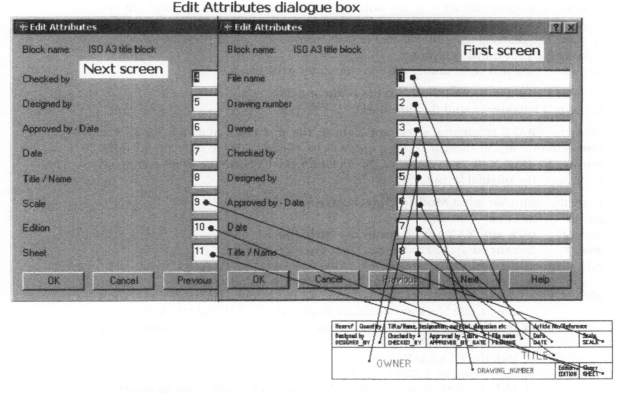

Figure 44.2 Identifying the 11 title block attributes.

Using an AutoCAD opened template file – exercise 1

1   The opened Iso A3 template file should still be displayed.

2   *a*) left-click on PAPER from the status bar to enter model space and note the 'thick black outline'. This is the drawing area
   b) draw a line from 0,0 to 420,290, i.e. A3 paper available
   c) erase the line.

3   Rather than start a new drawing we will insert a previously created drawing, so menu bar with **Insert-Block** and:
   prompt   Insert dialogue box
   respond **pick Browse**
   prompt   Select Drawing File dialogue box
   respond *a*) scroll and pick the C:\BEGIN folder
              b) scroll and pick a saved activity, e.g. act23
              c) pick Open
   prompt   Insert dialogue box
   with   Name: ACT23
   respond *a*) activate Explode
              b) de-activate the on-screen prompts
              c) enter the Insertion point as X: 0; Y: 0; Z: 0
              d) enter Uniform Scale as 1
              e) enter the Rotation angle as 0
              f) pick OK.

4   The selected drawing (file or Wblock?) will be inserted into the ISO A3 template drawing.

5 a) erase unwanted objects (e.g. the border) which have been inserted and are not required
   b) move the drawing to a suitable area of the screen
   c) optimise your drawing layout
   d) make a new text style, name: sta, with text font: Arial Black.

6 a) enter paper space with PS <R> at the command line
   b) Zoom a window around the title block.

7 With the ERASE command, select any item in the title block and right-click. The complete title block, sheet markings, etc. will be erased to leave only the inserted activity drawing. Note the 'shape' of the viewport then UNDO the erase command to 'restore' the original layout.

8 From the menu bar select **Modify-Object-Attribute-Single** and:
   prompt   Select a block
   respond **pick any XXX text item in the 'title block'**
   prompt   Enhanced Attribute Editor dialogue box
   with   a) Tab options: Attribute, Text Options, Properties
            b) Attribute tab active with: Tag, Prompt and Value details
   respond a) resize the dialogue box by dragging the lower edge downwards until the 11 title block attributes are displayed – i.e. FILENAME–SHEET
            b) drag the tag division line to the right to display the complete tag names as Figure 44.3(a)
            c) pick the FILENAME line
            d) alter value to: R2005/AB/01
            e) pick the DRAWING_NUMBER line (and FILENAME entry changes)
            f) alter value to: 456-BDF
            g) continue to pick the tag lines and alter the values as follows:

Tag	Value
OWNER	FARCAD
CHECKED_BY	RMF
DESIGNED_BY	HTC
APPROVED_BY_DATE	enter to suit
DATE	02-03-04

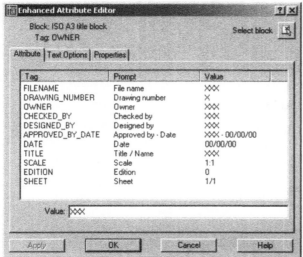

(a) The default values

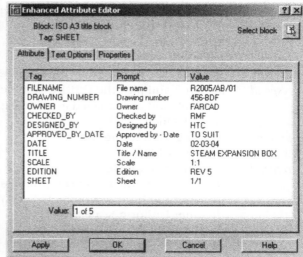

(b) Our entered values

Figure 44.3 The Enhanced Attribute Editor dialogue box for the title block with ACT23.

TITLE	STEAM EXPANSION BOX
SCALE	1:1
EDITION	REV 5
SHEET	1 of 5

and dialogue box at this stage as Figure 44.3(b)
now   *a)* pick the OWNER tag – highlights
        b) pick the Text Options tab
prompt   `Text Options dialogue box`
respond *a)* Text Style: scroll and pick sta
        b) Height: alter to 10
        c) Oblique Angle: alter to 10
        d) pick the Properties tab
prompt   `Properties dialogue box`
respond *a)* scroll color and pick Red
        b) pick Apply then OK.

9. The drawing screen will be returned and the attribute values entered in step 8 will be displayed. The original XXX OWNER attribute will have been replaced by the value FARCAD in red at a height of 10 and with a 10 obliquing angle.

10. Using the sequence **Modify-Object-Attribute-Single**, and your imagination, modify some of the other entered attributes for colour, height, text style, etc.

11. When you have completed the attribute 'editing', zoom previous and return to model space. Your drawing should now resemble Figure 44.4.

12. Finally, right-click on ISO A3 Title Block from the Layout tabs and:
    prompt   `Shortcut menu`
    respond **pick Rename**
    prompt   `Rename Layout dialogue box`
    respond *a)* alter name to MYLAYOUT-1
            b) pick OK.

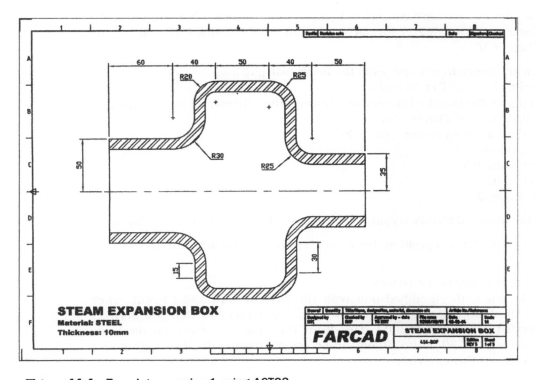

Figure 44.4 Template exercise 1 using ACT23.

13 Menu bar with **File-Save As** and:
   a) Save Drawing As dialogue box
   b) File type extension is *.dwg
   c) scroll to your C:\BEGIN folder and enter any suitable drawing name.

14 *Notes*
   a) we began the exercise with a template file – extension .dwt
   b) we saved the drawing as a drawing file – extension .dwg
   c) the original template file is thus unchanged
   d) this is surely useful to the user.

15 This first template exercise is now complete.

Template – exercise 2

1 Close any existing drawings then menu bar with **File-New** and:
   prompt   Create New Drawing dialogue box
   respond   **pick Use a Template**
   prompt   Use a Template dialogue box
   respond   a) scroll at Select a Template
               b) pick **Iso a0-named plot styles.dwt**
               c) note the preview and description
               d) pick OK.

2 The screen will return a paper space A0 drawing.

3 Check the layers – same six named layers as first example.

4 Move the cursor to lower left corner of white area and the co-ordinates will be approx 0,0. At the upper right corner of the white area the co-ordinates are approx 1180,840.

5 Enter model space with MS <R> or pick PAPER from status bar.

6 As with the previous exercise we will not create a new drawing but will insert the modified large-scale drawing of the factory layout with I beams – completed and saved in Chapter 38.

7 Menu bar with **Insert-Block** and using the Insert dialogue box:
   a) scroll and pick the C:\BEGIN folder
   b) scroll and pick the modified large-scale drawing from Chapter 38 (if not available, pick LARGESC from Chapter 36)
   c) activate the on-screen prompts (all tick)
   d) activate Explode
   e) insertion point: 0,0
   f) XYZ scale: 0.02
   g) rotation angle: 0.

8 Hopefully the modified factory layout will be displayed within the A0 sheet layout.

9 Erase the black border and position the factory layout for maximum effect.

10 *Task*
   a) Modify the entrance to the factory.
   b) If your layout is the modified one with the circular wall and I beams, then redefine the block BEAM to your own spec, but remember that the drawing was inserted at a scale of 0.02. This from the original factory layout being drawn at a scale of 50. Think about this before attempting the task. Figure 44.5 is my layout for this exercise.

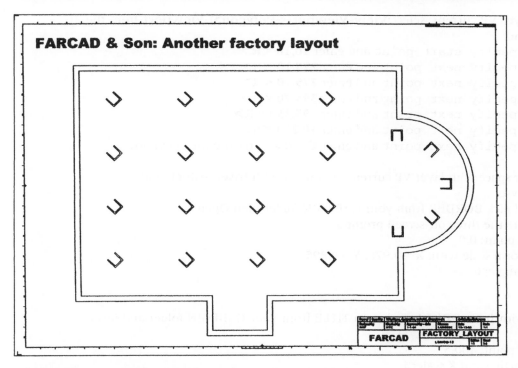

Figure 44.5 Template exercise 2 with an A0 template file.

11 Note that in my Figure 44.5 I have scaled the title box by a factor of 2. This was only to allow the user to 'see' the ISO A0 template file title block entries. They were originally quite small.

12 The exercise is now complete and can be saved as a .dwg file with a suitable name.

Creating our own A3 template file

We will now create our own A3-sized template file and will use our existing A3SHEET drawing file as it has layers, linetypes, text styles and dimension styles already created.

1 Close all existing drawings then open your A3SHEET standard sheet from your BEGIN folder.

2 *a*) erase the black border
 b) make a new layer named VP, continuous linetype, any colour
 c) make this new layer current.

3 Menu bar with **Tools-Wizards-Create Layout** and:
 prompt   Create Layout - Begin dialogue box
 respond   alter the various dialogue boxes as follows:
         a) Begin Layout name: MY A3 LAYOUT
         b) Printer: None
         c) Paper size: Millimeters ISO A3 (420 × 297)
         d) Orientation: Landscape
         e) Title Block: none for now
         f) Define viewports: none for now
         g) Finish: pick Finish.

4 Still with layer VP current, select **View-Viewports-Polygonal Viewport** from the menu bar and:
   prompt   Specify start point and enter **10,10 <R>**
   prompt   Specify next point and enter **275,10 <R>**
   prompt   Specify next point and enter **275,70 <R>**
   prompt   Specify next point and enter **395,70 <R>**
   prompt   Specify next point and enter **395,250 <R>**
   prompt   Specify next point and enter **10,250 <R>**
   prompt   Specify next point and enter **C <R>** – to close the viewport.

5 Still in paper space with layer VP current, menu bar with **Insert-Block** and:
   a) pick Browse
   b) scroll and pick BORDER from your C:\BEGIN folder then Open
   c) de-activate the three on-screen prompts
   d) insertion point: 0,0
   e) non-uniform scale with: X – 0.976; Y – 0.895
   f) rotation angle: 0
   g) pick OK.

6 Menu bar with **Insert-Block** and open TITLE from your C:\BEGIN folder and insert with:
   a) insertion point: 400,5
   b) explode active and X scale: 1
   c) rotation angle: 0
   d) pick OK.

7 *a*) enter model space with command line MS <R> and note the viewport outline
   b) make layer OUT current.

8 Menu bar with **File-Save As** and:
   prompt   Save Drawing As dialogue box
   respond   *a*) scroll at Files of type
               b) pick AutoCAD Drawing Template (*.dwt)
   prompt   Save Drawing As dialogue box
   with   Template file active (i.e. Save in)
   respond   *a*) File name: enter A3LAYOUT
               b) pick Save
   prompt   Template Description dialogue box
   enter   The following lines of text but NO RETURN
               This is my A3SHEET standard sheet layout created in paper space. My wblocks BORDER and TITLE have been inserted. The layout sheet has layers, text style, units and dimension style customised to my own requirements. Saved in AC TEMPLATE
   and   dialogue box as Figure 44.6
   respond   pick OK.

9 Your A3SHEET standard sheet will be saved as a template file and:
   a) added to the list of existing AutoCAD templates in the TEMPLATE folder
   b) be able to be 'opened' as a template
   c) will not be 'over-written' when a drawing is saved.

10 Repeat step 8, but save your A3LAYOUT template file in your C:\BEGIN folder. You should have the same template description dialogue box as before.

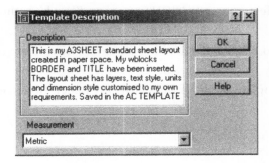

Figure 44.6 The Template Description dialogue box for the A3SHEET standard sheet.

Task

1 Close all existing files.

2 Menu bar with **File-New** and:
   prompt   Create New Drawing dialogue box
   respond **pick Use a Template**
   prompt   Select a Template list
   respond a) either   1. scroll and pick A3layout.dwt (should be at the top of the list)
              or   2. navigate your BEGIN folder and pick A3LAYOUT.dwt
              b) pick OK.

3 The screen will display your A3 layout template file.

4 a) in model space, insert any drawing (e.g. ACT33) full size with 0 rotation
   b) explode, modify and reposition the inserted drawing
   c) layout similar to Figure 44.7.

5 Menu bar with **File-Save As** and:
   a) File type: *.dwg?
   b) Save as any suitable name.

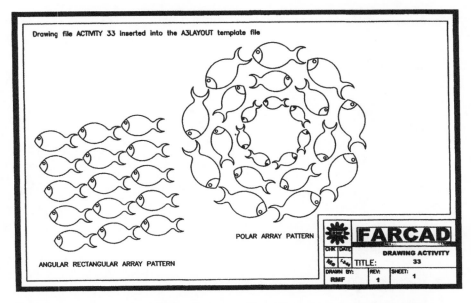

Figure 44.7 Using the A3LAYOUT template file with an inserted drawing file.

6 Close any existing file then re-select your A3LAYOUT template file which should be displayed 'as new'.

7 Thus a template file can be opened, used and a drawing file saved, leaving the original template file 'untouched'.

8 Try opening your A3LAYOUT template file from the C:\BEGIN folder. Does the same concept apply when saving?

Summary

1 A template file is a prototype drawing with various defaults set to user requirements.

2 AutoCAD has several template files conforming to different drawing standards, e.g. ANSI, ISO, etc.

3 Template files can be created and saved by the user.

4 Template files 'safeguard' the prototype drawing from being 'overwritten'.

5 Template files have the extension .dwt and can be saved in the AutoCAD Template folder or in user-defined folders.

Chapter 45
CAD standards

CAD standards allow the user to create a file that defines certain properties, these being layers, dimension styles, text styles and linetypes. This file is saved as a drawing standards file with the extension **.dws**. When a standards drawing has been saved it can be compared with other saved standards drawings and any differences in the four properties will be highlighted.

As all of our exercises have been saved with the .dwg extension, it will be necessary to create a standards drawing. We will use our A3SHEET standard sheet for this purpose, and to demonstrate how CAD standards drawings are used, we will:
a) save our original A3SHEET drawing as a .dws file
b) modify some of the properties in the A3SHEET drawing
c) save this modified drawing as another .dws file
d) compare the original standards file with the modified file.

Getting started

1 Close any existing drawings then menu bar with **File-Open** and from the Select File dialogue box:
 a) folder C:\BEGIN current
 b) pick A3SHEET then Open.

2 Immediately the file is displayed on the screen, menu bar with **File-Save As** and from the Save Drawing As dialogue box:
 a) File type: scroll-pick **AutoCAD Drawing Standards (*.dws)**
 b) File name: enter A3SHEET (if not already the default)
 c) save in: BEGIN folder
 d) pick Save.

3 The original A3SHEET drawing will still be displayed and will now be modified.

4 Menu bar with **Format-Layer** and from the Layers Properties Manager dialogue box, alter the following layer properties:

Layer name	Property to alter	New property
CONS	linetype	HIDDEN
TEXT	colour	green
DIMS	lineweight	0.3

5 Menu bar with **Format-Text Style** and using the Text Style dialogue box, alter the following:
Style name: Standard;
Font name: gothice.shx.

6 Menu bar with **Dimension-Style** and using the Dimension Style Manager dialogue box:
 a) highlight the A3DIM style
 b) right-click and pick rename

   c) alter name to DIMNEW
   d) pick Set Current then close the dialogue box.

7 Now that the modifications to the A3SHEET standard sheet are complete, menu bar with **File-Save As** and:
  a) File type: scroll-pick **AutoCAD Drawing Standards (*.dws)**
  b) File name: enter TEST
  c) Save in: scroll and pick BEGIN folder
  d) pick Save.

Comparing standards

1 Close all existing drawings.

2 Menu bar with **File-Open** and:
  prompt Select File dialogue box
  respond *a*) file type: Standards (*.dws)
      b) Look in: C:\BEGIN
      c) pick TEST then Open.

3 The standards file for TEST will be displayed (check that this file has the modified layers).

4 We now want to compare the properties of the open TEST standards file with those of our A3SHEET standard sheet, as this is the file which is our prototype, i.e. it is 'our standards file'.

5 Menu bar with **Tools-CAD Standards-Configure** and:
  prompt Configure Standards dialogue box
  with Two tabs: Standards and Plug-ins
  respond *a*) ensure Standards tab active
      b) pick +
  prompt Select Standards File dialogue box
  respond *a*) ensure your BEGIN folder current
      b) file type: Standard (*.dws)
      c) pick A3SHEET then Open
  prompt Configure Standards dialogue box
  with *a*) Standards tab active
      b) A3SHEET listed
      c) Description of File, Last Modified, Format
      d) dialogue box as Figure 45.1 (my display will differ from your display)
  respond **pick Check Standards**
  prompt Check Standards dialogue box
  with *a*) Problem
       Dimstyle 'DIMNEW'
       Name is non-standard
      b) Replace with
       1. A3DIM Dimstyle from A3SHEET Standards File
      2. perhaps other named dimension styles
  respond **pick A3DIM then the Fix tick**
      i.e. we are replacing the DIMNEW dimension style with A3DIM style from the A3SHEET standards file
  then Check Standards dialogue box
  with *a*) Problem
       Layer 'DIMS'
       Properties are non-standard

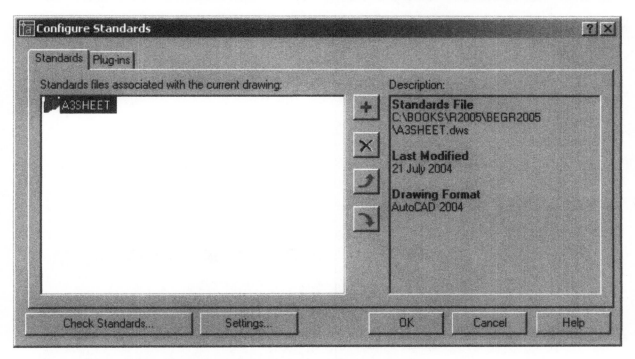

Figure 45.1 The Configure Standards dialogue box for A3SHEET.

	b) Replace with
	    1. DIMS layer from A3SHEET Standards file (highlighted and ticked)
	    2. perhaps other named dimension styles
	c) Preview of changes information – Figure 45.2(a)

respond	**pick DIMS then Fix tick** i.e. we are replacing the 0.30 mm lineweight value on layer DIMS with the default value from the A3SHEET standards file
then	`Check Standards dialogue box`

with	a) Problem
	    Layer 'TEXT': Properties are non-standard
	b) Replace with
	    TEXT layer from A3SHEET standards file (highlighted and active)
	c) Preview of changes information

respond	**pick the Fix tick** i.e. we are replacing the green colour on layer TEXT with the Blue default from A3SHEET
then	`Check Standards dialogue box`

with	a) Problem
	    Layer 'CONS': Properties are non-standard
	b) Replace with
	    CONS layer from A3SHEET standards file
	c) Preview of changes information

respond	**pick the Fix tick** i.e. we are replacing the HIDDEN linetype on layer CONS with the CONTINUOUS default from the A3SHEET standards file
then	`Check Standards dialogue box`

with	a) Problem
	    Textstyle 'Standard': Properties are non-standard
	b) Replace with
	    Standard text style from A3SHEET standards file
	c) Preview of changes information

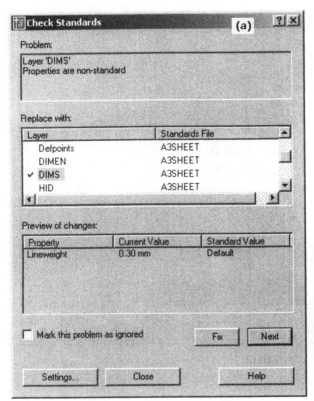
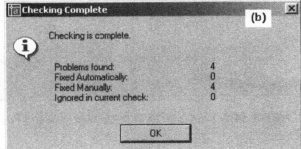

Figure 45.2 The (a) Check Standards dialogue box for layer DIMS and (b) the Checking Complete dialogue box.

respond	**pick the Fix tick** i.e. we are replacing the Standard text style with the gothice.shx font name with the Standard default from A3SHEET
then	`Checking Complete dialogue box`
with	details about the problems and fixes – Figure 45.2(b)
respond	a) read the dialogue box information b) pick OK then close.

This exercise is now complete. Do not save any changes.

Summary

1 CAD standards allow the user to 'check' layers, linetypes, dimension styles and text styles between several drawing files.

2 Standards files have the extension .dws and can be saved in an AutoCAD folder or a user-defined folder.

Chapter **46**

The AutoCAD Design Center

The AutoCAD Design Center is a drawing management system with several powerful advantages to the user, including:
a) the ability to browse and access different drawing sources
b) access drawing content including Xrefs, blocks, hatches and symbol libraries
c) view object definitions (e.g. layers) prior to inserting, attaching or copy and paste into the current drawing
d) search for specific drawing content
e) redefine a block definition
f) create shortcuts to drawings and folders
g) drag drawings, blocks and hatches to a tool palette.

In this chapter we will investigate several of these topics.

Getting started

1. Open your A3SHEET standard sheet and cancel any floating toolbars.

2. Menu bar with **Tools-Design Center** and:
   prompt Design Center window
   respond **study the layout and leave as displayed**.

3. The first time that the Design Center is activated, it may be docked at the side of the drawing screen area but it can be moved and resized by:
   a) left-click in the Design Center title bar and hold down the button
   b) drag into the drawing area
   c) resize to suit.

4. The Design Center window consists of the following:
   a) the Design Center title bar at the left
   b) a toolbar with several icon selections
   c) four tabs: Folders, Open Drawings, History, DC Online
   d) the tree side and hierarchy area on the left
   e) the content side with icons on the right
   f) an information bar below the tree view and palette
   g) an Auto-hide and Properties option.

5. The Design Center window is fully displayed in Figure 46.1(a). If Auto-hide is selected, the window 'collapses' and is displayed as Figure 46.1(b). To fully display the window, click Auto-hide again.

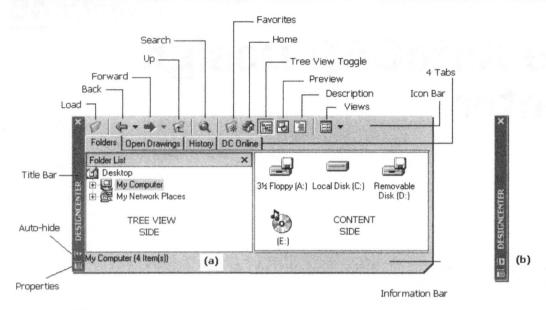

Figure 46.1 The Design Center window (a) fully displayed and (b) with Auto-hide on.

Using the tree view and hierarchy

1 The tree side of the Design Center window consists of a **hierarchy** of list of names with a (+) or a (−) beside them.

2 If a name in the tree side is selected (left-click) then the content side of the window will reveal the contents of the selected item.

3 If the (+) beside an item is selected then that item is '**explored**' and the tree side of the window will list the 'contents' of the selected item. The selected item's (+) will be replaced with a (−) indicating that it has been explored.

4 If the (−) is then selected, the explored effect is removed and the (+) is returned. The selection of the (−) is termed '**collapsing**'. The user is thus **navigating** with the Design Center.

5 *Note*: While 'explore' is the technical term associated with these activities, I prefer 'expand' as it compliments the 'collapse' term.

To demonstrate using the tree side, refer to Figure 46.2 and:

1 Note the Design Center layout as opened – Figure 46.2(a). Your display may originally differ from that shown but this is not important.

2 Explore My Computer (if available) by picking the (+) to give a tree expansion similar to Figure 46.2(b).

3 Explore the C: drive by selecting the (+) to give the folders on the C drive of your system – Figure 46.2(c). This will probably differ to what I have displayed.

4 Explore C:\BEGIN by selecting the (+) to give a listing of the files you have saved, similar to Figure 46.2(d).

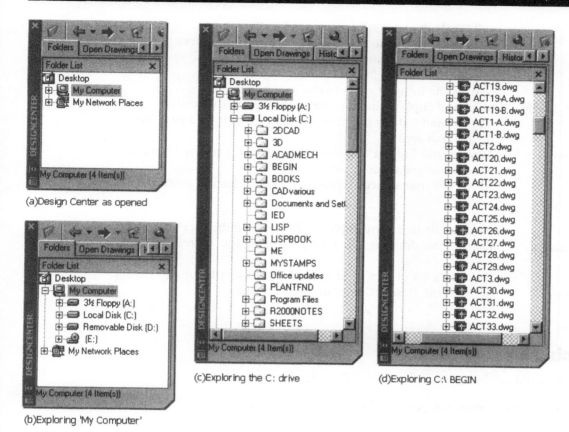

Figure 46.2 Exploring the tree side of the Design Center window.

5   Note that in Figure 46.2 I have only displayed the tree view side of the Design Center. This was to ensure all the displays 'fitted' into a single layout.

6   Move the pointer arrow onto BEGIN and right-click:
    prompt   Shortcut menu
    respond   **pick Explore**
    and   content side of the Design Center window will display icons of the contents of the named folder.

7   *Task*
    a) collapse your named folder by picking the (−) at the name
    b) collapse the C: drive
    c) collapse My computer and the Design Center should be displayed 'as opened'.

The Design Center menu bar and tabs

The menu bar of the Design Center allows the user access to the several icons and tabs. These are displayed and listed in Figure 46.1 and are:

1   *Icons*
    Load   allows the user to navigate to files and load into the contents area
    Back   returns to the most recent location in the history list
    Forward   returns to the next later location in the history list

	Up	will toggle the Design Center 'up a level'
	Search	allows the user to search for named drawings, blocks and non-graphical objects using specified criteria
	Favorites	lists the contents of favorites folder
	Home	returns the Design Center to the user home folder
	Tree View Toggle	hides and displays the tree side of the Design Center
	Preview	displays and hides a preview of a selected item in a pane below the content area
	Description	displays and hides a text description about a selected item in a pane below the content area
	Views	allows the user to access different displays for the content area.
2	*Tabs*	
	Folders	displays the hierarchy of files and folders on the user's computer
	Open Drawings	displays all drawings currently open in the current drawing session
	History	displays a list of files most recently opened with Design Center
	DC Online	accesses the Design Center online webpage.

Exercise 1

1 AutoCAD should still be active with A3SHEET displayed.

2 Activate the Design Center and position the window in centre of screen.

3 Expand the following:
   a) My Computer
   b) the C: drive
   c) the BEGIN (or your working) folder.

4 Right-click on BEGIN and pick Explore and the content side will display information about the selected folder. This may not be in the 'form' we want.

5 From the Design Center menu bar, scroll at Views and note the four options available:
   a) Large icon   thumbnail sketches of the files
   b) Small icon   names only with extension
   c) List   lists all contents of folder in alphabetical order
   d) Details   gives file name, size, type and when modified.

6 The four View options are displayed in Figure 46.3 for the content side of the Design Center window.

7 From the Design Center menu bar select Preview and Description and the content side will display two additional areas. These can be resized by dragging the lower 'border' up or down.

8 With large icons displayed, scroll and pick (left-click) any drawing icon from the content side and a preview and description (if applicable) will be displayed similar to Figure 46.4.

9 Scroll and pick another item (icon). If possible try and pick another type of file format. Figure 46.5 displays a BMP file of Figure 35.6 which was the Geometric Tolerance dialogue box. This dialogue box was 'screen dumped' from AutoCAD into a word-processing package for inclusion in the book.

The AutoCAD Design Center

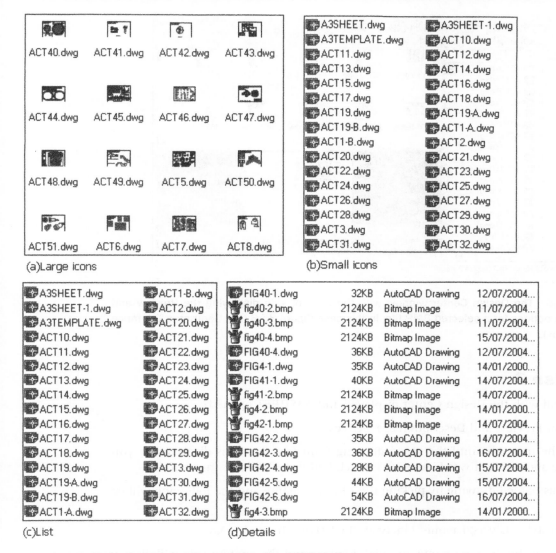

Figure 46.3 The Design Center menu bar VIEW options.

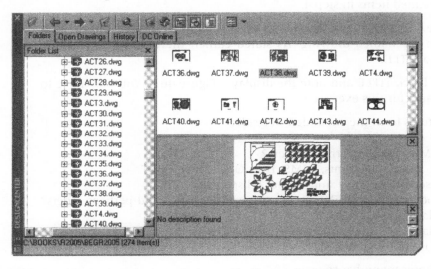

Figure 46.4 The Design Center window with the content side displaying the Preview and Description options for a selected item in the content side – Activity 38.

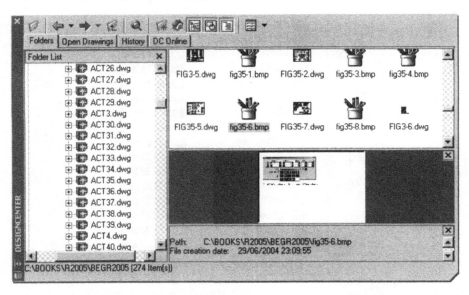

Figure 46.5 The Design Center window with the content side displaying the Preview and Description options for a selected BMP file, in this case Figure 35.6 (the Geometric Tolerance dialogue box).

Exercise 2

1. Are you still with the Design Center displayed and A3SHEET opened?

2. Ensure that Preview and Description are 'active'.

3. Scroll in the tree side until the saved drawing from Chapter 39 is displayed. If you are unsure about this, it was Figure 39.3 which had three Wblocks inserted.

4. Explore this named drawing by picking the (+) at its name and the following will be displayed:

    Blocks, Dimstyles, Layers, Layouts, Linetypes, Tablestyles, Textstyles, Xrefs.

5. Right-click on the drawing name and pick Explore and the content side will display icons for the eight named items in step 4.

6. In the content side, right-click on Blocks and pick explore to display information about the blocks in the named drawing. As stated earlier, there should be three – BORDER, PLIST and TITLE.

7. Pick (left-click) the block TITLE and note the display – Figure 46.6. You may have to resize the preview area for this exercise.

8. Explore the other items from the expanded tree side:
Selection	Result
Dimstyles	A3DIM listed and perhaps others?
Layers	0, CL, CONS, Defpoints, DIMS, HID, OUT, SECT, TEXT
Layouts	not used but Layout1 and Layout2 displayed?
Linetypes	ByBlock, ByLayer, CENTER, Continuous, HIDDEN, and perhaps others?
Tablestyles	Standard only listed?
Textstyles	Standard, ST1 and perhaps ST2, ST3 and ST4?
Xrefs	none listed (obviously?)

9. From the Design Center menu bar select:
    a) Up: displays contents of explored drawing, i.e. Figure 39.3
    b) Up: displays contents of the BEGIN folder

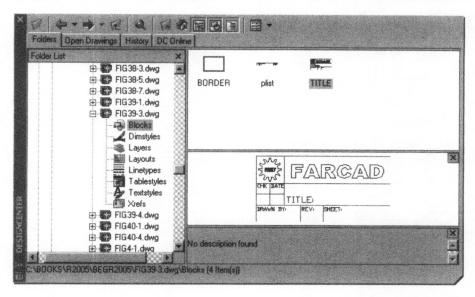

Figure 46.6 The Design Center window with the Preview and Description options for Figure 39.3. The blocks have been explored to display the three Wblocks used in this drawing.

    c) Up: displays the C: drive expansion
    d) Up: displays the My Computer icons.

Note: You may have explored other folders/files during this exercise and will need to select the 'Up' menu bar option more than the four times listed.

Exercise 3

1   Do you still have the Design Center displayed from exercise 2?

2   From the Design Center menu bar pick the **Search icon** and:
    prompt   Search dialogue box
    respond   *a*) scroll at Look for
                 b) pick Xrefs
                 c) scroll at In
                 d) pick Local disk (C:)
                 e) pick Browse
    prompt   Browse for Folder dialogue box
    respond   *a*) explore the C: drive and pick BEGIN
                 b) pick OK
    prompt   Search dialogue box
    with   named folder name displayed
    respond   *a*) at Search for the name: enter*
                 b) pick **Search Now**
    and   the search will begin
    then
    with   Search dialogue box
               a list of the 'found' Xrefs similar to Figure 46.7.

3   *Notes*:
    a) there should be two items found, both XREFEX in drawing files XREFLAY1 and XREFLAY2
    b) my Figure 46.7 has the folder name as C:\BOOKS\R2005\BEGR2005.

4   Now cancel the Find dialogue box and cancel the Design Center to leave the blank A3SHEET standard sheet as opened.

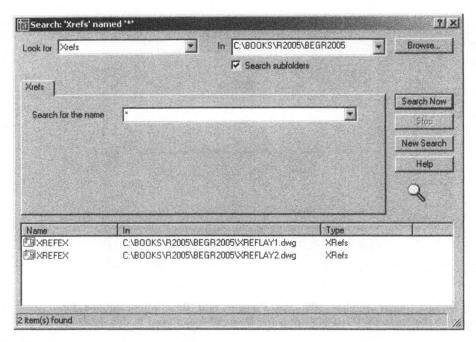

Figure 46.7 The Design Center Search facility. The selected item was 'XREFS' in the BEGR2005 folder.

Exercise 4

1 Close any existing drawings then menu bar with **File-New** and select Start from Scratch, Metric then OK.

2 Menu bar with **Format-Layer**, **Format-Text Style**, **Format-Dimension Style** and note the layers, text styles and dimension styles in this current drawing. These features (unless altered by the user) should be the AutoCAD defaults of:

   Layers: 0   Text Styles: Standard with txt.shx font   Dimension Style: ISO-25

3 Activate the Design Center and position to suit, i.e. docked or floating.

4 Explore the following by picking the (+) at the appropriate folder/file:
   a) My Computer
   b) the C: drive
   c) the BEGIN folder
   d) the A3SHEET drawing file.

5 From the explored A3SHEET tree hierarchy, explore Layers to display the layers from the A3SHEET drawing in the content side of the Design Center.

6 From the content side:
   a) left-click on the OUT layer
   b) hold down the left button
   c) drag into the drawing area
   d) repeat 'pick-and-drag' for layers HID and CL.

7 Explore Textstyles and 'pick-and-drag' all named text styles into the drawing area.

8 From the AutoCAD menu bar with **Format-Layer** and **Format-Text Style**, check that the selected layers and text styles are now available for use in the new drawing.

9 Collapse the explored A3SHEET in the tree side, i.e. pick the (−) at the drawing name.

10 Menu bar with **Insert-Block** and there are no block names listed in the current drawing – obviously?

11 *a*) from the tree side, explore the saved conference room drawing from Chapter 38 (Figure 38.5)
   b) explore Blocks
   c) left-click on each of the four blocks (SEAT, TABLE1, TABLE2 and TABLE3) and drag into the drawing area, positioning to suit. Figure 46.8 displays the Design Center window with TABLE2 selected
   d) menu bar with **Insert-Block** and the four 'pick-and-drag' blocks are listed.

12 This means that we have used the Design Center to 'insert' blocks from a previously saved drawing into the current opened drawing. This is contrary to the statement made in Chapter 38, i.e. **'blocks can only be used in the drawing in which they were saved'**.

13 Thus the Design Center allows ANY blocks, Wblocks, layers, linetypes, text styles, dimension styles, layouts and Xrefs to be 'inserted' from any drawing into the current drawing. This is a very powerful aid to the CAD user.

14 This exercise is complete and does not need to be saved.

15 Collapse all the expanded tree hierarchies and close the Design Center. Your screen should still display the four blocks inserted from Figure 38.5 with the Design Center.

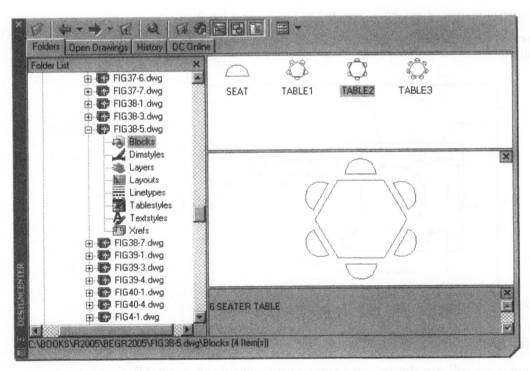

Figure 46.8 The Design Center window for the explored Figure 38.5 drawing with Blocks explored.

Exercise 5

1. Close any existing drawings then menu bar with **File-New** and select Start from Scratch, Metric then OK.

2. Draw the following:
   a) a 100 square as four lines
   b) a 30-radius circle at the square centre
   c) fillet a corner with radius 20 and chamfer a corner with distances of 20.

3. With the menu bar sequence **Dimension-Style**, note the current dimension style. I had ISO-25.

4. Use this dimension style and add the following dimensions, denoted in Figure 46.9 by (a):
   a) baseline linear
   b) circular – both diameter and radius
   c) an angular and a leader.

5. Activate the Design Center and explore:
   a) My Computer
   b) the C: drive
   c) your named folder (BEGIN)
   d) the A3SHEET standard sheet.

6. a) Right-click and explore Dimstyles, and A3DIM should be displayed in content side
   b) Drag-and-drop A3DIM into the current drawing.

7. Menu bar with **Dimension-Style** and
   a) the Dimension Style Manager dialogue box should display the A3DIM style
   b) set A3DIM current
   c) close the dialogue box.

8. Now add similar dimensions as step 4, denoted in Figure 46.9 by (b).

9. This exercise has shown another use for the Design Center. The exercise is complete. Save if required.

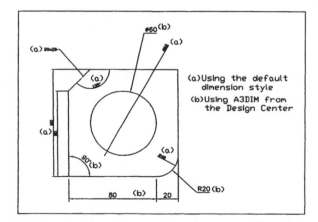

Figure 46.9 Using a dimension style from the Design Center.

Exercise 6

In this example we will use the Design Center to investigate the AutoCAD blocks available to the user in various drawings which come with the package.

1 Close all existing drawing then open your A3SHEET standard sheet with layer OUT current.

2 Activate the Design Center and position to suit.

3 Explore the following trees:
   a) C: drive
   b) Program Files – read notes**
   c) AutoCAD 2005
   d) Sample
   e) Design Center.

4 *Notes***
   a) Program Files is the folder into which my AutoCAD 2005 has been installed.
   b) Your system may use a different folder name.

5 a) explore Analog Integrated Circuits.dwg
   b) explore the blocks within this drawing
   c) Figure 46.10 should be the result of the various 'explore' operations.

6 Refer to Figure 46.11 and drag-and-drop into the opened A3SHEET the first named block from every drawing of the explored Sample–Design Center hierarchy. Remember that as they are blocks, you may have to scale after insertion. The following table is a list of these blocks:

Ref	File name	Block
(a)	Analog Integrated Circuits	10104 Analog IC
(b)	Basic Electronics	Battery
(c)	CMOS Integrated Circuits	14160 CMOS IC
(d)	Electrical Power	Alarm
(e)	Fasteners – Metric	Cross Pt Flathead Screw
(f)	Fasteners – US	Hex Bolt 1/2 inch side

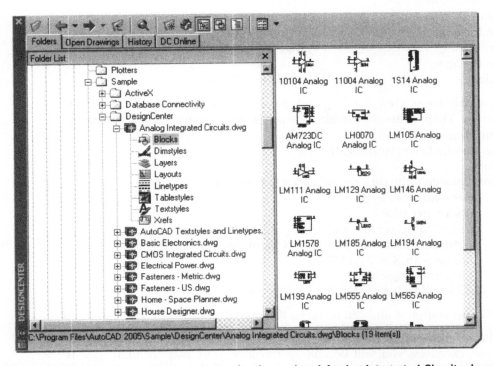

Figure 46.10 The Design Center window for the explored Analog Integrated Circuits drawing.

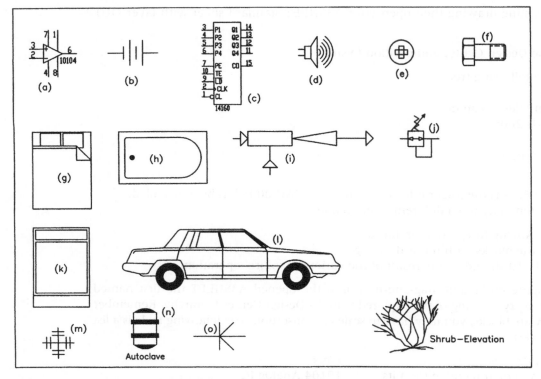

Figure 46.11 Inserting AutoCAD blocks using the Design Center.

(g)	Home – Space Planner	Bed – Queen
(h)	Home Designer	Bath Tub – 26 × 60
(i)	HVAC	Air Injector
(j)	Hydraulic-Pneumatic	Air Regulator
(k)	Kitchens	Base Cabinet
(l)	Landscaping	Car-Sedan side
(m)	Pipe Fittings	Cross-flanged
(n)	Plant Process	Autoclave
(o)	Welding	Double Bevel Weld

7 We have so far only considered the drag-and-drop method of inserting blocks from the Design Center. There is another method which is more accurate than that used. To demonstrate this:
   a) explore the AutoCAD Landscaping.dwg file
   b) explore Blocks
   c) from the content side right-click on Shrub-Elevation and:

prompt	Shortcut menu
respond	**pick Insert Block**
prompt	Insert dialogue box
respond	1. Insertion at X = 340 and Y = 15 – or own values
	2. Scale with X = 2 and Y = 2
	3. Rotation angle: 5

   d) the selected block will be inserted at the desired point.

8 This exercise is now complete and can be saved if required.

This chapter has introduced the user to the AutoCAD 2005 Design Center. Hopefully having tried the exercises, you will be able to navigate your way through it. It is fairly easy, but takes some patience and practice.

Chapter **47**

Toolbars and tool palettes

Toolbars have been used since the first line and circle were drawn and most users will probably be more than satisfied with the existing toolbar icon buttons. It is possible to create toolbars to user requirements thereby improving drawing productivity. Tool palettes can also be created for user blocks and hatch patterns. In this chapter we will investigate how both these draughting tools can be customised.

Customising a toolbar

1 Open your A3SHEET file (drawing or template).
2 Select from the menu bar **View-Toolbars** and:
   prompt   Customize dialogue box
   with   a) five tab selections: Commands, Toolbars, Properties, Keyboard, Tool Palettes
              b) two 'sections': Toolbars and Menu Group
   and   a) Toolbar tab current
              b) several active toolbars denoted with a tick – e.g. Draw, Modify
              c) ACAD as the current menu group
   respond **pick New**
   and   possible ACAD menu group message displayed
   respond **pick Yes**
   prompt   New Toolbar dialogue box
   respond a) Toolbar name: enter MINE
              b) Menu Group: ACAD – Figure 47.1
              c) pick OK
   prompt   Customize dialogue box
   with   a) MINE listed and active (X)
              b) new toolbar displayed at top of screen as Figure 47.2(a)
   respond **pick Close**.

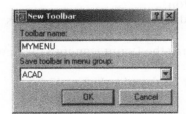

Figure 47.1   The New Toolbar dialogue box.

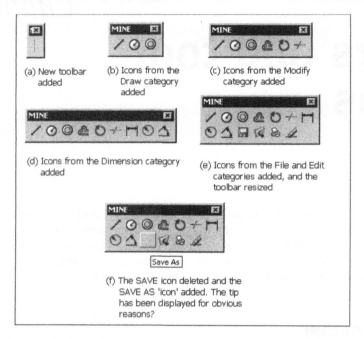

Figure 47.2 Customising a toolbar.

3   Move the new toolbar to the left-centre of the screen by:
    a) moving the cursor into the title bar area of the new toolbar
    b) hold down the left mouse button
    c) drag the new toolbar to desired position
    d) release the left mouse button.

4   Repeat the **View-Toolbars** menu bar selection and:
    prompt   Customize dialogue box
    respond   **pick the Commands tab**
    prompt   Commands tab display
    with   two sections:
                a) Categories
                b) Commands
    respond   **pick the Draw category**
    and   Command section displays all the draw commands with description and tip
    respond   a) pick the Line button form the Commands side (Figure 47.3)
                b) hold down the left mouse button
                c) drag the icon button into the new MINE toolbar
                d) release the left mouse button
                e) scroll at Commands and drag the Circle Center Radius icon into the new toolbar
                f) scroll at Commands and drag the Donut icon into the new toolbar
                g) pick Close
    and   the new MINE toolbar will be displayed as Figure 47.2(b).

5   Select **View-Toolbars** from the menu bar and:
    prompt   Customize dialogue box
    respond   **activate the Commands tab**
    and   list of Categories available

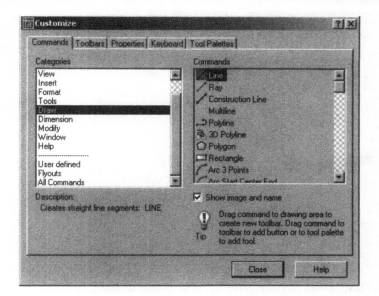

Figure 47.3 The Customize dialogue box with the Commands tab active and the Draw category selected.

    respond   *a)* pick the Modify category
                  b) scroll and pick the Offset icon and drag into the new MINE toolbar
                  c) scroll and pick the Rotate icon and drag into the new toolbar
                  d) scroll and pick the Trim icon and drag into the new toolbar
                  e) new MINE toolbar as Figure 47.2(c)
                  f) pick the Dimension category
                  g) pick and drag the following icon buttons into the new toolbar – Linear Dimensioning, Radius Dimensioning, Angular Dimensioning – Figure 47.2(d)
                  h) pick Close.

6 At the command line enter **TOOLBAR <R>** and:
    prompt   Customize dialogue box
    respond   *a)* activate the Commands tab and File category
    prompt   *b)* pick and drag the Save, Open and Plot icons into the new toolbar
                  c) activate the Edit category
                  d) pick and drag the Erase icon into the new toolbar
                  e) pick Close.

7 Your MINE toolbar is complete and can be positioned and resized to suit – Figure 47.2(e).

8 Now:
    a) use all the icons in your toolbar except the SAVE icon – why?
    b) activate the Customize dialogue box with the Commands tab current
    c) right-click the Save icon and from the shortcut menu pick Delete then OK from the message box
    d) now add the SAVE AS icon to your toolbar – no icon, just a blank space as Figure 47.2(f)
    e) close the dialogue box.

9 Your customised toolbar can be closed and opened at any time as it has now been added to the AutoCAD list of toolbars and can be activated with the menu bar selection **View-Toolbars** – Figure 47.4.

10 This section on creating a customised toolbar is now complete.

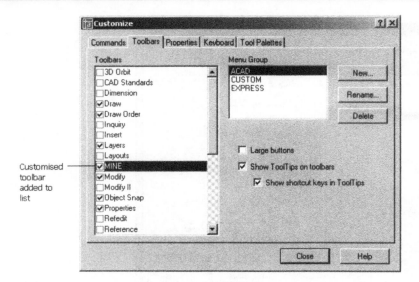

Figure 47.4 New toolbar MINE added to the current list of available toolbars.

Tool palettes

A tool palette is a draughting aid which can be customised by the user with the addition of drawings, blocks and hatch patterns. These AutoCAD objects can be dragged from the Design Center or cut and pasted from other tool palettes. To view the default tool palette:

1 A3SHEET standard sheet should still be opened.

2 Menu bar with **Tools-Tool Palettes Window** and:
prompt   Tool Palettes window
with   Four tab options (Figure 47.5):
            a) Sample office project
            b) Imperial Hatches
            c) ISO Hatches
            d) Command Tools
respond   study then cancel the window.

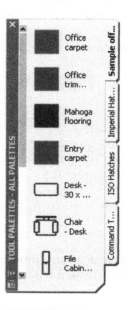

Figure 47.5 The default TOOL PALETTES window.

Creating a new tool palette

1 Ensure A3SHEET is opened and activate the Design Center and Tool Palettes windows.

2 Explore:
   a) your BEGIN folder on the tree side to display the folder contents with large icons
   b) your equivalent to Figure 39.3 which has the three blocks of border, plist and title
   c) explore Blocks from the tree side to display the three named blocks in the content side.

3 **Right-click in the content side** and:
   prompt   shortcut menu
   respond **pick Create Tool Palette**
   prompt   Tool Palette window
   with   *a*) the four default tabs
              b) a new tab (Figure 39.3 in my case)
              c) the three blocks added to this palette
   respond **right-click the new tab name**
   prompt   shortcut menu
   respond *a*) pick Rename Tool Palette
             b) alter name to **MYTRY** then <RETURN>.

4 The new tool palette will be renamed.

5 *a*) explore a saved drawing from Chapter 40
   b) explore the blocks from this drawing and TROPHY displayed in the content side?
   c) left-click on TROPHY and drag into the MYTRY tab – Figure 47.6.

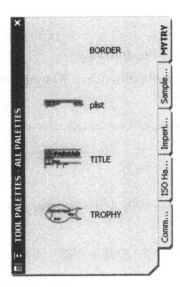

Figure 47.6 The new MYTRY tool palette with blocks added from previous drawings.

6 Right-click the MYTRY tab and from the shortcut menu:
   a) pick View Options
   b) select List View then OK.

7 Cancel the Tool Palette and Design Center windows to complete this exercise.

This chapter has been a brief introduction to customising toolbars and tool palettes. These are very powerful draughting aids if used correctly.

Chapter 48

Sheet sets

What is a sheet set?

1. Any major drawing project consists of a series of drawings/layouts and organising these into a 'usable order' can be very time-consuming. AutoCAD 2005 allows multiple drawing files to be organised into a sheet set, this being achieved with the Sheet Set Manager. New sheets can be created from 'scratch' and added to the existing sheet set and existing drawing files can be imported into the sheet set.

2. In simple terms:
   a) **a sheet set is a named collection of drawing sheets**
   b) the Sheet Set Manager allows the user to organise, display and manage sheet sets
   c) **each sheet in a sheet set is a layout in a drawing (DWG) file**.

The Sheet Set Manager

We will now create a sheet set from our existing drawings so:

1. Open your A3SHEET and menu bar with **Tools-Sheet Set Manager** and:

    prompt   Sheet Set Manager palette
    with   *a*) Sheet List control at top
              b) Several buttons which vary depending on tab selected
              c) Three tabs:
                  1. Sheet List
                  2. View List
                  3. Resource Drawing
                  4. Figure 48.1 displays the basic Sheet Set Manager palette
    respond   *a*) scroll at the Sheet List control
              b) pick New Sheet Set
    prompt   Create Sheet Set - Begin dialogue box
    respond   *a*) Existing drawings active
              b) pick Next
    prompt   Create Sheet Set - Sheet Set Details dialogue box
    respond   *a*) Name of new sheet set: enter **MINE-1**
              b) Description: enter to suit, e.g. MY FIRST ATTEMPT AT A SHEET SET USING DRAWINGS FROM MY NAMED FOLDER
              c) Store sheet set data file (.dst) here: ensure named folder is listed. It should be if you have opened your A3SHEET from your named folder. If it is not listed then use the navigate button and select it
              d) read the note, it may be useful
              e) pick Next

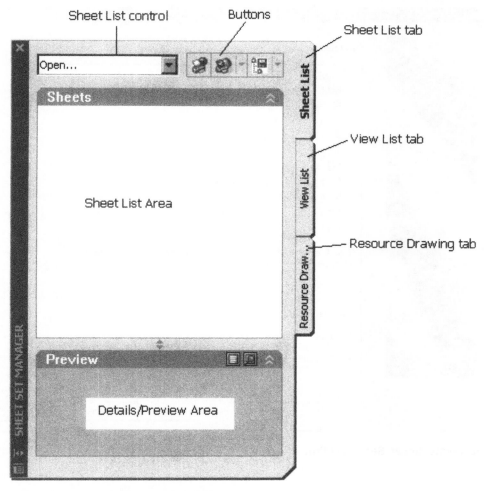

Figure 48.1 The SHEET SET MANAGER.

prompt	Create Sheet Set – Choose Layouts dialogue box
respond	**pick Import Options**
prompt	Import Options dialogue box
respond	**all options active then pick OK**
prompt	Create Sheet Set – Choose Layouts dialogue box
respond	**pick Browse**
prompt	Browse for Folder dialogue box
respond	*a)* navigate to your named folder
	b) pick OK
prompt	Create Sheet Set – Choose Layouts dialogue box displayed and all drawings having layouts will be displayed
respond	*a)* de-activate (i.e. remove drawing name tick) drawings which only list Layout1 and Layout2
	Notes
	1. you may have several drawings with Layout1/Layout2 listed. These are the layouts which have not been used by us. We only want layouts we have used
	2. I have several layouts which you may not have in your named folder
	b) pick Next

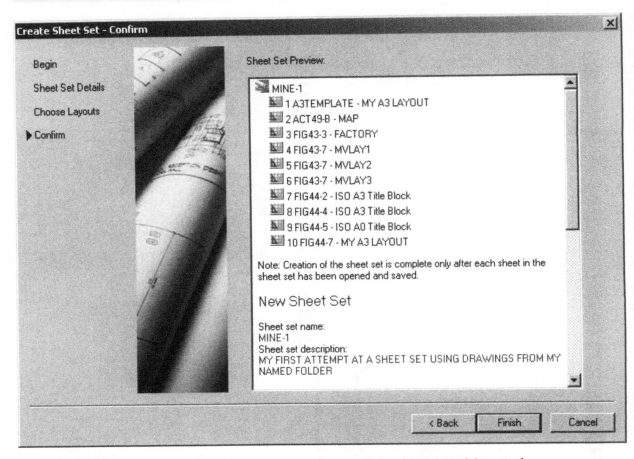

Figure 48.2 The Create Sheet Set – Confirm dialogue box with selected layouts from named folder.

    prompt   Create Sheet Set – Confirm dialogue box (similar to Figure 48.2)
    respond   **pick Finish**.

2   The sheet set palette will be displayed with the listed layout data.

3   Now cancel the Sheet Set Manager palette and close A3SHEET with no changes but with AutoCAD still active.

Adding a View List to the palette

1   AutoCAD still active, so menu bar with **File-Open Sheet Set** and:
    prompt   Open Sheet Set dialogue box
    respond   *a*) navigate to your named folder (if not current)
                b) pick Mine-1
                c) pick Open
    prompt   Sheet Set Manager palette with MINE-1 active
    respond   *a*) right-click on A3TEMPLATE – MY A3 TEMPLATE (or similar name)
                b) pick Open from the shortcut menu
    and   the template file is opened with appropriate layout tab active.

2   Ensure that paper space is active.

3   Pick the **Resource Drawings tab** from the palette and:
    prompt   Sheet List area is blank except for Add New Location
    respond   **pick Add New Location** (two left-clicks or right-click and select)
    prompt   Browse for Folder dialogue box
    respond   a) named folder active
               b) pick Open
    prompt   Sheet Set Manager palette displayed with list of draw-
               ings from named folder
    respond   a) scroll and select any activity drawing
               b) right-click
               c) from shortcut menu select Place on Sheet
    and   selected activity drawing displayed
    prompt   Specify insertion point at command line
    respond   move drawing to suitable position and let it settle
    prompt   tooltip displayed
    respond   right-click to display scales
    prompt   select suitable scale (e.g. 1:4) then position.

4   Repeat step 3 and place another four activity drawings on the A3TEMPLATE sheet, selecting suitable scales and insertion points. Figure 48.3 is my A3TEMPLATE sheet with the activities.

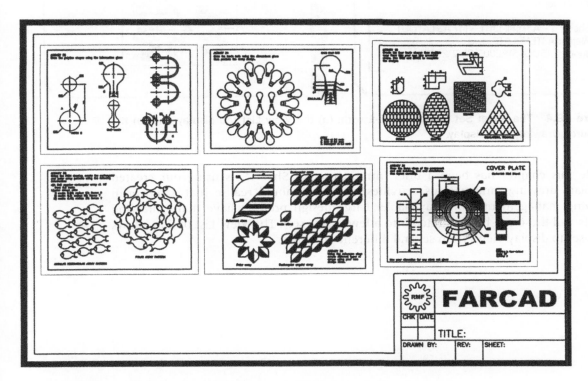

Figure 48.3   Adding activity drawings to the MINE-1 sheet set.

5   Activate the View List tab and the palette will be similar to Figure 48.4(a). With a right-click, the user can:
    a) rename and renumber the named drawing list
    b) display individual sheets
    c) Figure 48.4 also displays the Resources Drawings tab with the named folder 'opened'.

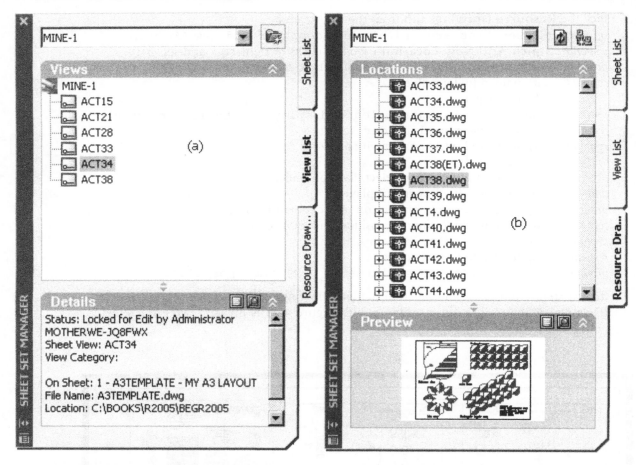

Figure 48.4 The Sheet Set palette MINE-1 with: (a) the View List tab display and (b) the Resource Drawings tab display.

While this chapter has been a brief introduction to Sheet Sets using our created template drawings and various activities, the user should realise that a powerful drawing management tool is available. Large design projects have many drawing sheets and these can be integrated into a few sheets sets. The View List option allows the user to display views from several different drawings.

Chapter **49**

'Electronic' AutoCAD

AutoCAD allows the user to generate electronic drawing files, these files being in Drawing Web Format (DWF). These DWF files can be opened, viewed or plotted by third-party persons having access to DWF Viewer. In this chapter, we will investigate several of these electronic generating devices.

Creating a DWF file

DWF files are 'at the heart' of AutoCAD electronic generation, and to demonstrate how these are created:

1. Close any existing drawing files.
2. Open any of your saved AutoCAD activities, e.g. ACT38 and:
   a) remove all text and dimensions
   b) with layer TEXT current, add a revision cloud and an item of text by referring to Figure 49.1.

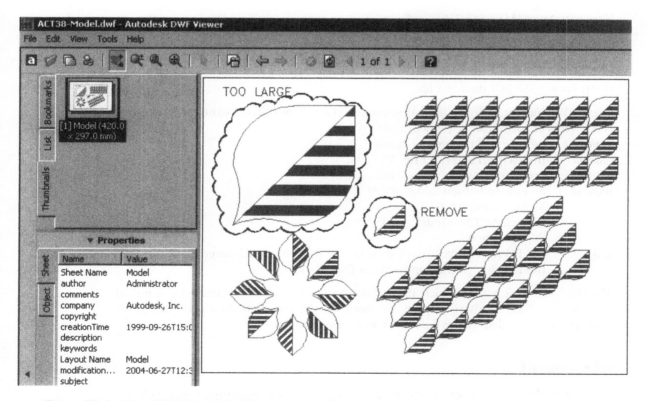

Figure 49.1 The ACT38-Model.dwf file opened in DWF Viewer.

3   Menu bar with **File-Plot** and:
    prompt   Plot dialogue box
    respond a) Printer/plotter: scroll and select **DWF6ePlot.pc3**
               b) Paper size: scroll and select ISO full bleed A3 (297 × 420)
               c) Plot area: What to plot: Display
               d) Plot offset: Center the plot active
               e) Plot style table: None
               f) Plot scale: Fit to paper active (tick)
               g) Plot options: set to own requirements
               h) Drawing orientation: Landscape active
               i) pick Preview
    and   centred preview of activity drawing displayed
    respond **right-click**
    prompt   shortcut menu
    respond **pick Plot**
    prompt   Browse for Plot File dialogue box with file type *.dwf
    respond a) scroll to your named folder
               b) accept file name – ACT38-Model.dwf (or similar)
               c) pick Save
    prompt   Plot and Publish Job Complete with No errors or warnings
               message displayed
    respond **cancel the message**.

4   Menu bar with **File-Save As** and:
    a) alter file name to ACT38(ET) – for future use
    b) ensure saving as a *.dwg file
    c) pick Save.

5   Now exit AutoCAD.

Autodesk DWF viewer

1   Open your A3SHEET drawing file.

2   Open DWF Viewer with the taskbar sequence **Start-Programs-Autodesk-Autodesk DWF Viewer** (or similar sequence) and:
    prompt   Getting Started screen
    respond **pick Open a DWF File**
    prompt   Open File dialogue box
    respond a) ensure named folder active
               b) file types: DWF
               c) pick ACT38-Model
               d) pick Open
    and   DWF Viewer screen will display the ACT38-Model saved DWF file with the revision cloud comments as Figure 49.1.

3   The DWF viewer is a 'tool' which allows third parties to open, view and print AutoCAD drawings (in DWF format) without the need for the complete AutoCAD draughting package to be installed. This is a useful management facility.

eTransmit

eTransmit allows the user to create a set of AutoCAD drawings (only DWG or DWT formats) which can be posted on the Internet or sent to others as an e-mail attachment. The process generates a report file which allows the user to add notes and a password if required. The files to be transmitted can be stored by the user:

1. in a named folder
2. in a created self-executable or zip file.

We will demonstrate the concept by example, so:

1. Close any existing drawings then open ACT30(ET).
2. Menu bar with **File-eTransmit** and:
   - *prompt*   Create Transmittal dialogue box
   - *with*   Current Drawing information: tree and table
   - *respond*   **pick Transmittal setups**
   - *prompt*   Transmittal Setups dialogue box
   - *respond*   **pick New**
   - *prompt*   **New Transmittal Setup dialogue box**
   - *respond*
     - a) New Transmittal setup name: enter **MYTRIAL**
     - b) Based on: Standard
     - c) pick Continue
   - *prompt*   Modify Transmittal Setup dialogue box
   - *respond*
     - a) Transmittal package type: scroll and pick **Self-extracting executable (*.exe)**
     - b) Transmittal file folder: scroll and pick named folder if required
     - c) Transmittal file name: Prompt for a file name
     - d) Transmittal options: set as required – e.g. Prompt for password active
     - e) Transmittal setup description: add as required – e.g. My first attempt
     - f) pick OK
   - *prompt*   Transmittal Setups dialogue box with MYTRIAL added to list
   - *respond*   **pick Close**
   - *prompt*   Create Transmittal dialogue box with MYTRIAL added to transmittal setup
   - *respond*
     - a) add notes as required (Figure 49.2)
     - b) pick OK
   - *prompt*   Specify Self-extracting Executable File dialogue box
   - *respond*
     - a) scroll and pick named folder – should be current
     - b) file name should be ACT38(ET) - MYTRIAL.exe
     - c) pick Save
   - *prompt*   Transmittal - Set Password dialogue box (if password option activated)
   - *respond*
     - a) Password: enter to suit
     - b) Confirm: enter same password
     - c) pick OK.

3. The command line will return the message:

   Updating fields
   Transmittal created: named folder path and file name

4. Now exit AutoCAD and open Windows Explorer and:
   a) navigate to your named folder
   b) arrange the icons by name

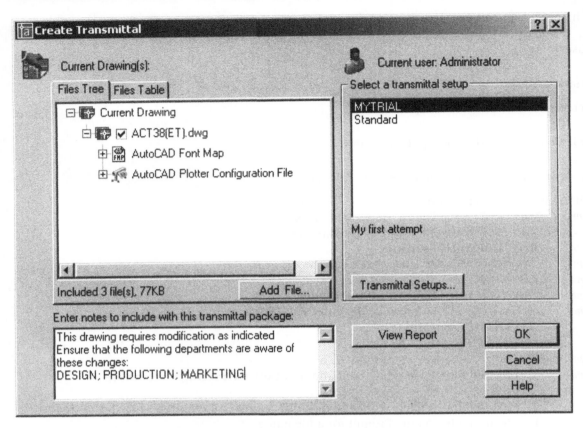

Figure 49.2 The Create Transmittal dialogue box for ACT38(ET).

    c) right-click on the new ACT38(ET)-MYTRIAL icon and pick Properties from the displayed Shortcut menu
    d) note the dialogue box display similar to Figure 49.3 then Cancel.

5 To demonstrate how an eTransmit file is 'unpacked', Windows Explorer should still be active, so double left-click on the ACT30(ET) executable file and:
    prompt   Transmittal dialogue box
    respond   *a)* pick Browse
                  b) enter a new path for extracting the file
                  c) pick OK
    prompt   Transmittal password dialogue box
    respond   enter your password then pick OK
    prompt   Transmittal message with 'All items were successfully extracted'
    respond   pick OK
    respond   Explorer screen returned.

6 *a)* From Explorer, scroll and pick the named folder to which the eTransmit was extracted and the following files should be listed:
       1. acad.fmp – an AutoCAD font map file
       2. ACT38(ET) – the extracted AutoCAD drawing
       3. ACT38(ET).txt – a text document.
    b) Double left-click on the ACT30(ET) text file and Notepad will display the Transmittal report. Cancel Notepad.
    c) Double left-click on the ACT30(ET) drawing file and AutoCAD will be opened and display the transmitted file.

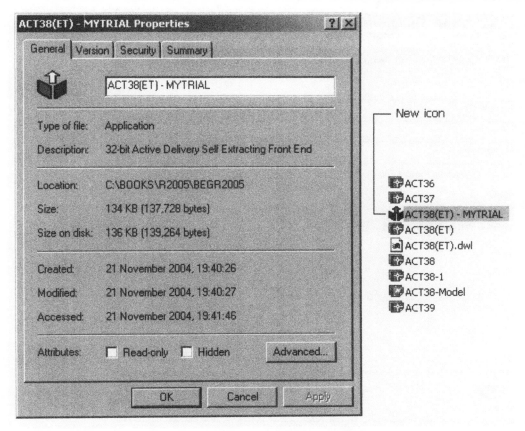

Figure 49.3 Properties dialogue box for ACT38(ET) - MYTRIAL.

7   AutoCAD files can be attached and send as an e-mail, and Figure 49.4 is a typical e-mail screen with two attachments:
    a) the ACT30(ET).dwg file
    b) the ACT30(ET).exe file.

8   While an AutoCAD drawing file can be sent as an attachment, the file may be very large and this can cause 'problems' when it is being 'opened' by the receiver. An exe (or zipped) file may be easier to send electronically.

Publishing to the Web

AutoCAD allows the user to create Web pages of existing drawings. To demonstrate the concept:

1   Start a new metric drawing from scratch to display the typical AutoCAD blank screen.
2   Menu bar with **File-Publish to Web** and:
    prompt   Publish to Web (Begin) dialogue box (similar to the Layout dialogue box)
    respond a) pick Create New Web Page
               b) pick New
    prompt   Publish to Web - Create Web Page dialogue box
    respond a) Web page name: enter MYPAGE
               b) Ensure parent directory is your named folder
               c) Add any suitable description – Figure 49.5
               d) pick Next

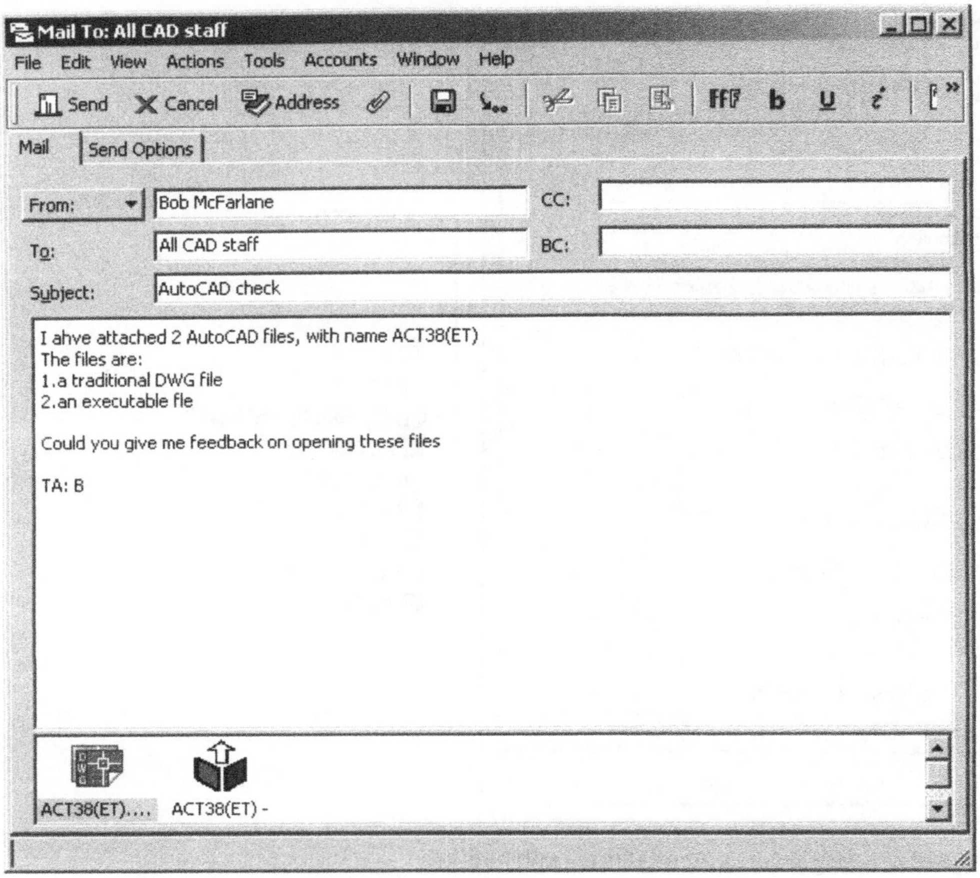

Figure 49.4 Sending AutoCAD files as attachments by e-mail.

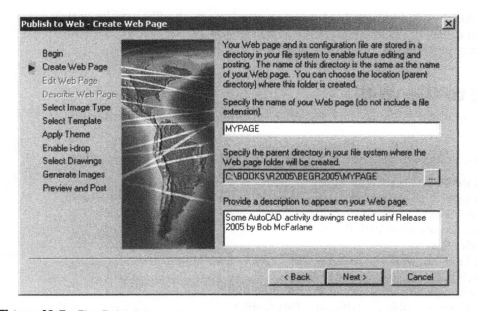

Figure 49.5 The Publish to Web – Create Web page dialogue box.

prompt	Publish to Web - Select Image Type dialogue box
respond	*a)* select type from list: DWF
	b) pick Next
prompt	Publish to Web - Select Template dialogue box
respond	*a)* select a template type – e.g. Array plus Summary
	b) pick Next
prompt	Publish to Web - Apply Theme dialogue box
respond	*a)* scroll and select an element, e.g. Classic – Figure 49.6
	b) pick Next
prompt	Publish to Web - Enable i-drop dialogue box
respond	activate i-drop then pick Next
prompt	Publish to Web - Select Drawings dialogue box
respond	at Drawing, select the (. . .) navigate button
prompt	Publish to Web dialogue box
respond	*a)* scroll to your named folder
	b) select an activity – e.g. ACT36
	c) pick Open
prompt	Publish to Web - Select Drawings dialogue box
with	ACT36 listed
respond	*a)* Layout: select Model – probably active
	b) Label: alter to GEAR (or similar)
	c) Description: enter to suit
	d) pick Add->
and	GEAR added to Image list
now	select several drawings from your list and:
	a) Layout: Model
	b) add a suitable label
	c) add a suitable description
	d) pick Add->
then	pick Next
prompt	Publish to Web - Generate Images dialogue box

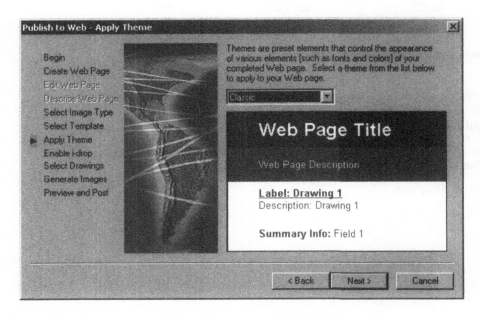

Figure 49.6 The Publish to Web – Apply Theme dialogue box.

respond	*a*) Regenerate all images
	b) pick Next
prompt	Plot progress information displayed
then	`Publish to Web - Preview and Post` dialogue box
respond	pick Preview
and	Internet Explorer with Images of drawings – Figure 49.7
respond	*a*) view your images
	b) close the Internet Explorer to return to AutoCAD
and	select Finish.

MYPAGE

Some AutoCAD activity drawings created using Release 2005 by Bob McFarlane

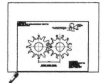

GEAR
SPUR GEARS created using the array command

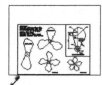

PROPELLOR
Various lobed propellors created from a single blade design

LIGHT BULB
Pattern design

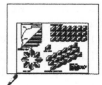

LEAF
Leaf designs created using the array command

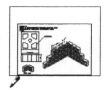

WALL
Isometric wall created from a single brick design

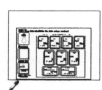

STAMPS
Stamp layout with attributes

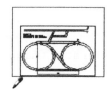

RAILWAY
Railway track layout using multilines

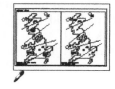

MAP
Weather map created with xrefs/blocks

Figure 49.7 Preview of the created Web page.

3 You have now created a Web page which can be:
  a) edited to your requirements
  b) posted as required.

4 *Note*: I hope that in this chapter the user has realised that AutoCAD has uses other than drawing. The Web page creation is very useful and relatively simple.

This exercise is now complete.

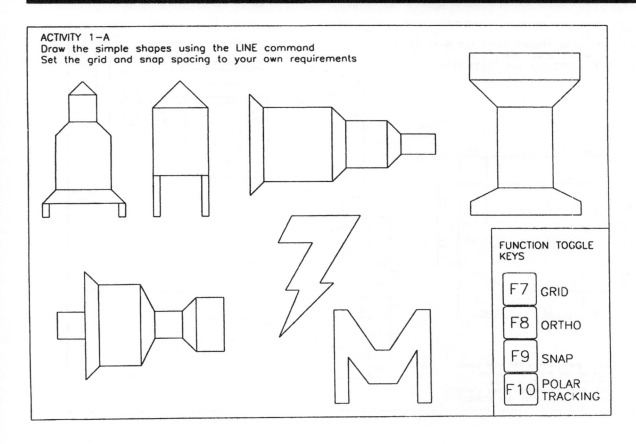

ACTIVITY 1-A
Draw the simple shapes using the LINE command
Set the grid and snap spacing to your own requirements

FUNCTION TOGGLE KEYS
- F7 GRID
- F8 ORTHO
- F9 SNAP
- F10 POLAR TRACKING

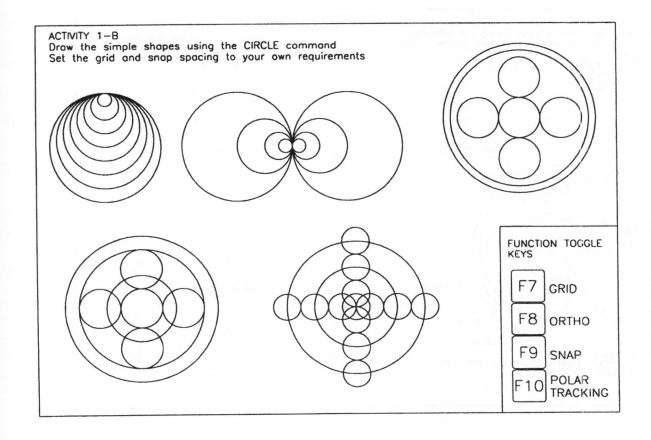

ACTIVITY 1-B
Draw the simple shapes using the CIRCLE command
Set the grid and snap spacing to your own requirements

FUNCTION TOGGLE KEYS
- F7 GRID
- F8 ORTHO
- F9 SNAP
- F10 POLAR TRACKING

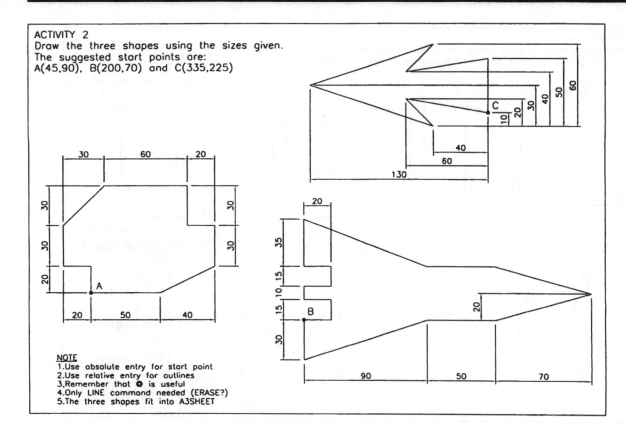
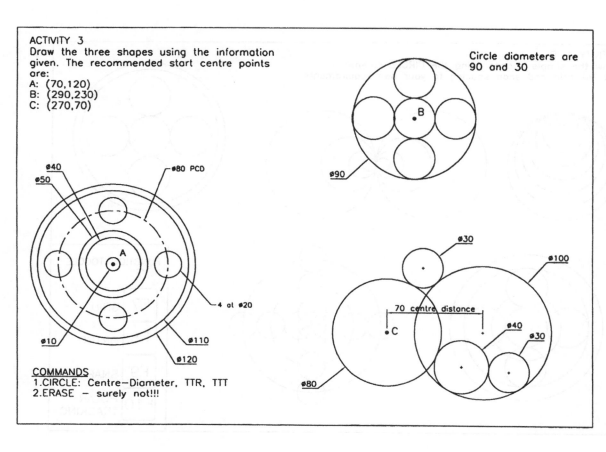

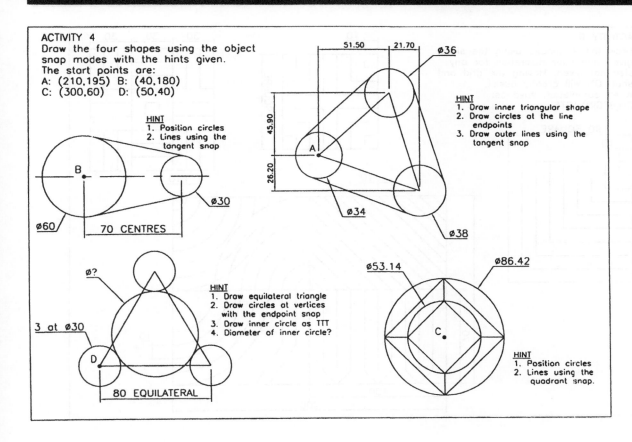

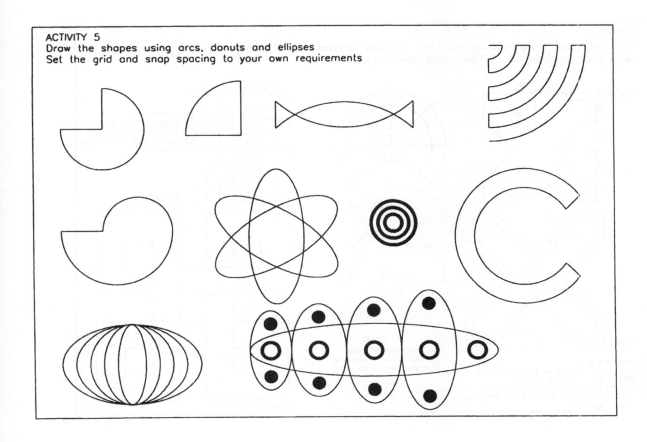

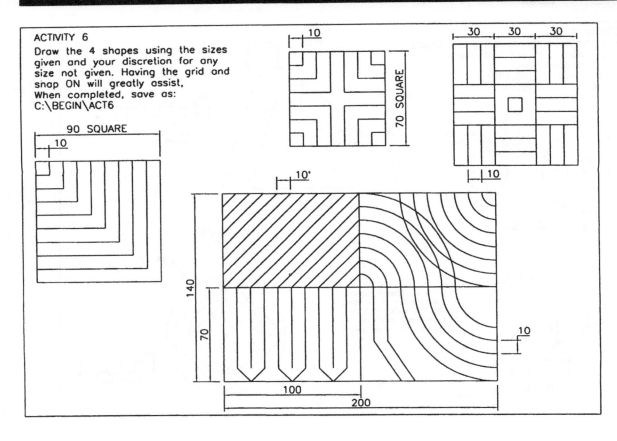
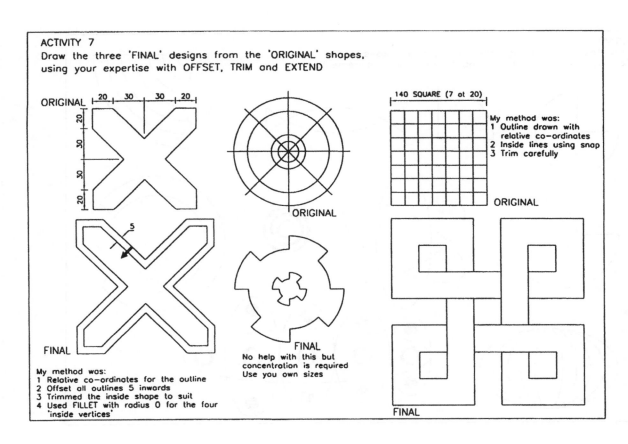

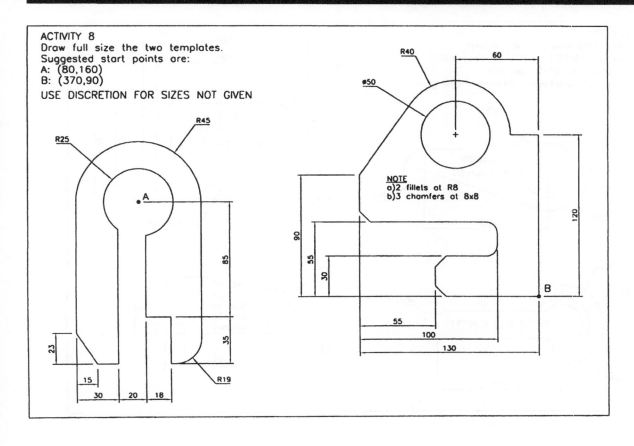

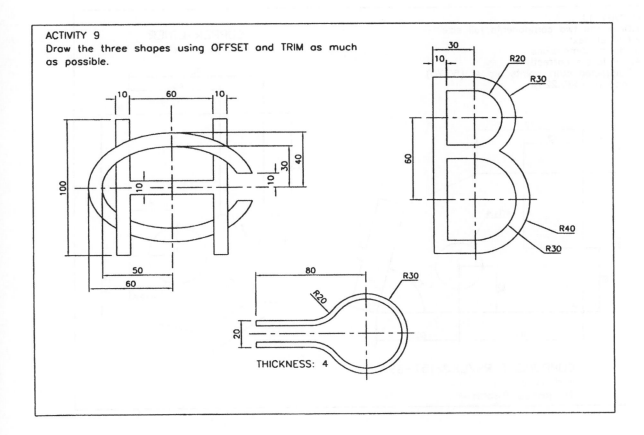

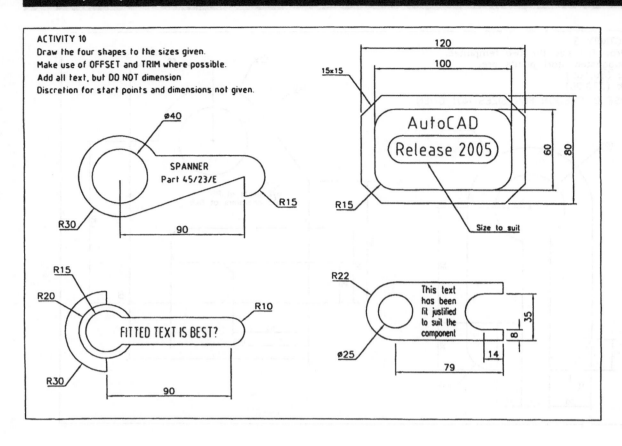

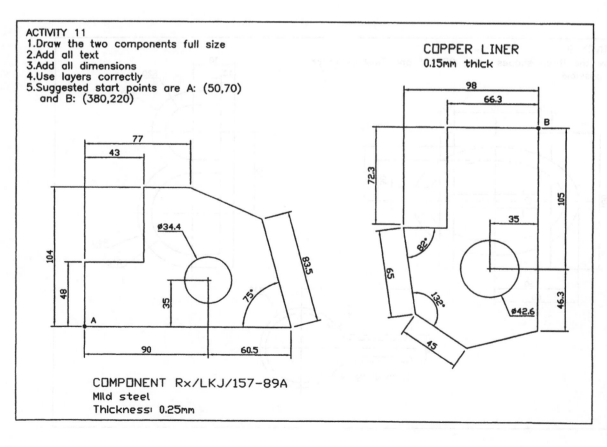

ACTIVITY 12
Draw, dimension and add text. Use layers correctly.

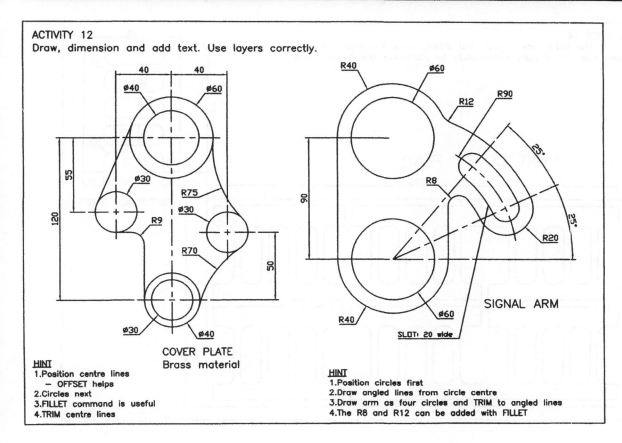

COVER PLATE
Brass material

HINT
1. Position centre lines
   – OFFSET helps
2. Circles next
3. FILLET command is useful
4. TRIM centre lines

SIGNAL ARM

HINT
1. Position circles first
2. Draw angled lines from circle centre
3. Draw arm as four circles and TRIM to angled lines
4. The R8 and R12 can be added with FILLET

ACTIVITY 13
Draw, fully dimension and add text
Easier than you think!!!

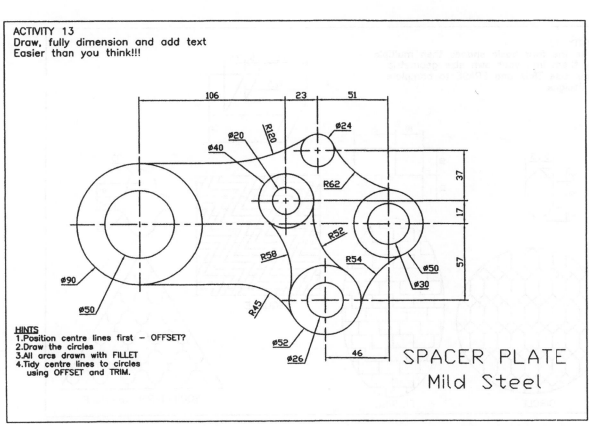

SPACER PLATE
Mild Steel

HINTS
1. Position centre lines first – OFFSET?
2. Draw the circles
3. All arcs drawn with FILLET
4. Tidy centre lines to circles
   using OFFSET and TRIM.

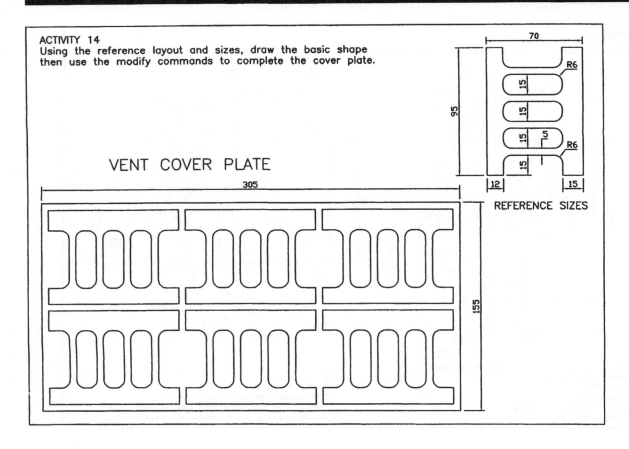

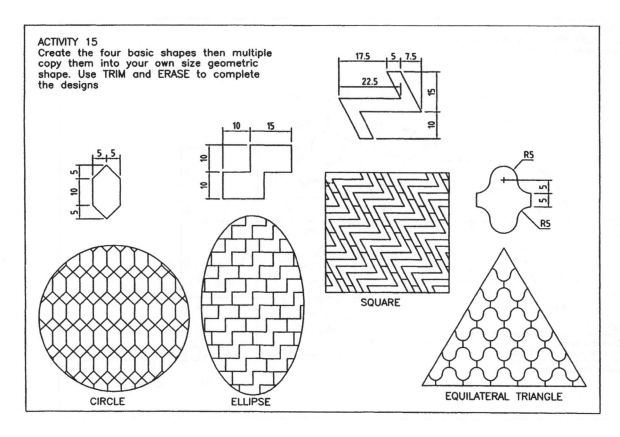

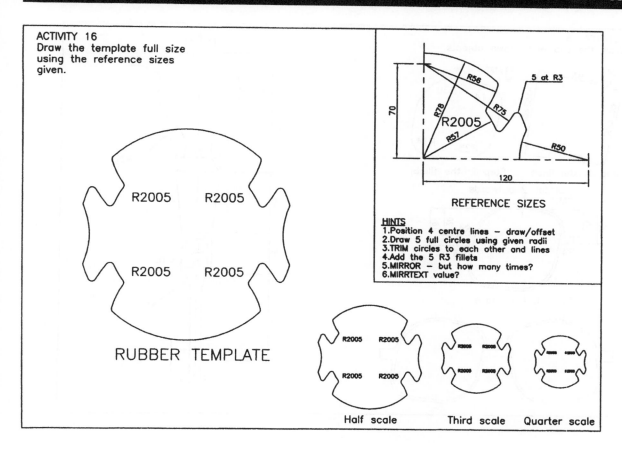

ACTIVITY 16
Draw the template full size using the reference sizes given.

HINTS
1. Position 4 centre lines – draw/offset
2. Draw 5 full circles using given radii
3. TRIM circles to each other and lines
4. Add the 5 R3 fillets
5. MIRROR – but how many times?
6. MIRRTEXT value?

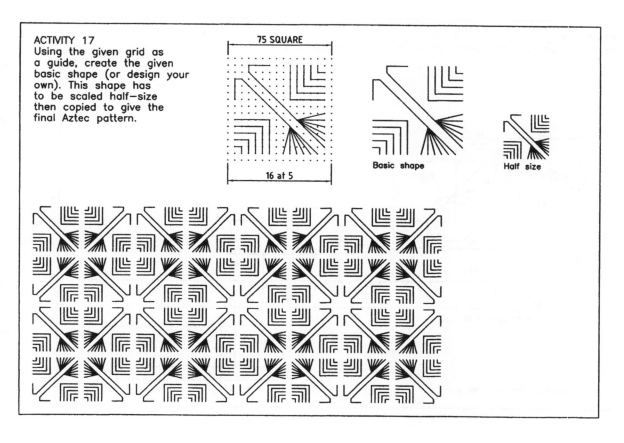

ACTIVITY 17
Using the given grid as a guide, create the given basic shape (or design your own). This shape has to be scaled half-size then copied to give the final Aztec pattern.

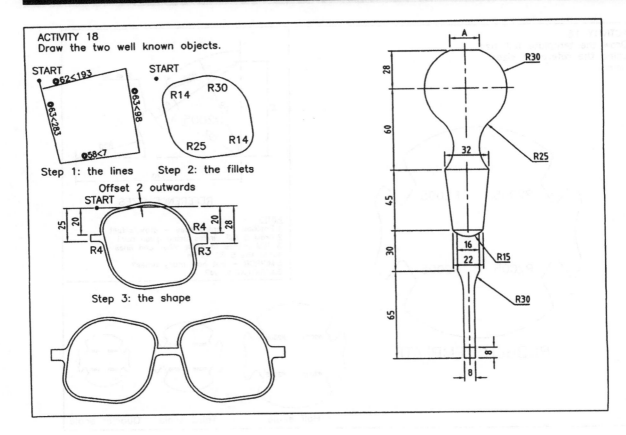
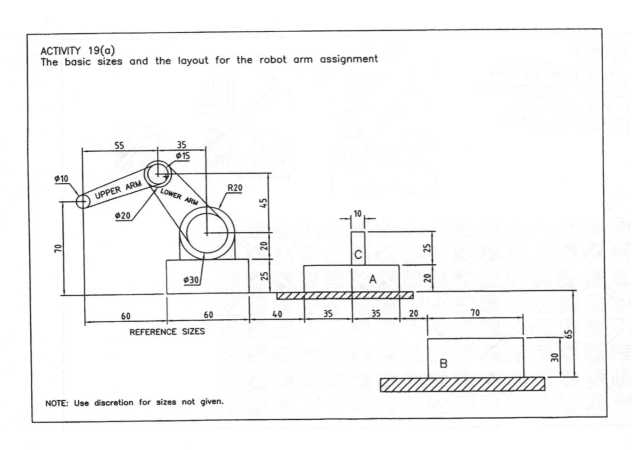

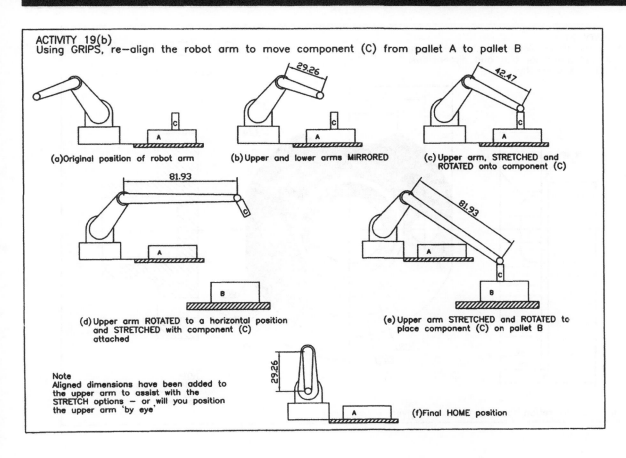

ACTIVITY 19(b)
Using GRIPS, re-align the robot arm to move component (C) from pallet A to pallet B

(a) Original position of robot arm
(b) Upper and lower arms MIRRORED
(c) Upper arm, STRETCHED and ROTATED onto component (C)
(d) Upper arm ROTATED to a horizontal position and STRETCHED with component (C) attached
(e) Upper arm STRETCHED and ROTATED to place component (C) on pallet B
(f) Final HOME position

Note
Aligned dimensions have been added to the upper arm to assist with the STRETCH options – or will you position the upper arm 'by eye'

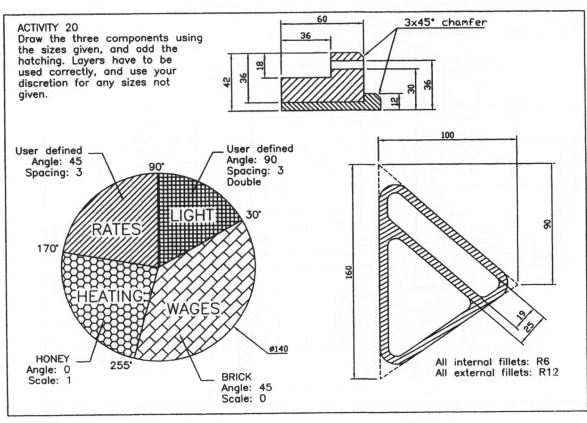

ACTIVITY 20
Draw the three components using the sizes given, and add the hatching. Layers have to be used correctly, and use your discretion for any sizes not given.

All internal fillets: R6
All external fillets: R12

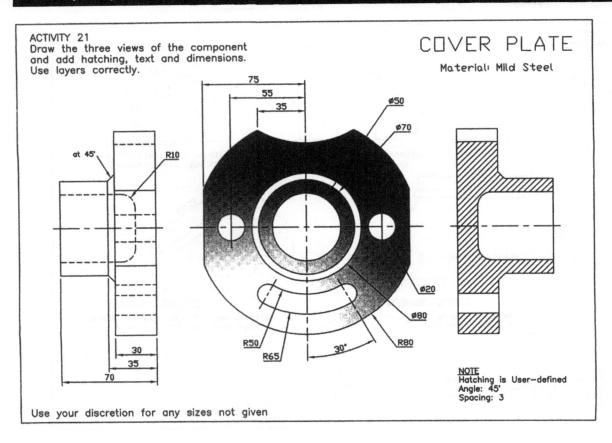

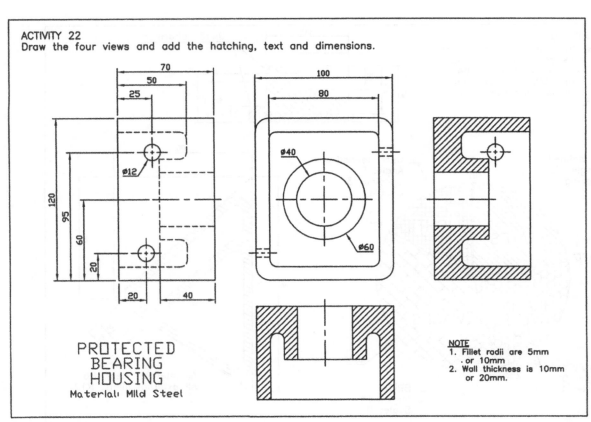

ACTIVITY 23
Draw the component and add text, hatching and dimensions.

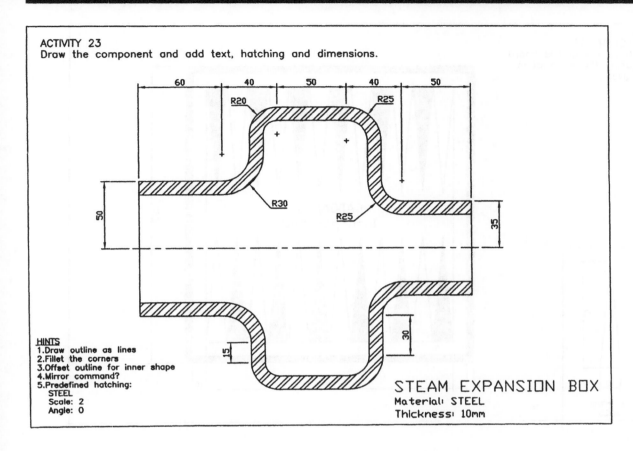

HINTS
1. Draw outline as lines
2. Fillet the corners
3. Offset outline for inner shape
4. Mirror command?
5. Predefined hatching:
   STEEL
   Scale: 2
   Angle: 0

STEAM EXPANSION BOX
Material: STEEL
Thickness: 10mm

ACTIVITY 24
Draw full size, adding the hatching.

GASKET COVER

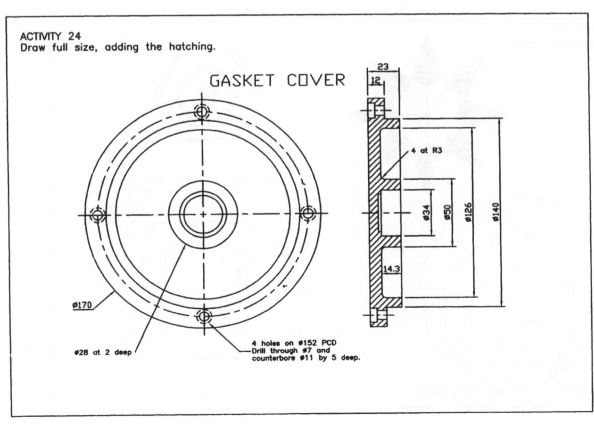

Ø170
Ø28 at 2 deep
4 holes on Ø152 PCD
Drill through Ø7 and
counterbore Ø11 by 5 deep.

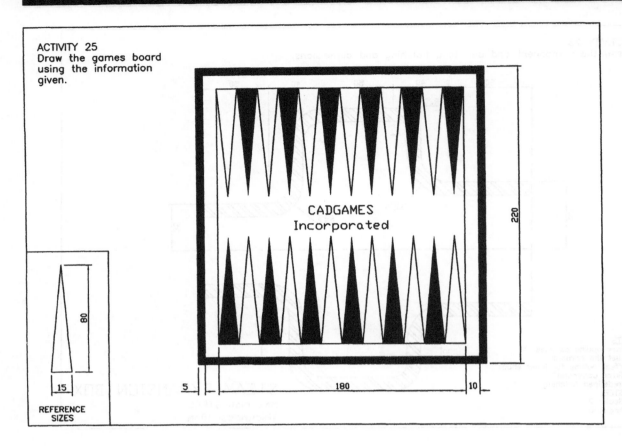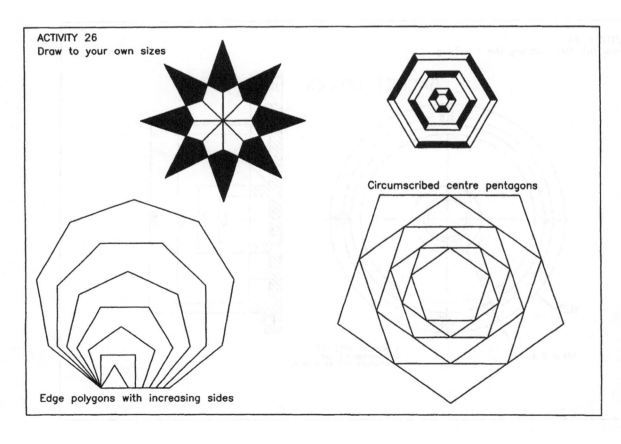

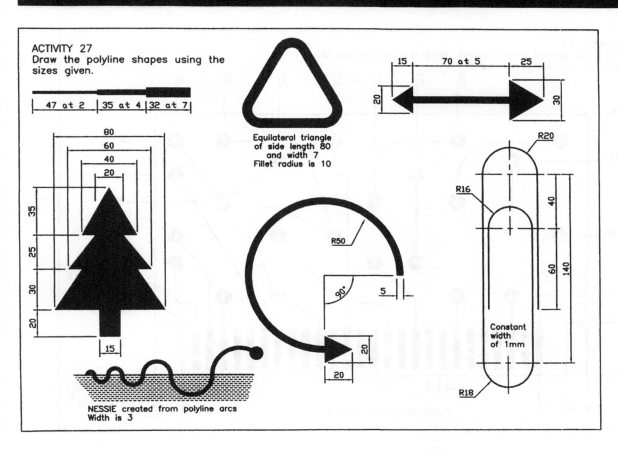
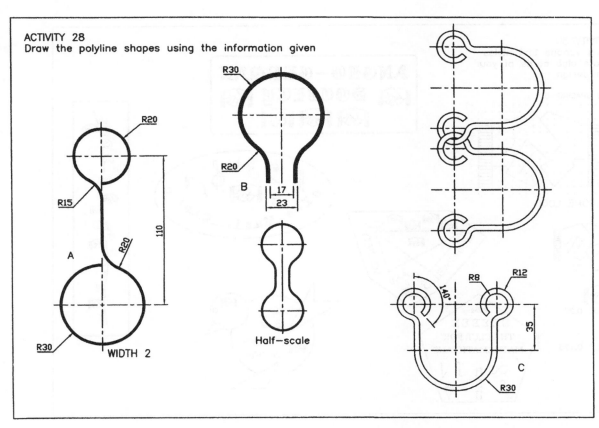

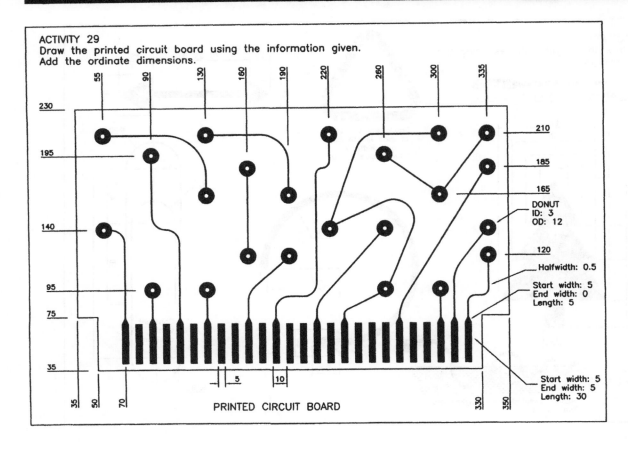

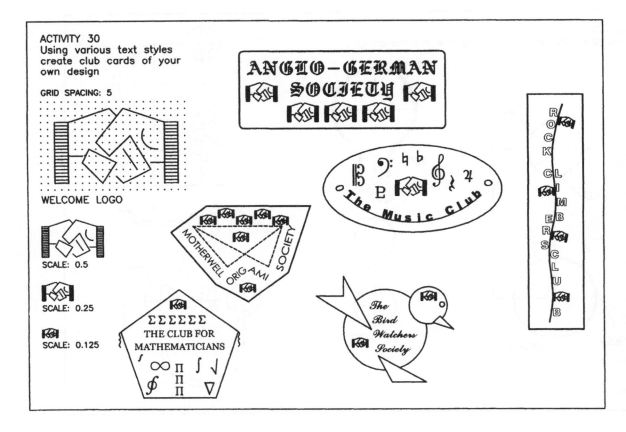

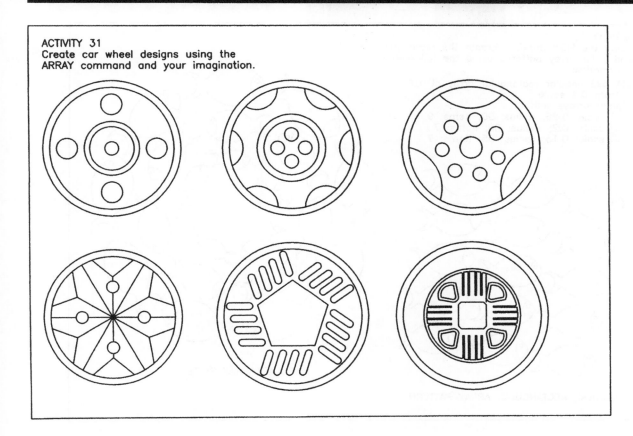

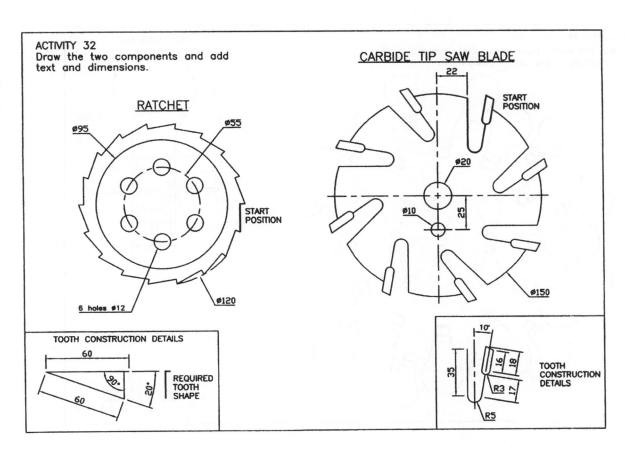

ACTIVITY 33
Using the FISH drawing, create the rectangular and polar array patterns using the following information:
a) A 5x3 angular rectangular array at 10° from 0.2 scale
b) polar arrays with:
   i) scale: 0.25; radius: 90; items: 9
   ii) scale: 0.2; radius: 60; items: 7
   iii) scale: 0.15; radius: 35; items: 7

POLAR ARRAY PATTERN

ANGULAR RECTANGULAR ARRAY PATTERN

ACTIVITY 34
Draw the basic bulb using the dimensions given then produce the array design.

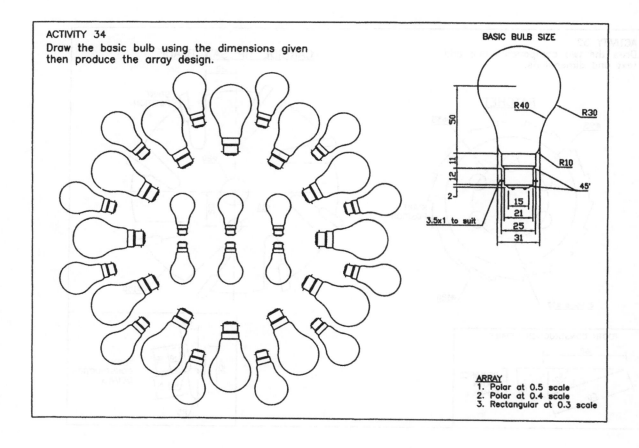

BASIC BULB SIZE

ARRAY
1. Polar at 0.5 scale
2. Polar at 0.4 scale
3. Rectangular at 0.3 scale

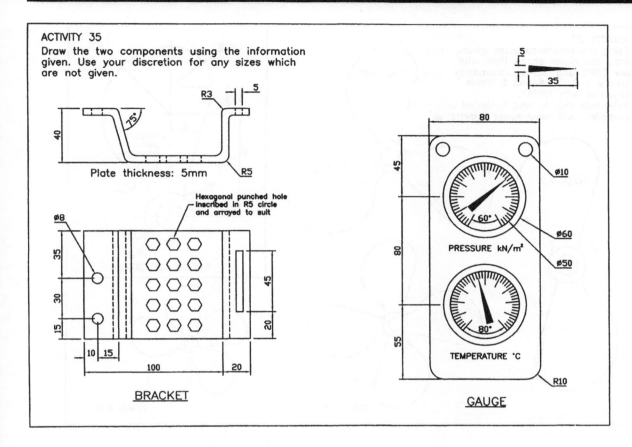

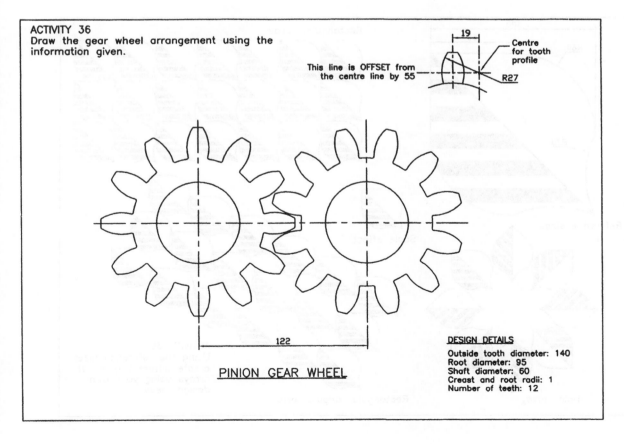

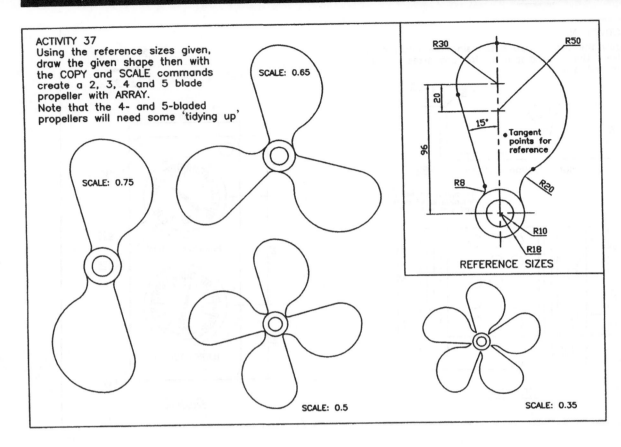
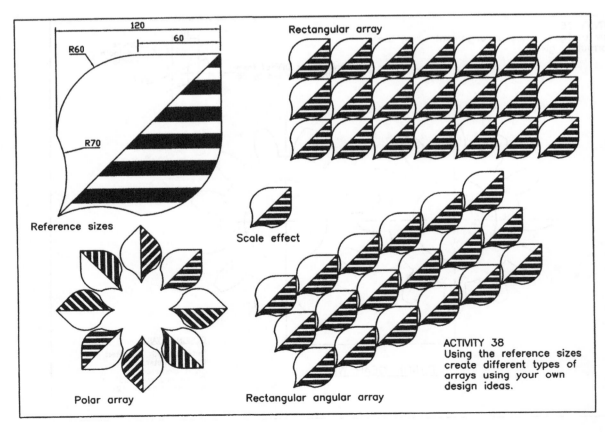

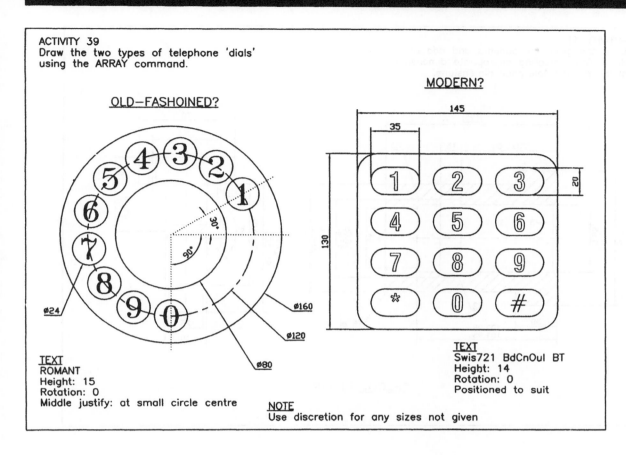

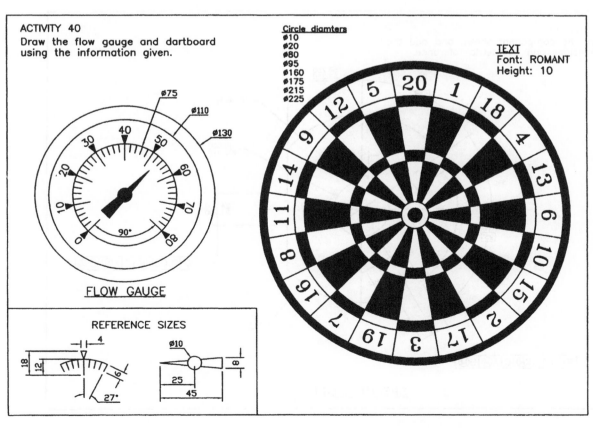

ACTIVITY 41
Draw the given components and add all dimensions, creating appropriate dimension styles for the tolerance dimensions.

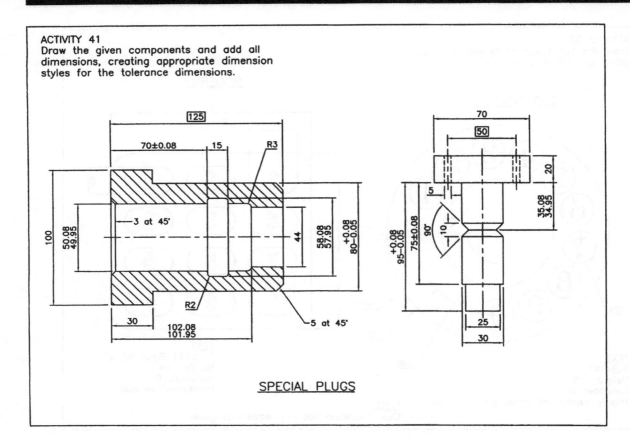

SPECIAL PLUGS

ACTIVITY 42
Draw the component shown and add the dimensions and geometric tolerance.

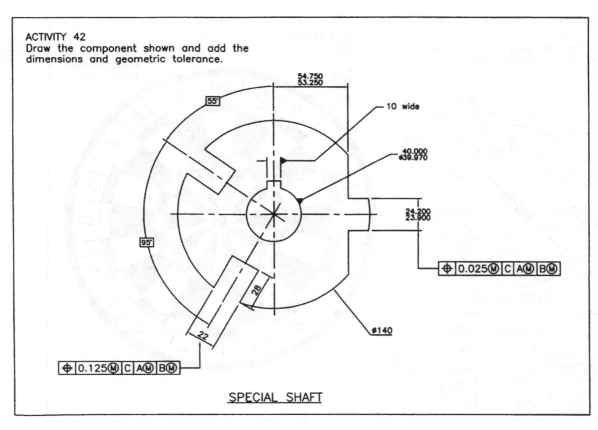

SPECIAL SHAFT

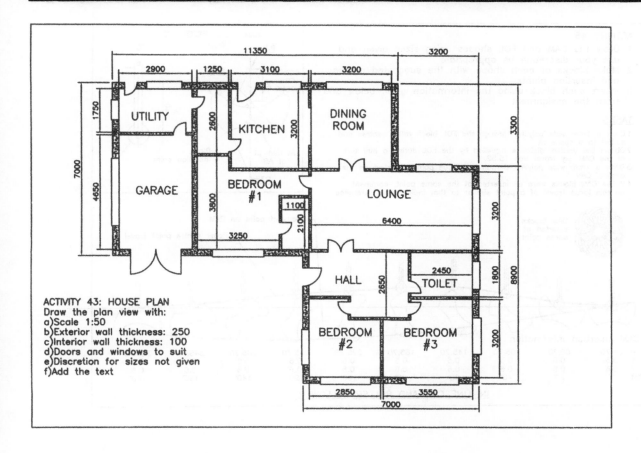

ACTIVITY 43: HOUSE PLAN
Draw the plan view with:
a) Scale 1:50
b) Exterior wall thickness: 250
c) Interior wall thickness: 100
d) Doors and windows to suit
e) Discretion for sizes not given
f) Add the text

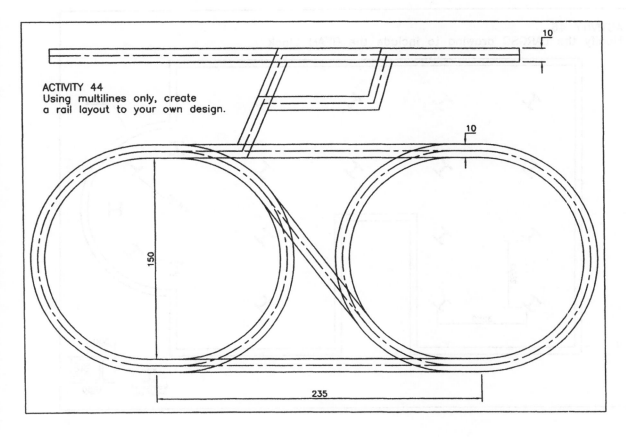

ACTIVITY 44
Using multilines only, create a rail layout to your own design.

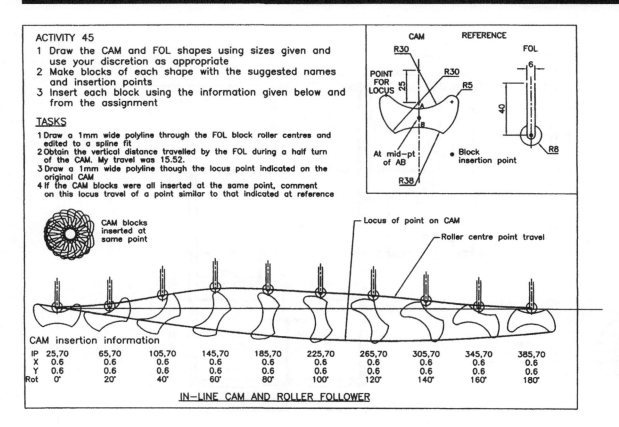

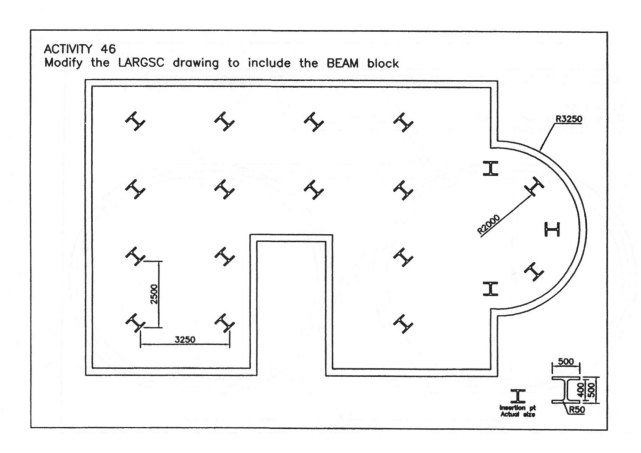

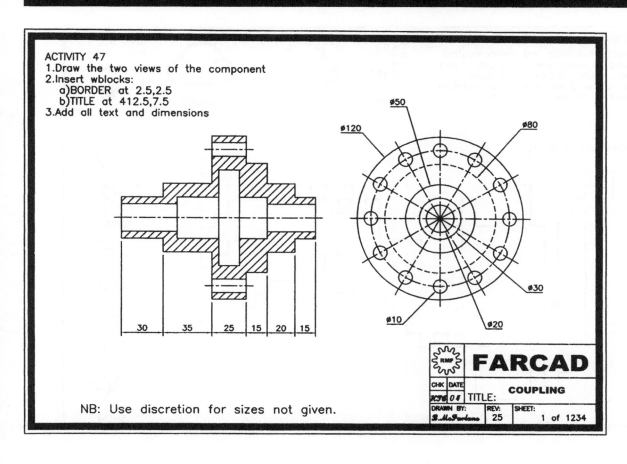

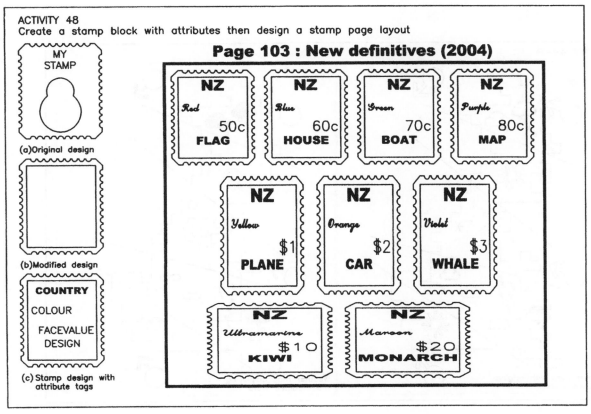

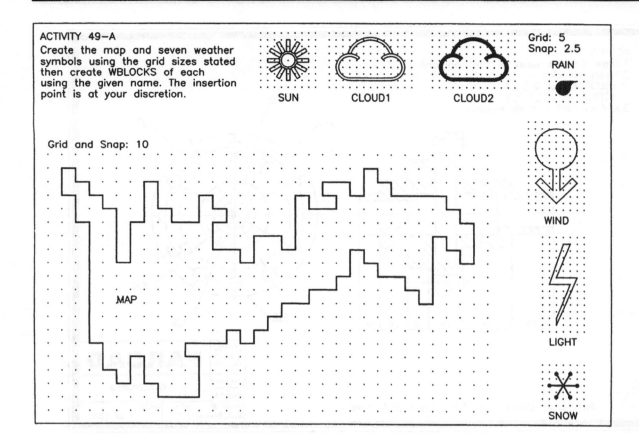
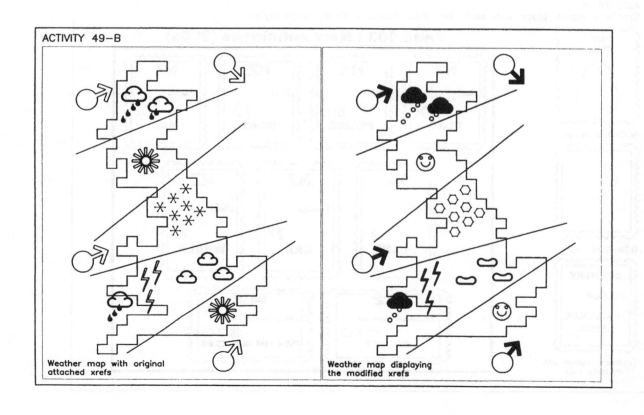

ACTIVITY 50
Draw the two isometric views using the sizes given then construct both an isometric and oblique pictorial.

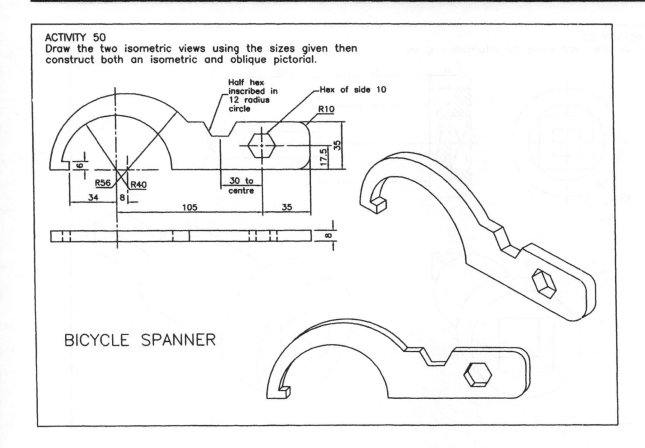

BICYCLE SPANNER

ACTIVITY 51
Design an isometric garden wall block from the basic 100 square x 40 sizes, scale by 0.25 and create a wall.

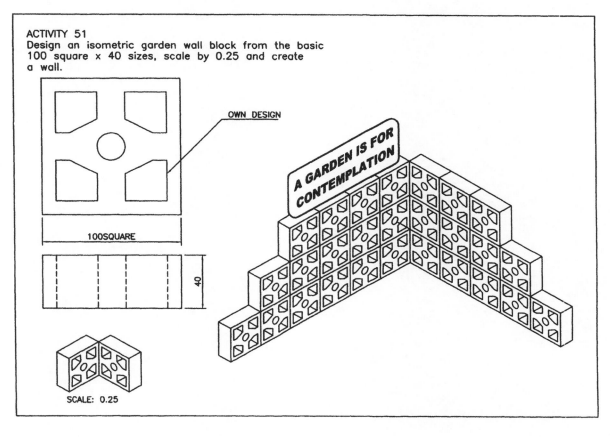

OWN DESIGN

100 SQUARE

SCALE: 0.25

A GARDEN IS FOR CONTEMPLATION

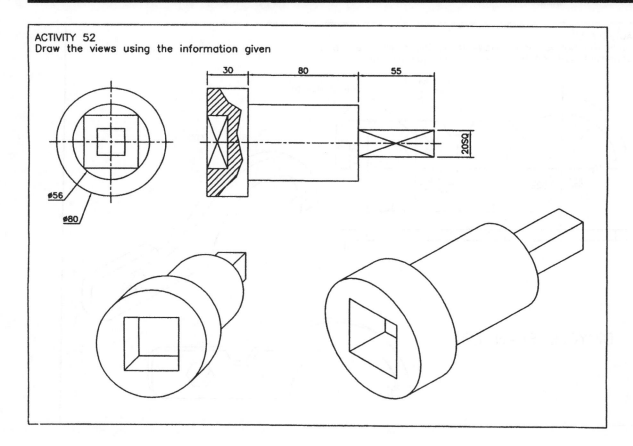

ACTIVITY 52
Draw the views using the information given

Index

2D Solid 177
Absolute co-ordinates 39
Align 198
Aligned dimensions 111
Angular dimensions 111
Arc text 215
Arcs 61
Area 204
Associative hatching 167
Autodesk DWF viewer 370
AutoSnap 55
AutoTrack 55

Break 195

Cabinet projection 319
Calculator 206
Cascade menu 6
Cavalier projection 319
Chamfer 86
CHANGE 242
Closing files 32
Colour 70
Command 5
Command prompt 4
Comparing dimensions 256
Comparing standards 344
Complex linetypes 275
Construction lines 147
Conventions 42
Co-ordinate icon 4
Copy 126
Current layer 74
Cursor 5
Customising toolbars 359

Default 6
Defining attributes 301
Delete layers 81
Dialogue boxes 7
Diameter dimensions 110
DIMASSOC 330
Dimension options 123
Dimension terminology 112
Dimension variables 257
Dimensions: MS and PS 329
Direct entry 41
Distance 203
Divide 194

Donut 62
DWF file 369

Edit vertex (polyline) 189
Editing text 102
Ellipse 62
Erasing 17
eTransmit 370
Exploding blocks 284
Exporting table data 230
Express menu 214
Extend 91

Feature control frames 259
Fields 231
Fillet 85
Fly-out menu 9
Function keys 10

Geometric tolerance 259
Gradient fill 170
Grid/snap 24, 41
Grip selection 141
Groups 277

Hatch modify 168
Hatch styles 162

Imperial drawing 264
Importing text 224
Isocircles 316
Isometric drawing 314
Isoplanes 315

Join (polyline) 188

Large scale drawing 266
Layer 0 291
Layer Properties Manager 66
Layer states 75
Layers toolbar 79
Layout tab 5
Leader dimension 111
Lengthen 196
Linear dimension 109
Linetypes 68
Lineweight 244
List 203
LTSCALE 95, 243

Match properties 96
Measure 194
Menu 5
Menu bar 3
Mirror 132
MIRRTEXT 133
Move 128
Multiple copy 131

New layers 71

Object snap modes 52
Object snap tracking 56
Objects 6
Oblique drawing 318
Offset 90
OOPS 19
Ordinate dimension 120
Origin move 42

Pan 151
Pickfirst 244
Planometric drawing 320
Point 178
Point filter 145
Point identify 203
Polar array 234
Polar co-ordinates 40
Polar tracking 26
Polygon 176
Polyline options 182
Polyline: line and arcs 183
Predefined hatch patterns 164
Properties toolbar 80
Publishing to Web 373

Quick dimension 122

Rays 149
Rectangles 41
Rectangular array 234
Relative co-ordinates 39
Rename layers 81
Revision cloud 216

Rotate 129
Running object snap 53

Save and save as 33
Scale 129
Selection set 20
Sheet set manager 364
Single line text 101
Small-scale drawing 268
Snap from object snap 55
Solid 177
Solid fill 63
Spellcheck 220
Standard toolbar 4
Starting AutoCAD 2
Status bar 4
Stretch 199

Template 9
Text and hatching 169
Text control codes 212
Text justification 104
Text modifications 222
Text tables 226
Time 205
Title bar 2
Toggle 9, 26
Tolerance dimension 258
Tool palette 5, 9, 362
Toolbar 5, 7
Transparent zoom 155
Tree view 348
Trim 93
TTR and TTT 47, 48

User-defined hatching 158

View List 366
Viewports 325

Windows buttons 3
Window/crossing 19
Wipeout 225
Wizard 9, 14

Zoom 152

For Product Safety Concerns and Information please contact our EU representative GPSR@taylorandfrancis.com Taylor & Francis Verlag GmbH, Kaufingerstraße 24, 80331 München, Germany

Printed and bound by CPI Group (UK) Ltd, Croydon, CR0 4YY
08/06/2025
01897007-0019